THE
ARCH OF
KNOWLEDGE

By the same author

*Darwinian Impacts: An Introduction
to the Darwinian Revolution*

THE ARCH OF KNOWLEDGE
AN INTRODUCTORY STUDY
OF THE HISTORY OF THE
PHILOSOPHY AND METHODOLOGY
OF SCIENCE

DAVID OLDROYD

Methuen

New York and London

First published in the USA in 1986 by
Methuen & Co.
in association with Methuen, Inc.
29 West 35th Street, New York, NY 10001

Published in Great Britain by
Methuen & Co. Ltd
11 New Fetter Lane, London EC4P 4EE

Printed in Singapore

Library of Congress Cataloging in Publication Data
Oldroyd, D.R. (David Roger)
    The arch of knowledge.

    Bibliography: p.
    Includes index.
    1. Science — Philosophy — History. 2. Science —
    Methodology — History.
    I. Title.
Q174.8.043 1986   501   85-21758
ISBN 0-416-01331-7
ISBN 0-416-01341-4 (pbk.)

British Library Cataloguing in Publication Data
Oldroyd, D.R.
    The arch of knowledge: an introductory study of the
    history of the philosophy and methodology of science.
    1. Science — Philosophy
    I. Title
    501   Q175

    ISBN 0-416-01331-7
    ISBN 0-416-01341-4 Pbk

CONTENTS

PREFACE

This book attempts a formidable, and in a sense impossible, task. It seeks to give an introductory account of what I take to be the leading ideas that have been put forward in the Western intellectual tradition about the nature of scientific knowledge and the ways it may be acquired. Needless to say, the subject is so vast that it cannot be treated comprehensively within the compass of a few hundred pages in a manner that will give general satisfaction. Nevertheless, the magnitude, and perhaps the foolhardy character, of the undertaking does not make it any the less worthwhile. On the contrary, there is, I believe, a constant need for clear introductory expositions of the history of thought, for with the endless proliferation of human inquiry it becomes ever more difficult to set one's feet securely on the first rungs of the ladder of intellectual history. The purpose of this book, then, is to provide the reader with some assistance towards locating these lower rungs, so that he or she may climb thereafter with greater ease and comfort, not becoming exhausted and frustrated by fruitless attempts to find the way of ascent in the early stages.

Given such overall intentions, some remarks may be in order here about the why and the wherefore of the mode of exposition that has been adopted. Many courses of instruction in philosophy treat the subject from a quasi-historical perspective. One begins a program of study with an examination of the dialogues of Plato, the works of Locke, Berkeley and Hume, or some other notable philosophical writer. But usually the classical texts are chosen for study because they raise questions that are of perennial philosophical interest — such as the nature of knowledge, the mind–body problem, the existence of God, the nature of goodness, and so on. The texts themselves, and more particularly the manner in which they fit snugly together as beads on the chain of intellectual history, are often treated as if they are of secondary significance or interest, compared with the philosophical problems that they raise. The consequence of this is often, I believe, a curiously distorted or incomplete view of the history of philosophy, even though the subject is ostensibly examined through its past. So one frequently encounters some rather extraordinary historical leap-frogging in the course of a training in philosophy: Hume leads to Plato, who is followed by Descartes, Kant, Wittgenstein, Aristotle, Hegel, and all stations to Feyerabend. To be sure, such pedagogical anarchy can serve certain purposes perfectly well. The detailed examination and critique of a particular author can provide an excellent training for the mind. But it may also lead to historical confusion, if no indication is given of the historical connections between the various writers that are examined. And the 'ahistorical' examination of past philosophers may lead one to see their problems, not as they were perceived in their own time, but as they appear of importance to us. This is not always conducive to clear understanding, even when the explicit intention is to use philosophical texts of former times to raise contemporary philosophical problems.

In the foregoing remarks, I may have given the impression that this book is yet another history of philosophy. But this would be misleading. What we shall be doing here is examining the history of the philosophy and methodology of *science* — not the whole of philosophy. The scientific movement is made up of a community of people who seek to gain knowledge of the world by the use of various kinds of observational and experimental procedures, not merely by thinking or talking about problems as do philosophers — though obviously a deal of thought and talk is involved along the way in scientific work. The scientist seeks to discover the regularities of nature and the laws describing these regularities; and theoretical explanations are proposed to account for such laws. We are interested here in the history of theories *about* science. Needless to say, philosophers have a good deal more to do than think about science, but it is remarkable how much of their time is spent in this activity. So while our field is narrower than the history of philosophy as a whole, the area of our concern is still dauntingly large.

The thought and talk about science is often referred to as 'metascience'; so the philosopher of science or 'metascientist' is often described as functioning at the 'meta-level' with respect to science itself. (In philosophical jargon, a 'metalanguage' is a language or system of symbols used to refer to, describe, discuss, or talk about another language.) So the metascientist is not a scientist *per se*: he talks about, discusses, appraises or evaluates what the scientist or the scientific community is doing or saying. But although science and metascience are not one and the same, the metascientist and the scientist need not be separate persons, and the distinction between the two, and the division of labour between the scientist and the metascientist, are in large measure products of the twentieth century. So while it is important to recognise that science and metascience are distinct enterprises, it is also the case that certain scientists have made important contributions to philosophy of science; and some philosophers have made significant contributions to science or mathematics. However, the present tendency is towards specialisation in philosophy of science, and not many contemporary scientists are making front-rank contributions to philosophy of science — or vice versa.

Our task, then, is to trace the broad history of metascience – or the history of the philosophy and methodology of science – from antiquity almost to the present. To this end, we shall seek to give an exposition of the ideas that have been put forward over the centuries about the nature and forms of knowledge, how knowledge is acquired, and the relationships between our ideas about the world and the nature of the world itself. (That is, we shall be concerned, among other things, with the history of *epistemology*.) This task will involve an examination of the roles, in the prosecution of scientific inquiry, of observation, of experimental methods, of logic and language, of classificatory procedures, of analogical thinking and the deployment of models in explanation, of social formations, and much else besides. We shall also be concerned with the history of attempts to understand the nature of being, of 'what is' — the 'stuff' or 'substance' of things. (That is, we shall also be concerned with the history of *ontology*.) So, although there is certainly a difference between philosophy and philosophy of science, the latter is really but a subset of the former, and much of what is discussed in this book could also find a place in a general history of philosophy. However, the features of the history of philosophy that will be examined — chiefly questions epistemological — have been picked out because of their particular relevance to the philosophy of science.

Besides remembering the relationship between philosophy and philosophy of science, one should also recognise the important differences between science and philosophy. An epistemologist or ontologist will seek to determine the nature of knowledge, or of being, by introspection, discussion, examination of the structure of language, or by some other strictly 'philosophical' procedure. The scientist, too, may be interested in knowledge and how it is acquired, and in the nature of being. (For example, the problems of the epistemologist and the psychologist, and the ontologist and the chemist overlap to a considerable degree; and in a sense the scientist is always a philosopher of a kind, though often unwittingly so.) But the methods of the scientist customarily involve experimentation of one sort or another, while the philosopher (in so far as he is strictly following his trade) does not engage in experimental work, though he may notice and comment on such work carried out by practical scientists.

In considering the history of metascience, then, we have to operate on several overlapping fronts, so to speak. Chiefly, our concern is with the history of ideas about science. But also we must examine some issues that lie more within the province of general philosophy. And some points of the history of science itself will command attention. The treatment adopted will be essentially expository, rather than critical. But while the account is intended to be clear and uncluttered by discussion of secondary texts, the accompanying notes and bibliography should assist the reader towards locating more advanced works, and controversies in the secondary literature. In a few places, however, attention will be drawn in the main body of the text to some of the differences of opinion and interpretation that abound in the secondary sources.

I do not suppose that any apology is needed for taking an historical approach to the study of metascience, though in fact rather few introductory texts have been written where this has been the guiding principle. A knowledge of the history of ideas concerning the nature and methods of science is a valuable component of philosophical education. By the same token, knowledge of the history of general philosophy can assist greatly towards an understanding of the ways in which scientific inquiries are carried out and the kind of thinking that is characteristically employed by scientists in their investigations. Thus a study of the history of metascience — or the history of the philosophy and methodology of science — certainly offers advantage for the novice in both science and philosophy. And for the would-be student of metascience, all is likely to be confusion if there is no familiarity with the background to current controversies and areas of inquiry.

But over and above these claimed educational desiderata for students of science and of philosophy, there is the important consideration that, in examining the history of metascience, one is looking at a major and essential component of the intellectual history of the Western world. Consequently, in examining this history, one is better able to comprehend the world we presently inhabit and the ways we try to 'make a living' in it. Moreover, quite apart from questions of 'self-knowledge', the study of intellectual history has, I believe, its own inherent interest, and it would do so even if we were not heirs to this history.

But what route shall we choose, as we try to cut our way through the vast jungle of primary and secondary sources relevant to our theme? One major clue to the journey that we shall follow is to be found in an admirable book by C.M.Turbayne, entitled *The Myth of Metaphor* (Yale University Press, New Haven, 1962), and specifically in its section on 'Analysis and Synthesis'. This discusses a venerable tradition of a two-fold pathway for the establishment of knowledge — from an examination of observable phenomena to general rational 'first principles' ('analysis'); and from such 'first principles' back again to observables, which are thereby explained in terms of the principles from which they are held to be deducible ('synthesis'). Following the history of this methodological program (the 'shape' of which gives the clue to my title — *The Arch of Knowledge*), we find an invaluable thread that will help guide us through the course of our inquiry.

Another topic that we shall pursue with some determination is concerned with the attempts that have been made to give a satisfactory account of the relationship between our thoughts about the world and the nature of the real world itself, and to establish some kind of correspondence between thoughts and things. This problem — rather more diffuse than the story of the development of ideas about the methodological pathways of analysis and synthesis — provides another broad area for our consideration. Many other issues will be encountered during the course of our exposition, but I shall not seek to identify them at this juncture. The reader will meet them at their due time and place.

I should, however, make special remark that one of the 'growth areas' in contemporary philosophy of science is concerned with the social dimension of knowledge, where ideas of the exponents of what is called 'sociology of knowledge' are presently having a considerable impact on metascientific studies. Since this area is currently the subject of intense philosophical debate, it might be thought inadvisable to include consideration of such matters in an introductory text. However, I have some sympathy with sociology of knowledge doctrines, which are, I believe, of considerable interest and importance, and I have seen fit, therefore, to include some consideration of them in my penultimate chapter. But the reader should be appraised that this is an active field of inquiry, and that the views I offer on this may not command universal assent. The same is true, of course, throughout the text, and indeed where all historical interpretations are involved. Even so, one should be aware of one area, at least, where there is very little unanimity of philosophical opinion.

In concluding these prefatory remarks, I should like to record my warm thanks to a number of people who have assisted in the preparation of this text in various ways: John Clendinnen, John Forge, Guy Freeland, Douglas Howie, Jane Oldroyd, Jonathan Powers, Diane Quick, Margaret Schlink,

John Schuster and Alan Walker. No doubt some errors and philosophical solecisms have remained undetected or uncorrected, and I must myself be held responsible for any such lapses. But I am most grateful to these people for their encouragement and their considerable assistance, either theoretical or practical. I trust that they will not regret their association with my undertaking.

David Oldroyd
Sydney, 1985

CHAPTER 1

THE ANCIENT
TRADITION

Plato and the foundation of the tradition of the 'arch of knowledge'

Modern science had its origins in Ancient Greece[1] when men began to speculate about the nature of the cosmos, write down their thoughts, and subsequently teach them formally in schools or academies. In the Greek system of education,[2] as described by Plato (428/7–348/7BC) in his *Republic*[3] and in the *Laws*,[4] pupils were first trained in gymnastics (with dancing) and music; also in reading and writing. Later came elementary studies of arithmetic, astronomy and music; and then arithmetic, geometry and astronomy at a more advanced level. According to Plato's ideal program of training for the 'guardians' — who were eventually to become rulers of the state — education should carry on through adult life, becoming ever more abstract and philosophical, with special emphasis on mathematics and 'dialectic'.[5] Abstract mathematics was considered to be of fundamental significance as a means for training the mind. And the study of music would eventually turn from producing and listening to actual sounds to the contemplation of the simple numerical ratios that govern musical harmony.[6] So also in astronomy. Elementary education would enable the student to know how to use the calendar, for example — a purely practical concern. The practical astronomer would make observations of the heavenly bodies; but the aim of 'philosophical' astronomy was to find the geometrical relationships according to which the heavenly bodies moved — and ultimately the task was to discover the form of the calculations according to which the cosmos had originally been created.[7]

The philosopher who successfully accomplished Plato's educational program would eschew practical considerations. Not for him the mundane task of measurement in order to construct a building. This would be the task of the slave. The educated free man, by contrast, would have his mind focussed on what we today would call the theoretical aspects of the problem. Thus in Plato's time there was a distinction between theoretical and practical knowledge and between theoretical and practical modes of inquiry, corresponding to the social division of the time.

All this may give us some inkling of the general nature of Greek science. The world revealed by our senses is one that is constantly in flux and change, though there does seem to be some constancy 'lying behind' changing appearances. To account for the phenomena of both continuity and change, the pre-Socratic philosophers such as Thales, Anaximander and Anaximenes had supposed that there was some underlying substratum for the material cosmos, which by taking on a variety of forms could give rise to the changing world of appearances.[8] Plato's approach was more intellectualist. He envisaged an unchanging reality existing in the domain of 'Ideas', rather than in some material substrate.

Plato's celebrated doctrine of Ideas (also called 'forms', 'exemplars' or 'paradigms') is not easy to come to terms with in a few words,[9] and over the centuries it has given rise to a truly gargantuan expository and critical literature[10] to which we shall not seek to do justice here. One difficulty in interpreting Plato's doctrine is that it was gradually developed during the course of his lifetime, so that there are significant differences between the early and late dialogues;[11] and the problem is compounded by the fact that the dialogues' exact order of composition is uncertain in some degree. We shall not, however, feel obliged to enter into the scholarly controversies that have flowed from these circumstances. Let us proceed at once to draw out the matters that are specifically relevant to our inquiry, discussing first a simple, homely example.

In everyday language we are accustomed to distinguish between proper names (such as 'Jane', the 'Sydney Opera House', the '*Titanic*', or the '*Eroica Symphony*'), which refer to particular individual entities, and names that refer to classes of things (such as 'wife', 'concert hall', 'ship' or 'musical composition'). These latter general terms refer to or signify *classes* of similar entities — or mental concepts thereof. Philosophers call them 'universals', distinguishing them from 'particulars' such as 'Jane', the 'Sydney Opera House', etc.

The question immediately arises in philosophy as to whether such universals or general concepts have any 'real existence'. If I think of my wife, Jane, it seems that there is, as it were, a correspondence between the (real, physically existing) person concerned and the idea that I may have of her, the word Jane acting as a mediating link between the two. But the situation seems to be different with respect to *general* notions such as 'wife'; there is no physically existing entity, 'wife', as such — only a collection, or class, of somewhat similar beings, each of which can meaningfully be referred to by the term 'wife'. But one may ask: How can one form the general ideas of 'universals' when one never actually encounters them in a physical way with one's sense organs? For with universals there can be no 'correspondence' between object, word, and ideas. The question seems to have greater force when one considers more abstract terms such as 'love', 'beauty' or 'justice', rather than 'wife' or 'concert hall'. So what is the philosophical status of 'universals'?

It was to questions of this kind that Plato's doctrine of 'Ideas' or 'forms' was addressed. In the nineteenth century, it was quite commonly supposed that he was thinking of what he called 'Ideas' as no more than mental concepts. But this interpretation, though useful in some measure, does not do proper justice to Plato's philosophical system, either in his early or his late dialogues. Plato supposed that for every idea or mental concept that we may have, there are corresponding, *objectively existing*

Ideas (forms, exemplars or paradigms). And when we are acquiring knowledge we are gaining an understanding of (or apprehending) these Ideas.[12]

In Plato's early dialogues, these Ideas were supposed to be immanent[13] in particular things. (For example, one might say that the 'Idea' of roundness is immanent in a sphere.) And in the early dialogues it is the *particulars* that are claimed to be real. But as Plato's thought developed he began to invest the *Ideas* with objective reality and to suppose that they existed in some manner independently of real objects. Finally, in the late *Timaeus*, the Ideas were portrayed as wholly transcendent,[14] as if they somehow existed in a divine realm, quite separate from the physical world of everyday life.[15] This doctrine, though not perhaps so very plausible to us today, might help to account for the similarities and differences of the objects we see around us. The horses that we see, for example, are all generally similar, but have individual differences. Perhaps such 'natural kinds' might be accounted for if each physically existing object were a manifestation of some correspondent transcendent 'form' or 'Idea', being always an imperfect realisation of the 'Ideal'.

Plato's *Republic* is usually regarded as his greatest work. In it he set out his vision of a utopian social system and form of government, with philosophically trained 'guardians' acting as the rulers. The guardians would undergo a life-long education, so that only by the age of about fifty would they be deemed ready to rule. And the function of their educational process, about which I made just a few remarks above, was to enable them to grasp or apprehend the forms or Ideas — and particularly the Idea of justice, so needful for a just ruler.

The information that we can acquire through our sense organs — knowledge of appearances — is deceptive and illusory, according to Plato.[16] Also, though one may happen to form a 'true judgement' correctly, this can be fortuitous. So merely being 'right' about something doesn't mean that one has *knowledge*. To have knowledge, Plato argued, one must apprehend the forms or Ideas of things. But how is this to be done?

Plato gave several answers to this in his various dialogues. In the *Symposium*[17] he considered how one comes to apprehend the Idea of beauty. One starts by inspecting an object that can generally be recognised as beautiful. Interestingly, Plato suggested a beautiful boy as a suitable object of contemplation. Then one finds several such boys and looks for the common pattern of beauty in all of them. Thereafter, one moves on to look for the beauty that is manifest in learning, in customs and laws, and so on. Finally one should come to apprehend the general abstract Idea of beauty in the world of forms, so that one may put aside the contemplation of slave boys of specially pleasing appearance. It is interesting that Plato, with his customary rhetorical skill, suggested a starting point where — within a given culture — there might indeed be a reasonable degree of agreement as to what is or is not beautiful. For the erotic urge is common to most of us, even though we may not all have a penchant for pederasty. But the proposal that the method might somehow reveal the common 'form' of beautiful boys, buildings, legal institutions, and everything else, is implausible. Indeed we may dispute the claim that such disparate 'kinds' have anything much in common in the way of beauty.

In another well-known passage, in the dialogue *Meno*,[18] Plato suggested that one might come to know forms by the fact that they were originally known to the mind in some previous incarnation. (In effect he was supposing that we have some inborn or *a priori* knowledge — an assumption that was an anathema to subsequent 'empiricist' philosophers.) There is a discussion between Socrates and a slave boy in which the teacher manages to elicit from the boy a statement of a mathematical theorem that he had never encountered before. Plato suggested that the child was really possessed of the knowledge all the time, but only subconsciously so to speak. So it was the teacher's task to 'draw out' this knowledge of forms — and hence we get the origin of the modern word 'education'.[19]

Neither the argument of the *Symposium* nor that of the *Meno*, in relation to the discovery of forms, need be taken very seriously (which is not to say that Plato thought them trivial), but the account to be found in the *Republic* is of surpassing importance, and really provides us with the starting point for our whole history of the philosophy and methodology of science.[20] I shall, therefore, quote the relevant passage in full. As in most of Plato's extant writings, the text is in the form of a dialogue, the speakers here supposedly being Socrates (Plato's own teacher) and Glaucon, a student of philosophy.

'Then, take a line cut in two unequal segments, the one for the class that is seen, the other for the class that is intellected — and go on and cut each segment in the same proportion. Now, in terms of relative clarity and obscurity, you'll have one segment in the visible part for images. I mean by images first shadows, then appearances produced in water and in all close-grained, smooth, bright things, and everything of the sort, if you understand.'

'I do understand.'

'Then in the other segment put that of which this first is the likeness — the animals around us, and everything that grows, and the whole class of artifacts.'

'I put them there,' he said.

'And would you also be willing,' I said, 'to say with respect to truth or lack of it, as the opinable is distinguished from the knowable, so the likeness is distinguished from that of which it is the likeness?'

'I would indeed,' he said.

'Now, in its turn, consider how the intelligible section should be cut.'

'How?'

'Like this: in one part of it a soul, using as images the things that were previously imitated, is compelled to investigate on the basis of hypotheses and makes its way not to a beginning but to an end; while in the other part it makes its way to a beginning that is free from hypotheses; starting out from hypothesis and without the images used in the other part, by means of forms themselves it makes its inquiry through them.'

'I don't,' he said, 'sufficiently understand what you mean here.'

'Let's try again,' I said. 'You'll understand more easily after this introduction. I suppose you know that the men who work in geometry, calculation, and the like treat as known the odd and the even, the figures, three forms of angles, and other things akin to these in each kind of inquiry. These things they make hypotheses and don't think it worthwhile to give any further account of them to themselves or others, as though they were clear to all. Beginning from them, they go ahead with their exposition of what remains and end consistently at the object toward which their investigation was directed.'

'Most certainly, I know that,' he said.

'Don't you also know that they use visible forms besides and make their arguments about them, not thinking about them but about those others that they are like? They make the arguments for the sake of the square itself and the diagonal itself, not for the sake of the diagonal they draw, and likewise with the rest. These things themselves that they mold and draw, of which there are shadows and images in water, they now use as images, seeking to see those things themselves, that one can see in no other way than with thought.'

'What you say is true,' he said.

'Well, then, this is the form I said was intelligible. However, a soul in investigating it is compelled to use hypotheses, and does not go to a beginning because it is unable to step out above the hypotheses. And it uses as images those very things of which images are made by the things below, and in comparison with which they are opined to be clear and are given honor.'

'I understand,' he said, 'that you mean what falls under geometry and its kindred arts.'

'Well, then, go on to understand that by the other segment of the intelligible I mean that which argument itself grasps with the power of dialectic, making the hypotheses not beginnings but really hypotheses — that is, stepping-stones and springboards — in order to reach what is free from hypothesis at the beginning of the whole. When it has grasped this, argument now depends on that which depends on this beginning and in such fashion goes back down again to an end; making no use of anything sensed in any way, but using forms themselves, going through forms to forms, it ends in forms too.'

'I understand,' he said 'although not adequately — for in my opinion it's an enormous task you speak of — that you wish to distinguish that part of what is and is intelligible contemplated by the knowledge of dialectic as being clearer than that part contemplated by what are called the arts. The beginnings in the arts are hypotheses; and although those who behold their objects are compelled to do so with the thought and not the senses, these men — because they don't consider them by going up to a beginning, but rather on the basis of hypotheses — these men, in my opinion, don't possess intelligence with respect to the objects, even though they are, given a beginning, intelligible; and you seem to me to call the habit of geometers and their likes thought and not intelligence, indicating that thought is something between opinion and intelligence.'

'You have made a most adequate exposition,' I said. 'And, along with me, take these four affections arising in the soul in relation to the four segments: intellection in relation to the highest one, and thought in relation to the second; to the third assign trust, and to the last imagination. Arrange them in a proportion, and believe that as the segments to which they correspond participate in truth, so they participate in clarity.'

'I understand,' he said. 'And I agree and arrange them as you say.'[21]

Let us try to interpret this important and well-known passage. As required by Plato's account, we may divide a line into four segments and allot to each a corresponding epistemological 'level', and also a corresponding ontological (or 'ontic') level,[22] as in Figure 1. (We shall also introduce some lettering into the diagram to facilitate discussion.)

The main division (*AC* and *CE*) is intended to indicate the distinction between the realms of mere 'opinion' (*doxa*) and 'knowledge' (*episteme*). Obviously, in Plato's book it is the philosopher's task to transcend the mere mundane world and its phenomenal appearances and somehow ascend to the world of forms — to the domain of knowledge. First, however, we need to grasp what the nature of this world of forms might be. We are of course, familiar with the world of physical objects (designated *BC*, above); and we are aware that there may be images, replicas of these, etc., which are somehow less real than the objects themselves. (A mirror image is 'less real' than the object that is mirrored, on this view; a waxwork of Winston Churchill is 'less real' than Winnie himself.) Now, grasping this point easily enough, the pupil is asked to suppose that the whole upper level of the line (*CE*) bears the same relation (in respect of reality) to the lower level (*AC*) as does the upper level (*BC*) of the lower main segment (*AC*) to its lower level (*AB*). That is, Plato is suggesting that *BC/AB* = *CE/AC*. (Also, we may presume, *BC/AB* = *DE/CD*, though this is not stated explicitly.) Thus, the upper level of the whole (*CE*) represents the world of forms where knowledge is to be sought; the lower level (*AC*) is merely the realm of mundane objects and appearances.

Figure 1

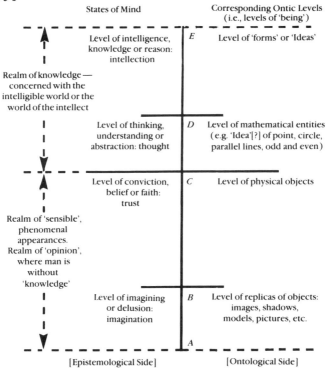

	States of Mind		Corresponding Ontic Levels (i.e., levels of 'being')
Realm of knowledge — concerned with the intelligible world or the world of the intellect	Level of intelligence, knowledge or reason: intellection	E	Level of 'forms' or 'Ideas'
	Level of thinking, understanding or abstraction: thought	D	Level of mathematical entities (e.g. 'Idea'[?] of point, circle, parallel lines, odd and even)
Realm of 'sensible', phenomenal appearances. Realm of 'opinion', where man is without 'knowledge'	Level of conviction, belief or faith: trust	C	Level of physical objects
	Level of imagining or delusion: imagination	B	Level of replicas of objects: images, shadows, models, pictures, etc.
		A	

[Epistemological Side]   [Ontological Side]

We must now notice the division of the upper part of the line into the two segments: *CD* and *DE*. Here, the whole methodological aspect of Plato's inquiry into knowledge becomes manifest. He was familiar with the work of the Greek geometers.[23] In geometry, one starts with certain definitions, axioms or postulates (which we may collectively call 'first principles') and proceeds from them by a series of logical moves to deduce certain theorems. If the 'first principles' are correct, and one makes no mistakes in the reasoning, then one should be able to have full confidence in the truth and adequacy of the theorems. Note that although a geometer may use some drawings, cut-out pictures, ruler and compass or whatever, to assist him in his inquiry, the truth or otherwise of his operations in no way depends upon the precision with which he does his drawings or sketches, or makes his cut-outs. So far as the mathematics proper is concerned, the geometrical reasoning is carried out in the realm of ideas, through purely intellectual activity.[24] It makes no difference whether the triangle is drawn well or ill. This, I suggest, was one of the main considerations that led Plato to put forward his doctrine of Ideas, for though in a physically drawn triangle the interior angles may not exactly add up to 180 degrees, they certainly do so in the ideal and perfect world of 'forms'. That is really what geometry, as a theoretical enterprise, is all about.

But why should the segment, *CD*, be distinguished from the highest ontic and epistemic level, *DE*? The answer, as Plato informs us, was that the geometers failed to give reasons as to why their 'first principles' were true and acceptable, even though they lay within the general 'intellectual' segment of the line, rather than in the domain of mundane physical objects, or replicas thereof. Here, of course, lies the rub. Deduction seems to be possible, and relatively unproblematical; at least it appeared so within the framework of Greek geometry. But how could one have true knowledge if one simply guessed the starting points of the deductive chains – if they were plucked out of the air so to speak (or perhaps nailed into it)?

Plato's answer was that there could be a fourth segment of the line — at the ontic level of the forms or Ideas, where the mind (if it could succeed in apprehending the forms or Ideas) would finally attain true knowledge. To do this, concepts should be subjected to the process of *dialectic*.[25] This was an important and still commonly used philosophical procedure which, I take it, was to exercise the minds of the would-be guardians to a large degree in the later years of their educational program. The object of the dialectical exercise was to clarify concepts. (In fact, most of the *Republic* was devoted to attempting to clarify the concept of justice.) A concept was selected and a provisional definition proposed, which was then subjected to general criticism and comment amongst the persons taking part in the philosophical discussion. Objections might be raised against the provisional definition, which would then need to be modified in the light of criticism. So the process would go on, back and forth, until general agreement had been reached concerning the definition. Then, according to Plato, the relevant form would be apprehended, and — unlike that which obtains in the case of the hypothetical first principles of the geometers — true knowledge would have been acquired because the concepts[26] were subjected to a rigorous process of dialectical discussion.

It is worth commenting that Plato himself came to believe in the real existence of a transcendent realm, inhabited by 'Ideas' or 'forms' — which were very much more than the ideas or thought in our minds. Consequently, there was, in his view, a proper warrant for the security and structural stability of human knowledge. However, with hindsight (and here I am making an anachronistic comment, such as good historians are supposed to eschew — that is, my comment is made from the perspective of one who does not subscribe to the Platonic doctrine of Ideas and forms), it would seem that Plato's dialectical method consisted, in effect, of the attempted clarification of concepts and the establishment of agreed definitions of things. So, I suggest, an Idea was, for practical purposes, no more than an agreed verbal definition, no matter what Plato himself may have thought. Moreover, the philosophers' dialectical discussions would have to be based on knowledge already available to them. They would not, in themselves, generate new empirical information.

But, quite apart from what we may think about the real existence of the Platonic world of Ideas or forms, and even if we think that the 'method of dialectic' does little more than lead to the discovery of what one already knows (with some beneficial conceptual clarification, no doubt), there is much that is of the most profound importance for the history of Western thought (and for the history of Western metascience) in Plato's metaphor of the divided line. To understand why this is so, it is necessary to focus our attention on an important feature: the 'shape' of the supposed process for the discovery of forms.

This shape is, as Turbayne has pointed out, rather like an arch. There is supposedly an 'upward' movement from the information concerning 'particulars' received by the senses, to general concepts and first principles (of mathematics, or perhaps some other science). And also there is a 'downward' deductive 'pathway', supposedly carried out in the realm of Ideas in Plato, but not necessarily so in later writers — who, however, were pleased to employ the same general model for the 'structure' of knowledge and its method of acquisition and deployment. For in geometry the 'downward' deductive descent will terminate in 'theorems', still in the domain of Ideas. But in a science (in the modern sense of the word) one may arrive at predictions that may be capable of being tested experimentally. (In later chapters, we shall see how attempts were made to justify the 'movement' or 'translation' from the language in which the theoretical deductions of the science are conducted to the language of common everyday observation, or the language of 'pointer readings' in the laboratory. This may sound simple, but in fact in raises difficult philosophical problems, and some philosophers regard it as a misguided enterprise. Perhaps the belief that there is some fundamental difference between the language of theory and of observation is a vestige of the Platonic account of knowledge.)

Now any arch, to stand up firmly, needs two well-made legs and a well-shaped coping-stone, firmly placed at the top. The same might be said for the 'arch of knowledge', as Plato envisaged it.[27] The trouble with the geometers was that they built downwards from the air, so to speak, leaving their first principles pretty much unsupported (except in so far as they might lead to theorems that seemed to be true when tested with 'ruler and pencil'). The geometers neglected the other 'ascending' side of the arch. So for Plato they were mere 'technicians', not offering true knowledge.

Plato regarded dialectical discussion as the coping stone of his arch. But we should like to know much more about the process of ascent, which *itself* seems to involve the dialectical procedure. On the one hand we can see that knowledge of the 'way up', if discovered, will necessarily give one knowledge of the 'way down', for — to pursue our spatial metaphor just a little further than is really warranted — if you can find a pathway up a mountain, you should at one and the same time be finding the way down it. But this still evades the problem. How can one, starting from observations of the world, find the theoretical principles of a science? This is a difficult philosophical problem indeed – and difficult from a practical standpoint also — and really this book can be construed as an attempt to give an historical account of how men have wrestled with such questions.

Some notable modern philosophers such as Karl Popper (see page 301) have considered the question of how to 'ascend' the arch by suggesting, in effect, that you simply take a run and jump. That is to say, you formulate hypotheses as it were by guesswork and test by experiment the conclusions that flow deductively from these. This is the method of 'conjecture and refutation'; but it virtually ignores the 'ascending' side of the arch by saying that it refers to a psychological rather than a logical process, and consequently is not the concern of the philosopher of science. Plato would not, I think, have taken kindly to Popper's description of scientific methodology.[28] Another approach, which we find advocated in the writings of Sir Francis Bacon,[29] involved an attempted scaling of the arch by a very slow and steady ascent, with a vast mass of empirical information employed in order to serve as hand-holds at each step in the climb.[30] This method, too, as we shall see, does not really provide a satisfactory answer to our problem. And in fact it may be that the problem has been wrongly posed. But because of the extraordinary power of Plato's writings (and the fact that the equally influential Aristotle continued on the methodological trail that Plato had blazed, and for many other reasons besides), his particular way of picturing the structure of an adequate knowledge system — namely as an ascending pathway leading to first principles, which are somehow 'guaranteed' or 'made secure' by the power of dialectic, and a descending, deductive, explanatory procedure – has exercised a quite extraordinary influence over the philosophy of science in its historical development, even though many people seem to have used the spatial metaphor of 'ascent' to, or 'descent' from, first principles quite unwittingly.

A few further points may be worth discussion. Plato supposed that the geometers were only at the third segment of his divided line, CD, because they used hypotheses, which, as we have said, were not guaranteed by the activity of dialectic. Presumably, if the hypotheses were so guaranteed they would be converted to true first principles, belonging in the fourth segment, DE. Plato's hope that this might be achieved was, I believe, illusory; but the attempts to guarantee the truth of certain fixed starting points for ratiocination (deduction) provide a recurring theme in the history of the philosophy and methodology of science — one that we shall encounter many times in our story.

Secondly, we should note that the problem of how to scale the left-hand side of the arch is, in a sense, equivalent to the 'problem of induction', which we shall be discussing elsewhere in this book.[31] Induction, as defined by Aristotle, is a 'progress from particulars to universals'.[32] Or, we may say, it is a process in which general statements about classes of things are made on the basis of knowledge concerning particular constituent members of those classes; more loosely, it is arguing from the particular to the general. It is, unfortunately, logically invalid to do this,[33] and this is the thorn in the flesh[34] of all philosophy of science, and for scientists' attempts to gain true and certain knowledge of the world. Plato did not see his problem — in relation to the hypothetical character of the first principles of geometry — in terms of the so-called 'problem of induction'; but it is, in fact, closely related. If one could *deduce* general laws or principles on the basis of knowledge of particular items of empirical information, then the pathway to the apex of the arch would present no great problem to the would-be climber. But unfortunately this is not possible; and in fact a certain upward path would render the downward path largely superfluous.

A third point we must make relates to the so-called procedures of *analysis* and *synthesis*.[35] We do not find these terms used as such in the passage from the *Republic* that we have quoted above. But in subsequent writers the 'downward' 'deductive' path became known as synthesis and the 'upward' 'inductive' path was called analysis (or *composition* and *resolution*, respectively).[36] We shall, in this book, be much interested in the subsequent histories of these terms.[37]

Finally, before we proceed, we should note that although the process of deduction from accepted premises to conclusions may seem unproblematical, whereas one may look askance at the legitimacy of induction, in fact the true sceptic can be, and sometimes is, as doubtful about deductive moves as inductive ones.[38] However, for the present purpose, if we think of deduction as the converse of induction — that is, if we think of it as a process of arguing from the general to the particular — then there need not be too much distress involved in thinking of deduction as a self-evidently justified procedure.[39]

Plato himself was not what we would call a scientist, though he took a keen interest in mathematics and some interest in cosmology. In general, he downgraded interest in observations of physical phenomena. The Platonic philosopher's mind was to be directed 'upwards' to the domain of Ideas, where 2 plus 2 *really* equalled 4, where the angle sum of a triangle *was* exactly and perfectly equal to two right angles. He was not concerned with mundane particulars. Further, despite Plato's interest in mathematics and deduction, little had been written in his day about logic and about which forms of reasoning were to be taken as valid and which invalid. For establishing the definitions of things, Plato's most advanced technique was the so-called 'method of division'.[40] So Plato was hardly in a position to make a thoroughgoing job of working out a satisfactory philosophy of science. He did, however, develop an interesting and highly influential epistemology. The fact that it was conflated with ontology — a doctrine of a *hierarchy* of levels of being, which consequently had certain 'moral' overtones — was, perhaps, unfortunate.[41] But Plato's influence has been lasting and profound. This is why we have looked at one aspect of his work in some detail, even if it was but a small portion of the whole and was a contribution to 'philosophy of science' made at a time when science, as we understand it, could hardly be said to have existed.

Aristotle

Let us now turn our attention to Aristotle (384–322BC),[42] who as a young man was Plato's pupil at the Academy in Athens. In his early works Aristotle tended to sympathise with Platonic philosophy in some measure, but as his thought matured he gradually turned away from the system of his master, taking particular exception to the doctrine of forms or Ideas as real entities in some transcendent realm of the Ideal.[43]

Aristotle wrote extensively on a wide range of philosophical topics; but we may also with some justification refer to him as a scientist in that he made and recorded numerous empirical observations, particularly in marine biology, and he was concerned to offer *explanations* of phenomena by means of a wide range of explanatory devices. Thus, for example, he suggested that there were particular physical vapours arising from the interior of the Earth which would account for mineral deposits; or there was an 'unmoved mover' which would, in the final analysis, account for all the motions of the cosmos. We shall not be concerned to say very much about Aristotle's 'scientific' works,[44] but their existence should certainly be noted for his scientific interests had a profound influence on his philosophy. Aristotle also wrote extensively and influentially on such matters as ethics, politics and rhetoric.

Most of what Aristotle had to say on the method of acquiring knowledge is to be found in a group of works concerned with logic, collectively known as the *Organon*.[45] The works were: *Categories*; *On Interpretation*; *Prior Analytics*; *Posterior Analytics*; *Topics*; and *On Sophistical Refutations*.[46] Also important for our purposes are the *Physics*; and many of the ideas we shall discuss also appear in the *Metaphysics*.[47] The *Prior Analytics* was concerned with the processes of deduction, using the syllogism. The *Posterior Analytics* was concerned with the application of syllogistic reasoning, and how to form the premises for use in syllogisms. That is, it was concerned with inductive procedures.

We may notice first Aristotle's doctrine of 'categories', which though it may not have been of paramount importance in the Organon, evolved, as we shall see, into a doctrine of profound significance in the writings of Kant.[48] The word 'category' though wrapped round with a certain mystical aura in Kant, really means no more than a type of 'predicate'. According to Aristotle, there are just ten categories — ten kinds of things that may be said about (or predicated of) a thing. It has been suggested[49] that Aristotle arrived at his list of categories simply by imagining a man standing in front of him in the Lyceum, and asking himself what *kinds* of questions might be put respecting the man. Be this as it may, the proposed list of categories was:

what (or *Substance*),
how large (that is *Quantity*),
what sort of thing (that is, *Quality*),
related to what (or *Relation*),
where (that is, *Place*),
when (or *Time*),
in what attitude (Posture, *Position*),
how circumstanced (*State* or condition),
how active, what doing (or *Action*),
how passive, what suffering (*Affection*).[50]

For example, one could ask, concerning a thing before one, *what* 'substance' it was — and receive the answer 'man'; one could ask *where* it is — and be told 'in the Lyceum'; and so on. The categories, then, represented (according to Aristotle) the fundamental descriptive classes into which existent or real things, or attributes of things, might be placed (categorised).

Whether or not Aristotle did in fact specifically ask himself questions about a man standing before him in the Lyceum is not really important. It is, in fact, more likely that he was guided towards his list of ten categories by examining the characteristic features of Greek *language*, as much as the world of physical objects.[51] But chiefly he seems to have been engaged in an ontological inquiry as much as a logical one, and his purpose was probably to combat Plato's doctrine of real and existent Ideas, the philosophical/linguistic inquiry providing the means to this end. Aristotle was concerned to show that being resides in individual objects perceived by the senses — not in transcendent universals. However, one also needs to have some notion of categories (or possible predicates) in order to proceed to the syllogistic reasoning of the *Prior Analytics*, which was of such major significance in Aristotle's work as a whole.

Aristotle believed that the same general principles of reasoning held in all the sciences,[52] but that each science had its own particular set of first principles.[53] The principles of reasoning were displayed with consummate skill in the *Prior Analytics*, where the rules of syllogistic logic were set out at length. In fact, the logical device of the syllogism (of classes) was, it is believed, largely Aristotle's own invention.

A syllogism involves two assumed propositions (premisses) and a conclusion. Just to give some simple examples:

All Australians are salacious (Major premiss)
This perspicacious author is Australian (Minor premiss)
Therefore:
This perspicacious author is salacious (Conclusion)
No females are impregnable
All princesses are females
Therefore:
No princesses are impregnable

Both the foregoing arguments are logically correct, no matter whether one thinks the premisses happen to be true or false. But if we say:

Some men wear gaiters
All bishops are men
Therefore:
All bishops wear gaiters

we have an example of a logical fallacy — a sample of invalid reasoning. This is so, even though the conclusion may well be contingently correct. (Perhaps all bishops *do* wear gaiters!) On the other hand, one might have a syllogism that is perfectly acceptable, so far as a sample of reasoning by classes goes, even though the conclusion is absurd or patently false as a statement of fact. For example:

All bachelors are promiscuous
No promiscuous people are unhappy
Therefore:
No unhappy people are bachelors

The point, of course, is that syllogistic logic (and other kinds of logic) is concerned with forms of argument. It examines which kinds of reasoning are valid and which are invalid — not the truth or falsity of either premises or conclusions.

It is not our task here to examine the whole complex structure of categorical syllogistic reasoning.[54] Let it simply be said that so far as we know Aristotle virtually invented the process (or its formal presentation), and successfully codified the valid and invalid forms of syllogistic inference for the logic of classes.[55] Thus Aristotle did indeed provide an 'instrument' of reasoning for use in science, even if it was only of limited scope.[56]

But, the objection will very likely be raised: of what use is the syllogism if we have no knowledge of the premises of the argument? This is indeed a problem; and to pursue it further, in so far as it was treated by Aristotle, we must look at his 'theory of predicables', as presented in the *Topics*,[57] and his arguments about induction and arriving at first principles in the *Posterior Analytics*. This will give us some indication of Aristotle's theory of 'arch-climbing'.[58] He could, we may suppose, get down from the top, more or less safely, with the help of his syllogistic logic of the *Prior Analytics*. But how did he think one might make the ascent? Some thoughts on this are to be found in the volume entitled *Topics*, which although very likely composed after the *Posterior Analytics* may usefully be discussed first.

At the beginning of the *Topics* we are introduced to Aristotle's all-important doctrine of the five predicables. He supposed that there were just five ways in which a predicate might be related to a subject.[59] Some predicates were considered to be 'convertible' with their subjects;[60] others were not. If the relationship is one of convertibility, then if A is B anything that is B is also A; but in a case of non-convertible predication, if A is B then anything that is B is not necessarily A.[61] The 'convertible' predicates were held to be of two kinds: *definition* or *property*. If the predicate was non-convertible it was called an *accident*. The definition itself was made up of words stating the *genus* and *differentia* (or species). We must also introduce here the notion of essence, even though it has, unfortunately, had a somewhat baneful influence in the history of philosophy. 'A definition', wrote Aristotle,[62] 'is a phrase indicating the essence of something.' So the five predicate types considered in the 'theory' were genus, *differentia* (species), definition (essence), property, and accident.

Let us try to clarify this further with the help of some examples. Like Plato, Arisotle thought it was the philosopher's (or 'scientist's') task to search for the correct definitions of things (or more strictly of concepts or universals). To find definitions, some kind of classificatory process was called for. One had to find the genus to which a thing belonged and also its *differentia*, or species criterion. Together, these should give the fundamental characteristic qualities or attributes that were necessary and sufficient for a thing to be a thing of the kind or class to which it belonged. In Aristotle's mind (or at least in the system he bequeathed to the world), the doctrine of essences presupposed that the world was neatly divided and packaged into discrete 'natural kinds' or classes, with no fuzziness round the edges, so to speak. And each classificatory pigeon-hole would have its appropriate linguistic definition.[63]

It seems very likely that Aristotle's theory of definition, and his doctrine of essentialism,[64] arose from his interest in and knowledge of natural history and biology. In the animal and vegetable world, there do indeed appear to be discrete 'natural kinds'.[65] And in mathematics (particularly geometry) the essentialist approach works well. A circle, for example, can be defined as a plane figure (the genus), with every point of the figure equidistant from a fixed point (the *differentia*). Putting the genus and *differentia* together, we have the definition of the essence of a circle, the definition being convertible with the subject, circle. Plane figure is a *genus*, with several *species*, such as square, ellipse, pentagon, and so on.

The way in which essential definitions might be discovered is a question to which we shall return shortly. But leaving this point aside for the moment, I should emphasise that it was the search for such definitions that was the hallmark of Aristotle's overall methodological enterprise. For if one could find the defining essence of something, then (Aristotle supposed) it should be possible to deduce the *properties* of that thing from its definition. Thus, he wrote:

> A *property* is something which does not show the essence of a thing but belongs to it alone and is predicated convertibly of it. For example, it is a property of man to be capable of learning grammar; for if a certain being is a man, he is capable of learning grammar, and if he is capable of learning grammar, he is a man.[66]

We have here the germs of a rather attractive — or at least a not wholly implausible — methodology of science. One might envisage the scientist seeking for the essential definitions of things and then *deducing* the properties of them, rather as one can deduce theorems from axioms in a deductive system. Thus a large range of properties of things might be deduced from a relatively small number of defining essences. So there would be a considerable economy of thought, and many properties (or, more loosely, phenomena) might be *explained* by deducing them from defining essences. Thus would 'scientific' knowledge of the world be produced.

Unfortunately, however, the system was fundamentally unsatisfactory. Aristotle acknowledged, indeed emphasised, that there were certain attributes of things — namely *accidents* – that could not be predicated convertibly of subjects. An accident, he said:

> is something which can belong and not belong to any one particular thing; for example, 'a sitting position' can belong or not belong to some one particular thing.[67]

One can readily think of other 'accidental' predicates that cannot be deduced from a defining essence. For example, Aristotle's essential definition of man was that he was a 'rational animal'. A man may be tall, short, black, white, kind, unkind, etc., and these attributes obviously cannot be deduced from the defining formula: 'Man is a rational animal'. So 'accidents' would have to go unaccounted for according to Aristotle's theory. Well, perhaps one should not expect too much and Aristotle himself was quite willing to accept limitations to the scope of his method. But can one, as Aristotle supposed, deduce the 'property', 'capable of

learning grammar', from the essential definition, because of the convertibility of the property predicate? Aristotle might have thought one could.
But this the modern logician would strenuously deny. The notion of
educability to specific tasks is not contained with the notion of rationality.
One might acknowledge that any beings capable of learning grammar
would have to be rational. But there don't have to be any grammar learners
among the class of rational beings, unless one defines this to be so. In which
case, the point Aristotle was trying to make would be lost. It seems,
therefore, that the whole methodological enterprise foundered at this
point. So, recognising that there appears to be a fundamental difficulty in
Aristotle's method, we proceed nevertheless with our exposition. If we
identify reasoning from defining essences as movement downward from
the apex of the 'arch of knowledge', one might expect that the discovery
of definitions would constitute movement on the ascending side. What
account does Aristotle give of this? There is, to be sure, a section in the
Topics that attends to the methods for confirming definitions,[68] but it gives
one little guidance towards their discovery. Much more helpful is a
passage in the *Posterior Analytics*, which reads as follows:

> We must set about our search by looking out for a group of things which are alike in the
> sense of being specifically indifferent, and asking what they all have in common; then we
> must do the same with another group in the same genus and belonging to the same
> species as one another but to a species different from that of the first group. When we have
> discovered in the case of this second group what its members have in common, and
> similarly in the case of all the other groups, we must consider again whether the common
> features which we have established have any feature which is common to them all, until
> we reach a single expression. This is the required definition.[69]

Several points are worth noting here. First, we see that the suggested
method is dependent on sense observation; and the mind is evidently
considered to be capable of recognising similarities between some sorts of
particular objects; it can notice and apprehend classes. Likewise, it can
apparently notice similarities between some classes, and hence it can
mentally create genera. Perhaps one could, thereafter, proceed further to
more general groupings, until eventually one reached the most general
class, containing, as it were, 'all being'.[70] However, Aristotle did not
proceed thus far in the *Posterior Analytics*, and in fact he was only
concerned — in the hunt for essential definitions — to 'rise' from individuals, through species, to genera — not to higher groupings such as
families, orders or classes.[71]

A second important point, as I see it, is that Aristotle had no satisfactory
means for judging which attributes were essential and which were not.
One could, for example, recognise chimps and humans as 'natural kinds';
but then, was the possession of hair, milk, red blood, five fingers, a back-
bone, or what, the feature that should define the essence of the genus to
which both the human and chimp supposedly belonged? Clearly, the
correct choice would be a matter of luck as much as logic, so far as
Aristotle's 'method' was concerned. And indeed the subject of biological
taxonomy has been one of endless dispute ever since his time, with
particular controversy between those who thought that any system would
do, no matter how artificial — provided that clear criteria could be given

for the organisms in their various pigeon-holes in the taxonomic hierarchy
— and those who believed that there *was* a natural system, which might be
found by methodical examination of the data, and that it was the
naturalist's task to find it.[72]

In fact, of course, the discovery of definitions cannot be a logical
process, if one starts from ground level (so to speak) by examining
individual objects. If it were so, we could, I dare say, close this book at this
point. There would be no need for an 'arch of knowledge' — just a pillar
perhaps. And the distinction between deduction and induction would be
obliterated. Most metascientific problems would conveniently evaporate.
But needless to say, things are not so simple as this.

Aristotle himself was probably aware of the difficulty. To be sure, he
produced an admirable codification of the rules of syllogistic inference in
the *Prior Analytics*. These rules were to be put to use in the *Posterior
Analytics* and in the scientific treatises, so that Aristotelian science could
then be shown 'in action', so to speak. But in a sense the *Posterior
Analytics* is a great disappointment — though this should cause us no great
surprise, given the consideration that turning induction into deduction is
a lost cause. So although there is much talk, in the *Posterior Analytics*, of
obtaining 'knowledge by demonstration'[73] one is left in the greatest doubt
as to how the 'first principles' of the various sciences — or even, more
simply, the major premises for syllogisms — are to be obtained. Finally,
after skirting round the issue for page after page, Aristotle comes up with
the following obscure passage, right at the end of the treatise:

> Thus sense-perception gives rise to memory, as we hold; and repeated memories of the
> same thing give rise to experience; because the memories, though numerically many,
> constitute a single experience. And experience, that is the universal when established as
> a whole in the soul — the One that corresponds to the Many, the unity that is identically
> present in them all — provides the starting-point of art and science: art in the world of
> process and science in the world of facts. Thus these faculties are neither innate as
> determinate and fully developed, nor derived from other developed faculties on a higher
> plane of knowledge;[74] they arise from sense-perception, just as, when a retreat has
> occurred in battle, if one man halts so does another, and then another, until the original
> position is restored. The soul is so constituted that it is capable of the same sort of
> process... As soon as one individual percept has 'come to a halt' in the soul, this is the first
> beginning of the presence there of a universal (because although it is the particular that
> we perceive, the act of perception involves the universal, e.g. 'man', not 'a man, Callias').
> Then other halts occur among these proximate universals, until the indivisible genera[75]
> or ultimate universals are established. E.g. a particular species of animal leads to the genus
> 'animal', and so on. Clearly then it must be by induction that we acquire knowledge of the
> primary premises because this is also the way in which general concepts are conveyed
> to us by sense-perception.[76]

This interesting passage is worthy of comment. Aristotle is, I think,
describing some kind of psychological process, using the analogy of a
battle-ground in order to try to convey his meaning to his readers. One can
imagine a number of soldiers running away from a battle. Then, in various
places a few of the bolder spirits determine to make a stand (or 'halt')
against the pursuing enemy. As they do so, turning towards the enemy,
they gather around them a few other troops and successfully begin to beat
off their attackers. Then the various groups begin further to regroup (as
they say in military communiqués), and larger coalitions are formed.

Eventually, the soldiers in the previously retreating army are back in formation once again and successfully combine together to defeat their opposition.

The process of concept formation — of getting to know universals — is somewhat analogous, Aristotle would have us believe. Individual entities are perceived by the senses, and some of them are recognised as similar. Accordingly, they are grouped together as 'species'.[77] 'Higher' groupings may readily be formed, until the 'highest' grouping of things — relevant to a particular science — is reached. Aristotle believed that the human mind was specially endowed with a mental faculty (*nous*) that made possible the apprehension of general mental concepts, or universals. In this he was, in a sense, right, though today one would probably maintain that humans acquired this faculty over the long ages of evolutionary history, rather than being 'just made' that way.

But even if we grant this, it would hardly enable an Aristotelian methodology of science to be feasible (even though classificatory procedures certainly play an important role in modern science). The thing is that the first principles of a science are not expressed simply in terms of universal generalisations: 'All *P*s are *Q*s.' But this was the form of generalised statement that Aristotle's syllogistic logic was able to handle. And from the discussion of the 'battle' psychology that we have considered above, we can see that Aristotle envisaged the approach to the first principles of a science — and also the search for essential definitions – as involving a classificatory procedure and the formulation of universally generalised statements. But this would be insufficient as a general process of scientific empirical inquiry, except perhaps for biological taxonomy, no matter how much *nous* an Arisotelian scientist might be endowed with. And in fact, if one examines Aristotle's specifically scientific works, such as the *Meteorologica*, one does not find the methodology implicit in the *Organon* actually used to much practical purpose. However, this does not detract from Aristotle's contribution to logic in the *Prior Analytics*, or to the art of dialectical discourse in the *Topics*. And by his emphasis on the collection of empirical information, Aristotle was closer to modern science than was Plato, searching for transcendent forms or Ideas.

There are certain other aspects of Aristotle's scientific methodology, epistemology and ontology that we may consider somewhat more briefly. In the foregoing discussion, we have placed greatest emphasis on the discovery of essences and first principles, and wondered exactly what the relationship was of one to the other, and how they might be discovered.[78] But in the *Posterior Analytics* Aristotle mentioned that the recognition of the middle term in a syllogism was the cognitive process that required particular skill:

> Quickness of wit is a sort of flair for hitting upon the middle term without a moment's hesitation. A man sees that the moon always has its bright side facing the sun, and immediately realizes the reason: that it is because the moon derives its brightness from the sun... In [this case,] perception of the extreme terms enables him to recognize the cause or middle term. *A* stands for 'bright side facing the sun', *B* for 'deriving brightness from the sun', and *C* for 'moon'. Then *B*, 'deriving brightness from the sun', applies to *C*, the moon', and *A* 'having its bright side facing the source of its brightness', applies to *B*. Thus *A* applies to *C* through *B*.[79]

The argument here is set up in a quasi-logical fashion, as if the understanding of the logical relationships were the main thing in explaining phenomena. But as can be seen from the passage quoted, it is really some kind of *causal* agent or relationship that is necessary for understanding or explaining the phenomena. And the logic of the situation could not *in itself* reveal the cause. This needed to be known in order to set up the formal structure of the explanation in syllogistic style. However, the search for causal middle terms was a highly significant part of Aristotle's ideas on methodology. This is shown in an important section of the *Posterior Analytics*[80] where two forms of syllogistic 'demonstration' are described. As usual, Aristotle did not give his examples particularly clearly, but they may be reconstructed as follows:

1. Heavenly bodies that are near do not twinkle
   The planets are near
   Therefore: The planets do not twinkle
2. Heavenly bodies that are near do not twinkle
   The planets do not twinkle
   Therefore: The planets are near

Given the premisses, the two conclusions seem to follow reasonably satisfactorily, even though, in the second case, the argument is not formally valid as it stands. Aristotle did not regard this as an overwhelming objection, however, since he apparently thought it was legitimate to 'convert' the major premiss into the sentence: 'Heavenly bodies that do not twinkle are near'. (Actually, as Aristotle himself stated at the beginning of his *Prior Analytics*, it is really only logically permissible to 'convert' the sentence 'All A is B' into 'Some B is A'. And there was no empirical basis for the conversion.)

It will be noted that in the first syllogism one argues from a cause (the planets are near) to an effect (the planets do not twinkle). In the second case the argument runs from an effect (the planets do not twinkle) to a cause (the planets are near). In actual fact, neither argument could have been stated without the premiss (heavenly bodies that are near do not twinkle), which Aristotle himself acknowledged had to be assumed as known by induction or sense-perception. Nevertheless, the thought that one might educe causes from effects, as in the second argument, seemed to hold out a grand methodological promise to subsequent mediaeval and renaissance commentators on Aristotle; and as we shall see it exercised their attention very greatly, though it was really more an unfortunate distraction than anything.

The first form of demonstration above was known in Greek as '*Apodeixis tou dioti*'; in Latin it was called '*Demonstratio quia*'; in English we may render it as 'Demonstration that'. The second was '*Apodeixis tou hoti*' or '*Demonstratio propter quid*', which we may call 'Demonstration on account of which'. The great hope for the Aristotelian commentators was that somehow, by a suitable combination of these two forms of argument, one might produce a method for doing science which would break out of the methodological circle and construct the 'arch of knowledge' on a rational and certain basis, without the use of hypothetical reasoning.

This was, it now appears, a forlorn hope, and one that was not in fact actually sanctioned by the texts of Aristotle's *Organon*, taken as a whole. The commentators, however, linked the section in the *Posterior Analytics* on the heavenly bodies, etc. with some remarks in the opening section of the *Physics*, where Aristotle said that things were to be explained by causes and first principles, but that these could only be known from phenomena — 'from what is more immediately cognizable and clear to us'.[81] This text from the *Physics* seemed to imply a two-way methodological pathway, both inductive and deductive, as the tradition of the 'arch of knowledge' would sanction. But the Aristotelians hoped that somehow — most likely by worrying at the suggestions provided by the text of the *Posterior Analytics* that we have just been considering — a double pathway could be found that would be deductive in *both* directions. But it was not to be. And as we shall see, the collapse of this methodological phantasm coincided with the rise of the new science and new developments in ideas about method — in Galileo's mathematical/ experimental work, and in Bacon's emphasis on the need for new data as the basis of a great inductive pyramid.

I have alluded above to the notion of causes and effects as if they were fairly unproblematical. But Arisotle's actual theory of causation was quite complex, and it may be helpful to give some brief account of it here. The modern meaning of cause is narrower than that of Aristotle. Today, we usually call some thing or process which makes something happen, or an event occur, a cause. But Aristotle thought that this was only part of the story. To explain the existence and nature of a thing, or why an event occurs, one needed — according to Aristotle — to specify four kinds of cause.[82] These were the *'material'* cause, the *'motive'* or *'efficient'* cause, the *'formal'* cause, and the *'final'* cause. For example, to explain why a statue is as it is, one would need to specify: the material out of which it was made (marble); the tools that had shaped it, and how they acted on the marble; the pattern, formula or (in modern terms) the blueprint for the statue; and the idea or intention in the mind of the sculptor — the sculptor's purpose in making the work of art and the sculptor's intention to complete it the way he or she wanted it to be.[83]

The material and motive causes are virtually self-explanatory, being close to the modern notion of cause. Comment on the final and formal causes may be helpful, however. Most scientists today try to get along without the 'teleological' or 'final' cause (though it is particularly difficult to do so in embryology). The idea of some future state or condition 'causing' present processes is generally regarded as metaphysically repugnant, though certainly amongst religious people the idea of human history being a kind of enactment of God's divine purpose or plan — an essentially teleological notion — is widely esteemed. Aristotle himself envisaged all processes as the actualisation of inherent potentials. Thus, the acorn has within itself the potentiality to become an oak tree. So, in a sense, the oak tree is a 'final cause', somehow determining the acorn's growth and development. The adult tree is 'the sake for which' the process of acorn development takes place. Thus Aristotle's biological work, particularly his treatise on the *Generation of Animals*,[84] is teleological through and through. It may be noted that this aspect of Aristotle's philosophy was to prove congenial to subsequent Christian theologians, for one can readily talk about the realisation of the potential within a man, which will not be achieved if one is deflected from the straight and narrow pathway by sinful behaviour.

The 'formal cause' can be related back to our previous discussion of definitions and essences. It is difficult to give a clear-cut account of Aristotle's notion of formal cause. But it seems to have had at least three aspects. First, it might refer to the defining formula of a thing — the specification of its essence in terms of genus and species. In this sense, it was related to Plato's doctrine of forms and Ideas, discussed previously. In Aristotle (and Plato also, for all practical purposes), the 'form' was a verbal definition.

But the form (or formal cause) could also, as mentioned above, be thought of as a kind of blueprint. In this sense, it was rather like a representation of the shape of a thing, or even the very shape itself — and thus it meshed quite closely with the modern concept of form. At this point, Aristotle's theory of formal cause linked up with his matter theory, according to which all objects supposedly consisted of an underlying featureless, qualityless substrate — the '*hyle*'. Various 'qualities' could supposedly be impressed on this, yielding the four elements (earth, water, air and fire), which could be united in various combinations to produce different kinds of substance.[85] Or, by a slightly different approach, one could regard any substance as a kind of union of '*hyle*' and impressed 'form'. But the 'form' did not have to be restricted to shape. It could be temperature, texture, colour, and so on.

The third sense of form was closer to the modern notion of a mathematical formula describing some phenomenon. For example, the Pythagoreans discovered that simple whole-number ratios were linked with the musical consonances.[86] Thus, the phenomena of music might be accounted for in terms of the mathematical relationships found in the string lengths of consonant intervals. Clearly, there is a link between our word 'formula' and the ancient Aristotelian doctrine of formal cause.

This is as far as we need go in our exposition of the two great 'meta-scientists' of the ancient world, Plato and Aristotle. Plato established the notion of an 'upward' movement to first principles and a 'downward' deductive descent therefrom. Aristotle's concern was to deny the objective and transcendent existence of forms (or Ideas). Nevertheless, he made considerable use of the concept of immanent forms. He also assisted greatly in giving a detailed account of a least one kind of deductive reasoning (the categorical syllogism, in its various forms) and made some attempt to give an account of the way in which the first principles of a science might be attained. In this, he made use of ideas based upon classificatory systems and the relationships between classes of things. But the overall account of the upward movement to first principles was, I fear, quite unsatisfactory. On the one hand, forms were immanent essences; and on the other they were expressed in terms of verbal definitions. There was a most unfortunate conflation of logical and ontological properties. Language and reality were supposed somehow to be in a state of correspondence, one with another,[87] so that physical features of the world could be elucidated by linguistic analysis. Aristotle did, however, lay emphasis on the need for empirical inquiry. 'Facts' were certainly needed as a basis for all scientific investigation — not mere philosophical discussion. And in his scientific treatises he provided the world with accounts of remarkably astute observations that have impressed scholars many centuries later.[88]

Stoics, Sceptics and Neo-Platonists: Method, mathematics, logic and metaphysics conflated

I have dealt with a small selection of the writings of Plato and Aristotle in some detail, for they furnished the basis of a great deal that was to follow. But we shall only be able to treat the very long period between Classical Greece and seventeenth-century Europe very sketchily, touching on a few matters that I judge to be of particular interest or importance, and selecting particularly those points that relate to the history of the 'arch of knowledge'.

We shall not seek to give any general account of the Stoic philosophers, who, along with the Platonists, the Aristotelians, and the 'atomists', formed one of the main branches of ancient philosophy. But it is important to mention here that one of the Stoics, Chrysippus (c.280–206BC), is credited with devising the so-called 'hypothetical syllogism'[89] which could serve as important instruments of reasoning, over and above the 'categorical syllogisms' that one finds in the *Prior Analytics*. These hypothetical syllogisms, particularly the ones called *'modus ponens'* and *'modus tollens'* (set out in note 89) have proved of the highest importance in the history of the philosophy of science. For example, as we shall see in our discussions of the work of Sir Karl Popper in Chapter 8, the mode of reasoning nicknamed *modus tollens* (literally the 'taking-away style') is of fundamental significance to those who believe that the touchstone of scientific method is the attempted falsification of hypotheses.

In the matter of categories, the Stoics used a set of four (*substrate, quality, relation* and *relation-in-a-certain-way*), rather than Aristotle's ten. But this difference need not detain us here. Some of the later Stoics, notably the celebrated physician Galen of Pergamon (c.129–199AD), made important contributions to discussion of scientific method, which we shall refer to below. Here it may be mentioned that Galen wrote a treatise on logic that followed the ideas set forth by Chrysippus.[90]

We have already alluded to the 'upward' movement to first principles, called *analysis*, and the 'downward' deductive descent from first principles, called *synthesis*. However, the use of these terms was rather fluid, probably as a result of developments within the literature of mathematics. Of particular importance were a couple of sentences interpolated into one of the manuscripts of the *Elements of Geometry* of Euclid (c.280BC):

> *Analysis* is an assumption of that which is sought as if it were admitted and the passage through its consequences to something admitted (to be) true.[91]
> *Synthesis* is an assumption of that which is admitted and the passage through its consequences to the finishing or attainment of what is sought.[92]

These two passages were themselves discussed by the mathematician Pappus of Alexandria (*c*.300–350AD) in his commentary on the *Elements*. Pappus wrote:

> *Analysis*...takes that which is sought as if it were admitted and passes from it through its successive consequences to something which is admitted as the result of synthesis: for in analysis we assume that which is sought as if it were (already) done..., and we inquire what it is from which this results, and again what is the antecedent cause of the latter, and so on, until by so retracing our steps we come upon something already known or belonging to the class of first principles, and such a method we call analysis as being solution backwards...
>
> But in *synthesis*, reversing the process, we take as already done that which was last arrived at in the analysis and, by arranging in their natural order as consequences what were before antecedents, and successively connecting them one with another,[93] we arrive finally at the construction of what was sought; and this we call synthesis.[94]

By way of illustration of these procedures, we may quote Pappus's own example.[95] See Figure 2.

Figure 2

D   A   C   B

Let *AB* be divided in extreme and mean ratio at *C*, *AC* being the greater segment [i.e., *AB/AC* = *AC/BC*]; and let *AD* = ½*AB*.
I say that (square on *CD*) = 5(sq. on *AD*).

(Analysis)
For, since (sq. on *CD*) = 5(sq. on *AD*),
and (sq. on *CD*) = (sq. on *CA*) + (sq. on *AD*) + 2(rectangle *CA,AD*),
therefore (sq. on *CA*) + 2(rect. *CA,AD*) = 4(sq. on *AD*).
But (rect. *BA,AC*) = 2(rect. *CA,AD*),
and (sq. on *CA*) = (rect. *AB,BC*).
Therefore (rect. *BA,AC*) + (rect. *AB,BC*) = 4(sq. on *AD*),
or (sq. on *AB*) = 4(sq. on *AD*):
and this is true, since *AD* = ½*AB*.

(Synthesis)
Since (sq. on *AB*) = 4(sq. on *AD*)
and (sq. on *AB*) = (rect. *BA,AC*) + (rect. *AB,BC*),
therefore 4(sq. on *AD*) = 2(rect. *DA,AC*) + (sq. on *AC*).
Adding to each the square on *AD*, we have (sq. on *CD*) = 5(sq. on *AD*).

From this example, we can see how the geometrician might seek to confirm a result obtained through 'analysis' by means of a result obtained by 'synthesis'. In a sense, then, the methodology employed was analogous to the upward and downward movement constituting Plato's 'arch of knowledge', the one movement confirming, strengthening or supporting the other. But, in the geometrical example we have just considered, it should be noted that the procedures in the analytic and synthetic phases of the reasoning process were both *deductive*; neither was inductive. So we cannot — according to the geometrical tradition laid down here by Pappus — directly identify induction with analysis, and deduction with synthesis. Yet something like this identification did occur in subsequent authors. There arose, therefore, a mighty confusion in the terminology of metascience, which persisted well into the nineteenth century.[96] There were, in fact, no less than three traditions that became conflated one with another: the *geometrical* analysis and synthesis that we have just considered; the *methodological* tradition of analysis and synthesis (induction and deduction, or resolution and composition); and also the *rhetorical* procedures of thinking up ideas and then presenting them to one's auditors in a clear and logically coherent manner. For this last, the terms 'invention' and 'judgment' were commonly employed. Discussions on 'method' commonly treated rhetorical, scientific/empirical, and mathematical procedures together, without clearly differentiating one from another. I shall not, however, attempt to give an account of the history of rhetorical method in this text.[97]

Another possible source of confusion lay in the titles of Aristotle's two main logical works. Roughly speaking, one may associate the *Prior Analytics* with deductive logic and the *Posterior Analytics* with induction and the discovery of definitions, essences, first principles, and so on. Nevertheless, both treatises were entitled ... *Analytics*, employing Aristotle's word for 'logic'.

Then there was the felt need to reconcile the methodological remark at the beginning of Aristotle's *Physics* and the section of the *Posterior Analytics* concerning the search for middle terms (referred to in notes 79 and 80). And yet another possible source of confusion lay in the fact that Aristotelian definitions of essences might also be arrived at 'downwards', so to speak, using the Platonic 'method of division' rather than 'upwards' as described in the *Posterior Analytics*. Altogether, there was ample scope for misunderstanding.

Aristotle did not himself give a nice tidy example of the process of 'division', applied in a search for some definition, but many years later the Neo-Platonic commentator, Porphyry (232/3–c.304 AD), made good the deficiency in a beautiful example that has come to be known as the 'Tree of Porphyry'. In this,[98] by 'division', one could reach the definition of the essence of 'man' (see Figure 3).

The dichotomous division[99] went as far as 'man', below which level only different individuals could be recognised, 'man' being presumed to be the lowest level in the taxonomic hierarchy — the *'infima species'*.[100] The 'division' having been thus accomplished, 'man' could be defined as 'rational, sensitive, animate, corporeal substance'; or man's defining essence could be given by saying that he was a 'rational animal'. The process of dichotomous division, thus represented, might appear to be a kind of 'logical' process; but whether Porphyry in fact started from the bottom and worked up, or the top and worked down, we cannot now tell. He may, in fact, have done both; for, as we have said, finding your way up a mountain naturally involves finding your way down at the same time. (Even more likely, he simply drew on the work of some other author whose writings are no longer extant!)

But whatever Porphyry's actual procedure may have been, we can see another possible reason for the conflation of analysis and synthesis. For it is the 'downward' process of 'division' that involves analysis, in the sense that a process of breaking a whole into its parts is involved. And also there is a kind of 'logic' to this undertaking. For one selects a criterion for the division of a class and divides according to whether the members of the class *do* or *do not* possess the chosen characteristic. Of course, this isn't strictly a process of logical inference, but it seems to be a certain and secure process, once the decisions have been taken as to the criteria that are to apply at each level of the hierarchy. We assume, for example, that a body either *is* or *is not* animate. There is to be no ambiguous entity that is half-way between living and non-living. Consequently, in the 'taxonomic' tradition 'analysis' was associated with a quasi-deductive process, yet also with an inductive upward ascent, such as that employed by Aristotle in the *Posterior Analytics*.[101]

Figure 3

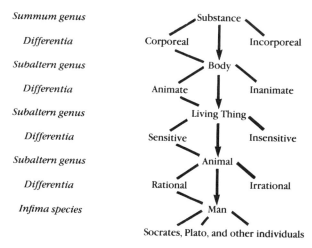

We have referred briefly above to the work of the physician, Galen, and it is worthy of note that there was a considerable input into theories of scientific method from the medical fraternity, of whom he was the leading personality of his day, continuing to exercise an influence for many hundreds of years. Amongst the physicians, there were those whom Galen referred to as belonging to the 'empiric' school and those who favoured the so-called 'dogmatic' approach.[103] The 'empirics' believed that theory was unnecessary in medicine. If a medicine worked, it should be used, even if its mode of action was quite unknown. The 'dogmatists', on the other hand, claimed that a physician should be possessed of some medical theory relevant to the problem in hand, and that appropriate medicines should be dispensed according to this theory. The disagreement between the two schools coincided very roughly with what was to become a long-lasting division in philosophy of science, namely that between empiricists and rationalists – between those who emphasised the value of a wide factual knowledge as a basis for scientific knowledge and those who emphasised the importance of theory and deductive reasoning. Putting the matter into the metaphorical terms that we have found congenial, the 'empirics' favoured the attainment of scientific knowledge by climbing the 'arch of knowledge' from the ground upwards; the 'dogmatists' thought it better to start at the top and make a steady deductive descent.

But Galen himself sensibly advocated the adoption of a middle way between the two extremes, or a combination of the two methods.[104] He also maintained that one should try to argue inductively from medical symptoms, or 'indications', to the causes of a disease. Then, knowing the causes, one should deduce which medicines would be appropriate and administer them accordingly in a 'rational' manner.[105] Galen, it seems, realised that an arch needs two legs if it is to stand firmly.

But in another passage Galen emphasised the importance of classification in the discovery of 'indemonstrable principles', whence by the method of division the 'absolute inquiry of everything' might be attained.[106] Evidently, the intellectual inheritance from Plato and Aristotle on matters epistemological and methodological was so confusing as to make it difficult for later writers — if they were to try to reconcile the statements of their leading predecessors — to produce wholly clear-cut and consistent accounts. Nevertheless, in Galen's idea of a double methodological pathway of analysis (resolution) *to* causes or first principles and synthesis (composition) *from* these in order to explain phenomena or deal with practical problems in a reasoned (rational) manner, there is perhaps a hint of what modern methodologists call hypothetico-deductivism,[107] though one must treat such a suggestion with caution for it is all too easy, with hindsight, to think one can see modern ideas in ancient writers. On the other hand, there seems to be little doubt that Galen clearly perceived the need for the search for knowledge to involve both rational and empirical elements.

Although the Aristotelian commentators and later Mediaeval methodologists did much to develop the work of Aristotle, there were some in antiquity who subjected his ideas to penetrating criticism. Such were the so-called 'sceptics', of the school founded by Pyrrho of Elis (c.360–270BC). Pyrrho taught that the human mind cannot penetrate to the inward essences of things, and that all knowledge is necessarily subject to some element of doubt. This position of sceptical doubt is sometimes referred to as Pyrrhonism, and it had a considerable influence on the history of Western thought, being one of the chief factors leading to Descartes' method of sceptical doubt which we discuss in Chapter 2.[108]

The physician, Sextus Empiricus (c.250AD), was an important disciple of Pyrrho, and provides us with our main source of information on the sceptical philosophy of antiquity. Sextus made what may seem to us to be the obvious criticism of the role of the syllogism in Aristotelian methodology, namely that in use it leads to a kind of circularity in reasoning.[109] To use Sextus's own example, we cannot say that *all* men are animals unless we know that Socrates is an animal. Consequently, the syllogism *per se* cannot help us one whit in our search for information about Socrates, and cannot in itself be the basis of scientific knowledge. The Aristotelians' point is, however, perhaps being lost in saying this. The significant feature is that, for the Aristotelian, natural kinds have essences; and the human mind is supposedly able to come to know them by means of its rationality. Thus in the celebrated 'scholastic' philosophy of St Thomas Aquinas (1225–1274), this was a significant component of the overall argument whereby he sought to forge a union of Aristotelian and Christian doctrine.[110] The human capacity to apprehend the essences of things was held to be a God-given gift to humanity, the only animal kind endowed with rationality. Incidentally, we may note here that Aquinas supposed that 'the human intellect knows by composition, division and reasoning',[111] a supposition that accorded well with the tradition of the 'arch of knowledge'.

We should also make here a few remarks about the doctrine of Neo-Platonism — which roughly speaking corresponds with the ideas of Plato, as refurbished and transmitted to the modern world by Plotinus (204–270AD). During the process of this transmission, the ideas of Plato and Plotinus became inextricably mixed up with Christian doctrines, for the fifth- or sixth-century Neo-Platonist Syrian scholar who wrote under the pseudonym of Dionysius the Areopagite was mistakenly thought to have been one of the disciples of St Paul (cf. *Acts*, Book 17, Verse 34). Thereby what were in fact Neo-Platonic ideas came to be thought of as part of the true Christian canon. Other important Neo-Platonic writers were Porphyry, Iamblichus (died *c*.330AD), Macrobius (born *c*.360AD), and Proclus (411–485).

We have seen in Plato's *Republic* that a hierarchy of intellectual states was envisaged, with a corresponding set of ontic levels; and the doctrine was skilfully illustrated with the aid of the metaphor of the line. However, the critic might well ask how the level of forms or Ideas could exist in some manner wholly independently of, and prior to, the world of material objects. In answer, we find Plato's doctrine restated by Plotinus in a considerably elaborated and emended version in a work entitled *Enneads*,[112] so called because each component book was divided into nine sections. Plotinus still supposed that the Ideas were transcendent to the material world, but they were to be construed as ideas in the mind of God. So God (also called by Plotinus the 'One') corresponded for the Neo-Platonists with the highest of Plato's forms — the 'Good'. Further, the 'One' was supposedly the source of all things — indeed of all 'being' in the universe. Just as the Sun, in a sense, gives 'being' to our planet through the radiation of heat, so also, according to the Neo-Platonists, did the 'One' give existence or being to the cosmos.[113] Thus the upper levels of the hierarchy of Ideas supposedly gave rise to the lower levels, in a process somewhat resembling the emanation or radiation of heat from the Sun. Consequently, a kind of 'causal' link was envisaged between the 'One' and all the lower levels of the hierarchy of Ideas. Material objects somehow acquired their material being by some mysterious emanatory process from on high — from the upper, non-material levels of the 'arch of knowledge'! In his *Celestial Hierarchy*, the pseudo-Dionysius described a hierarchy from God through the various ranks of angels, to man and beasts, right down to Beelzebub at the very nadir of spirituality.[114] So it was that Christianity and Neo-Platonism became conflated.

The Neo-Platonists and the Christians who came under the influence of Neo-Platonism via the writings of the pseudo-Dionysius thought that it was the task of men to try to raise themselves in the spiritual hierarchy by turning their minds towards the realm of intelligible ideas and away from concern with merely material objects. This required an appropriate cultivation of the intellect, and — as Plato himself would have approved — the study of mathematics and philosophy. It was for this reason, therefore, that in the writings of Proclus, for example, we find the study of mathematics exalted as an enterprise of a quasi-religious character.[115] Mathematics, the Neo-Platonist supposed, might indeed assist the philosopher to achieve the mystical or ecstatic union with God or the 'One' that was so earnestly sought. Thus, as with other writers in the long tradition of the 'arch of knowledge', there was an attempted securing of the apex by means of metaphysical arguments of a most fundamental kind.

The problem of universals

A good deal of the philosophical discussion that took place in the long interval between the collapse of ancient civilisation with the fall of the Roman Empire and the rise of the modern scientific movement was concerned with the question of the real existence — or otherwise — of 'universals'. Consider, for example, the taxonomic hierarchy of the animal kingdom. The species, man, belongs to a higher grouping of animals, the apes; these in turn belong to the primates, mammals, vertebrates, and to the set of animals as a whole. Any man that one may meet in the street is at one and the same time a man, an ape, a primate, and so on. So, to which universal category does he belong? If each object in the physical world has a corresponding Platonic (or Neo-Platonic) form or Idea, which is the appropriate one for a man: man, ape, mammal, or what? Or, if one is an Aristotelian rather than a Platonist, does this man that one encounters in the street have a single essence (as Aristotelian philosophy would seem to require) or simultaneously the essence of man, ape, mammal, etc.?

The problem was discussed quite early by the sixth-century scholar at the court of Theodoric the Goth, Boethius (480–524),[116] in a commentary on Porphyry's *Isagoge*.[117] Porphyry had wanted to know whether universals (general classes of things or qualities), as opposed to particulars (ie specific things to which proper nouns might appropriately be applied), could be said to have a real existence, or whether they were purely abstract entities. Boethius believed that universals were real features of sensible objects but were immaterial. He failed to answer Porphyry's question as to whether they might exist altogether independently of bodies. Leading spokesmen for the realist view, which attributed real existence to universals in a kind of Platonic sense, were Thomas Aquinas (1225–1274) and William of Champeaux (1068?–1122); the contrary position was powerfully stated by Peter Abelard (1079–1142) and by the 'nominalist', William of Ockham (c.1300–1349). The realists, who tended to dominate in the Middle Ages since their views were held to be theologically correct, believed that there were existing, real, essences or forms corresponding to each universal. This the nominalists simply denied, saying that universals were logically superfluous. Furthermore, humans could classify objects and qualities in any number of different ways. Objects or qualities are arbitrarily grouped (according to particular human interests, we would say today), and names are assigned to these groupings. There are no real essences, to which names correspond.

Although the nominalist account of the matter was stated so forcefully by Ockham in the fourteenth century,[118] it did not really prevail until the seventeenth century, which saw the final collapse of the Aristotelian system. It was John Locke (1632–1704) who then expounded the nominalist position in the most cogent manner, and, as we shall see in the next chapter, the philosophical shift from essentialism (and the belief in the real existence of universals in some form or other) to nominalism was associated with a number of important changes in scientific theory and practice. However, it should be emphasised that the nominalist stance was by no means Locke's particular invention. Nominalism was a significant component of the mediaeval philosophical tradition.[119]

Some mediaeval methodologists: Grosseteste, Duns Scotus and Ockham

Let us turn from this brief consideration of the ontological status of universals to look at the mediaeval developments of Aristotle's ideas on methods for the acquisition of scientific knowledge. That is, we are considering, once again, the methodology of science, though this can never be entirely divorced from epistemology and ontology. There are a great number of mediaeval writings that one might consider here as exemplars of the methodological writings of the period, but for reasons of space it will not be possible to embark on a full historical account. It must suffice to consider a particular example of special importance, namely the work of Robert Grosseteste (1168–1253), Bishop of Lincoln, a task that is greatly facilitated by the authoritative study of his work that was published some years ago by A.C.Crombie.[120]

By Grosseteste's (or Greathead's) day, most of the more important texts of Aristotle were available in Latin translation, and (as had previously occurred amongst the Hellenistic and Arabic scholars) there was a good deal of activity afoot in the preparation of exegetic commentaries on the master's texts. These commentaries were generally expository, but sometimes contained useful clarifications, or minor emendations and improvements upon the original texts. In the case of Grosseteste, Crombie claims that there occurred — within the mediaeval scholastic tradition, and in the context of a commentary on the *Posterior Analytics* – a distinct advance towards what is today called hypothetico-deductivism; that is, the view that scientific inquiry proceeds by the formulation of hypotheses and the experimental testing of their deductive consequences — the testing of the facts that would be true if the hypotheses were true. However, as I have said before, one must beware of reading modern conceptions into ancient texts.

In his commentary on the *Posterior Analytics*,[121] Grosseteste said that one may arrive at the definition of a thing by the method of division; this is the method of *composition*. But one may also arrive at the definition of something by starting with a group of things, subdividing it in some 'natural' way, and then looking at what the two groups have in common in order to determine the characteristics of the 'genus' (as opposed to the 'species' in the two subdivisions). This, said Grosseteste, is Aristotle's method of *resolution* used for the purpose of finding definitions. An actual example, offered by Grosseteste, consisted in the search for the 'common nature' of the colours associated with (for example) rainbows, prisms or iridescent feathers. The common nature was in fact given in the form of an hypothesis, namely that the colours were produced by a 'weakening' of the white light.

Thus, given the problem of finding a suitable 'definition' for 'colour of the spectrum', Grosseteste said that it was 'colours produced by weakening of white light'. Supposedly, then, the definition — thus found by *resolution* — could be used for the purpose of *demonstrating* the phenomena of colours.[122] Thus we see here that the method of analysis and synthesis was beginning to take on the appearance of hypothetico-deductivism, though it was doing so within the context of the Aristotelian framework of classification and definition hunting.

The location of the inquiry within the context of a taxonomic procedure appears more clearly in a second of Grosseteste's examples, which involved a search for the cause of some animals having horns while others were without. To do this, the following taxonomy was drawn up (actually in words, not by means of a diagram) as in Figure 4.

Figure 4

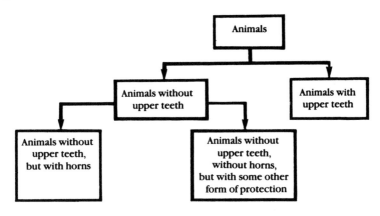

From this, Grosseteste was able to give the 'formal cause' of 'having horns' as: 'not having teeth in the upper mandible in those animals to which Nature does not give other means of preservation in place of horns'.[123]

But this statement of the 'formal cause' was insufficient for Grosseteste; he also wanted to know the 'efficient cause' responsible for the regularity revealed in this formal definition. He observed that there seemed to be a one-to-one correspondence between animals having only one row of teeth and more than one stomach. This, he suggested, arose because extra mastication in a second stomach was required when some teeth were absent. Further, he suggested that the material for the teeth might (in cases where teeth were lacking) go to make up the horns. Where the animals had alternative means of protection there might be no horns, but, as in the case of the camel,[124] there would be a hard horny pad in the upper jaw corresponding to the missing teeth.

Thus, by classifying animals according to certain observed regularities, Grosseteste could state certain 'formal causes', in the Aristotelian sense. But there was, it is clear, a logical gap between this and the proposed explanatory hypotheses — the 'efficient causes'. Grosseteste, however, was apparently well aware of this difficulty and suggested that the causes proposed with the aid of the 'method of resolution' should be subjected to observational testing — by looking at further cases to see whether the generalisation held good. By this means, false hypotheses might be eliminated.

Crombie has seen the whole procedure in a very modern light:

> He [Grosseteste] made deductions from rival theories, rejected those theories which contradicted either the facts of experience, and used those theories which were verified by experience to explain further phenomena.[125]

We should note that some of the examples that Grosseteste selected for discussion are to be found in Aristotle's *Posterior Analytics*. For example, we find the 'animal-horn' case;[126] but Aristotle only touched on the 'taxonomic' aspect of the problem and did not proceed further to propose possible hypothetical efficient causes to account for the observed regularities in the animal kingdom. Thus, in this instance we do seem to have a significant development of the original argument advanced by Aristotle. However, as we shall see below, some commentators since Crombie have been less inclined to see modern hypothetico-deductivism in the Mediaeval and Renaissance commentators on Aristotle.

We should notice here two further important points relating to 'method' that were developed in the fourteenth century, and can perhaps be regarded as precursors[127] of important later work. I refer to the 'method of agreement' and the 'method of difference' (as they were later called by John Stuart Mill[128]), which may be found in the writings of the Franciscan priest, John Duns Scotus (c.1265–1308), and William of Ockham respectively. Suppose one is trying to account for some particular effect. Then one may examine the accompanying circumstances each time the effect occurs, to see whether there is some common factor always present in these circumstances whenever the effect occurs. If such a common factor can be found, then this *may* be the cause of the effect.[129] The method is uncertain, of course, and Duns Scotus was apparently aware of this, and was not claiming more than that there was a natural propensity for a particular cause to be followed by a particular effect. However, he did seem to think that man possessed as an 'innate idea'[130] the principle that 'Whatever occurs in a great many instances by a cause that is not free,'[131] is the natural effect of that cause'.[132]

Ockham's contribution was to suggest that 'aptitudinal unions' (or natural propensities) may be displayed by looking at the circumstances accompanying an effect, to see if there may be some factor particularly associated with the effect. This is to say, the effect occurs if the accompanying factor is present but does not occur if it is absent.[133] However, like Duns Scotus, Ockham claimed no more than that his method could reveal 'aptitudinal unions' or propensities — not necessary connections.

The Paduan School

The gradual movement towards a hypothetico-deductive methodology of scientific inquiry was seemingly carried somewhat further in the writings of the so-called School of Padua, best known for its contributions to anatomy and medicine, but also important in the history of metascience. The Paduan School grew from the 'Averroist' movement of thirteenth-century Paris,[134] which was anti-clerical, and generally critical of St Thomas Aquinas's attempted synthesis of Christian and Aristotelian doctrines. Averroes (1126–1198) was perhaps the chief Arab commentator on Aristotle.

Amongst the Paduans, one can trace a reasonably direct line of writings on methodology of inquiry that culminated in the work of Galileo. However, their views were chiefly derived from Aristotle and his various ancient and mediaeval commentators. As behoved a medical school, there was also an admixture of ideas from Galen.

The Paduans, then, formed an important link between the ancient writers on method and the subsequent successful work of seventeenth-century natural philosophers such as Galileo. The Paduans sought to ring the changes on the old notions of resolution and composition thereby hoping somehow to produce a method for generating *certain* knowledge of empirical matters by a judicious combination of resolution and composition. In this they were not successful, as the subsequent history of metascience would lead us to expect; but this does not detract from their historical importance and interest. None of the Paduan methodologists can by regarded as a philosopher of outstanding achievement or renown. We shall, therefore, just refer briefly to a few of the persons concerned, and seek to indicate how their writings provided an important connection between the world of the 'ancients' and the 'moderns'.

In 1310, commenting on the *Posterior Analytics*,[135] Pietro d'Abano (1257–c.1315) distinguished between what he called 'demonstration *propter quid* [on account of which]' ('*doctrina compositiva*' or synthesis), and 'demonstration *quia* [that]' ('*doctrina resolutiva*' or analysis). Here we have the idea pursued that there might be a double way, the two components of which in combination would yield certain and demonstrative knowledge. As we saw above,[136] this prospect seemed to be indicated by a section of the *Posterior Analytics*, viewed in combination with a passage from the *Physics*. But it was a methodological mirage, and Pietro was unable to produce a solution to the problem in the way that was hoped.

Then drawing on the methodological tradition in medicine, Jacopo da Forli, (c.1360–1413) in a commentary on Galen's *Tegni* or *Ars Medica*, suggested that in trying to deal with fevers, the methodologically minded physician should first seek to 'resolve' the fever into its causes, and eventually determine the specific and distinct cause of the medical problem. Then, one should turn the procedure round, so to speak, and seek to infer the effects from their causes.[137] Clearly, we have here once again the 'arch of knowledge', this time in a medical context.

Another teacher of medicine at Padua, Hugh of Sienna (1376–1439), took a similar view to da Forli, saying:

> [T]he process of discovery in the case of demonstration through causes is resolutive, while that of setting forth the causes is compositive.[138]

This merely restated the traditional position, and does not eliminate the criticism that the 'double-pathway' method involved a circularity of argument. Yet the Paduans sought to deny this most strenuously. Paul of Venice (c.1370–1429) maintained that to know effects, and to know them causally, was not one and the same thing.[139]

A somewhat similar point of view was expressed by Agostino Nifo (1473–1538), in his commentary on Aristotle's *Physics* (1506), who stated that in order to explain phenomena they had to be examined by the intellect, and a cause was to be found that might serve as a middle term in a syllogism.[140] But for Nifo, the process of examination (or *'negotiatio'*) involved both composition and division. Somehow, it was to be this deployment of a *double* pathway that would prevent the whole methodological process from becoming hopelessly circular. It is uncertain exactly what Nifo had in mind, for he gave no example. But possibly (suggests Jardine[141]) he envisaged the used of Platonic division, as proposed by Pietro Pomponazzi (1462–1525) in relation to the discovery of the cause of an eclipse:

> There is an obstruction between the sun and the moon. It is either a body or a non-body. Not a non-body, because then it could not be an obstruction. It is either a transparent body or an opaque body. Not a transparent body, because then it could not obstruct. ... So it is [the] earth. And thus having acquired a more intense knowledge of that cause, I return to the effect [by demonstration *propter quid*].[142]

Yet the reader will very likely object that even with the use of division in this way, the question-begging of the procedure has not been eliminated: the argument is still circular, or ultimately dependent on an unsecured hypotheses. Interestingly, it appears that Nifo himself came to believe that the method, as an infallible means for the discovery of causes, was hopelessly circular, and he subsequently gave up the idea of '*negotiatio*' altogether, supposing that 'the discovery of the cause is syllogised only *coniecturale*'.[143]

It was chiefly on the basis of this passage that Randall was led to suppose that Nifo was recommending the use of hypothetical syllogism. That is, causes were to be proposed as hypotheses; in which case Nifo would have been well on the way to being an exponent of hypothetico-deductivism. This interpretation has, however, been queried by Jardine, who claims that the expression '*syllogismus coniecturalis*' referred not to conjectures or hypotheses, but was used by Nifo in the sense in which it had previously been used by Cicero — to refer to conclusive arguments drawn from true premisses. If this interpretation is correct, the claim that Nifo was anything like a modern hypothetico-deductivist may look shaky. On the other hand, one might say that in suggesting the Earth as the opaque body causing the eclipse a hypothesis has indeed been formulated, no matter what literary tradition Nifo may seem to belong to.

The culmination of the Paduan methodological writings appeared in the work of Giacomo Zabarella (1533–1589). He maintained that there were — despite the tangles and complexities of the methodological tradition — only two basic components to method: 'demonstration *quia*' and 'demonstration *propter quid*' (that is resolution and composition, or analysis and synthesis).[144] Again, we have reference back to the favourite passage in the *Posterior Analytics*. For Zabarella, however, resolution had two subdivisions: the attempt to reveal hidden causes (demonstration *a signo*) and induction. The account of induction was by no means free from difficulties. It appears that, in keeping with the Aristotelian tradition, it relied on an assumed capacity of the human mind to apprehend the essential natures of things — this time guided in the task by no less than the Holy Ghost. The essential natures, once discovered, could then serve as the premisses in the syllogisms of 'demonstration *quia*' and 'demonstration *propter quid*'.[145] In fact, however, when Zabarella gave an example of his method in operation — the discovery of the reasons for the generation and corruption of things — he turned directly to Aristotelian theory, coming up with the Aristotelian chestnut that matter is 'nothing in actuality but everything in potentiality'.[146]

So, despite all the years of discussion on method, nothing very much had been achieved by the methodologists in the way of providing new and worthwhile suggestions for the conduct of inquiry in the experimental way. Indeed, the whole exercise was chiefly concerned with the exegesis of Aristotle's texts, and those of later commentators. It was not based on any new kind of practice, nor did it suggest new ideas for such practice. There were, perhaps, some intimations of hypothetico-deductivism amongst the Paduans, but they were scarcely as well developed as those of Grosseteste in the Middle Ages. The Paduans' search for truth, certain demonstration, and indubitable first principles was incompatible with the free use of hypotheses, the mathematical analysis of phenomena, and the experimental testing that characterise scientific practice as we understand it. It should be emphasised, however, that even today there is considerable disagreement as to the nature of 'scientific method', and as we shall see in later chapters there are now some commentators who reject the suggestion that it is possible to give any coherent account of method. Certainly it is no longer thought, as was formerly quite often the case, that hypothetico-deductivism is the be-all-and-end-all of scientific method.

So whether the methodological writings of the Paduan School did or did not contain the germs of hypothetico-deductivism is not, perhaps, a matter to which we should devote too much attention. One thing is fairly certain, however. Within the texts of the sixteenth-century Paduan methodologists there were hints which Galileo — who with some justification can be regarded as one of the chief founders of 'modern' scientific method — found useful. When Galileo arrived in Padua in 1592, the University was still actively debating Zabarella's methodological pronouncements, and Galileo's own investigations were based on a methodology that had at least some of its roots in Aristotle's *Organon*. But by the time we reach Galileo, we are well and truly out of the ancient world. It is, therefore, time to take a pause, and reopen the discussion in a new chapter, where we consider some of the leading features of seventeenth-century metascience and metaphysics. This must suffice for our consideration of the ancient traditions.

Notes

1   For general surveys of Greek science, see, for example: M Clagett *Greek Science in Antiquity* Abelard-Schuman London 1957; B Farrington *Greek Science* Penguin Harmondsworth 1961; S Sambursky *The Physical World of the Greeks* Routledge & Kegan Paul London 1956; G E R Lloyd *Early Greek Science: Thales to Aristotle* Chatto & Windus London 1970; G E R Lloyd *Greek Science after Aristotle* Chatto & Windus London 1973

2   On the history of Greek education, see H I Marrou *A History of Education in Antiquity* Sheed & Ward London 1956

3   A good edition is *The Republic of Plato* (trans A Bloom) Basic Books New York & London 1968

4   *Plato in Twelve Volumes Laws* ... with an English translation by R G Bury 2 vols Harvard University Press Cambridge (Mass) and Heinemann London 1967–68

5   For the meaning of this troublesome word, 'dialectic', see p12

6   *Republic* 531b (Plato *op cit* [note 3] p210) (The pleasant-sounding musical intervals are produced by strings whose lengths bear simple ratios, one to another. For example: octave, 2:1; fifth, 3:2; fourth, 4:3, etc. The Pythagoreans had discovered these relationships and had sought to build a picture of the cosmos according to a musical model.)

7   *Republic* 529–530 (*ibid* pp208–9)

8   Thales supposed that the underlying substratum was water; Anaximander proposed an ineffable, indefinite entity called the 'boundless' (*apeiron*); and Anaximenes supposed that all things were different manifestations of air.

9   A good starting point for those who may wish to pursue the matter is W D Ross *Plato's Theory of Ideas* Clarendon Press Oxford 1951

10 The size of the literature on Plato is revealed by the fact that a recent bibliography for a fifteen-year period runs to 592 pages! (R D McKirhan *Plato and Socrates: A Comprehensive Bibliography, 1958–1973* Garland New York & London 1978)

11 Plato's texts were written in the form of discussions, usually with his master Socrates as one of the interlocutors. This literary form makes it very difficult to unravel Plato's *own* opinions.

12 We shall write 'idea' for a transitory thought occurring 'in our head', and 'Idea' for an objectively existent, and everlasting, form, as envisaged by Plato.

13 Immanent = inherent or 'indwelling'

14 Transcendent = existing apart from (and presumably excelling) the world, in a 'different plane' of existence. Some versions of theism represent God as transcendent with respect to the world; others represent God as immanent.

15 Plato *Timaeus* 51d (*Plato in Twelve Volumes IX Timaeus Critias Cleitophon Menexenus Epistles* with an English translation by R G Bury... Harvard University Press Cambridge [Mass] and Heinemann London 1975 p121)

16 Parallel lines may *appear* to converge; but in fact they don't. A coin may appear elliptical when viewed from a certain angle; but in fact it is round.

17 Plato *Symposium* 210a–211c (*Plato in Twelve Volumes III Lysis Symposium Gorgias* with an English translation by W R M Lamb... Harvard University Press Cambridge [Mass] and Heinemann London 1967 pp201–7)

18 Plato *Meno* 82b–85b (*Plato... Laches Protagoras Meno Euthydemus* with an English translation by W R M Lamb... Harvard University Press Cambridge [Mass] and Heinemann, London 1962 pp305–19)

19 Latin *educere* = to draw out

20 As pointed out in the Preface (p4), we take our starting point here, following the argument presented by C M Turbayne.

21 *Republic* 509d–511e (Plato *op cit* [note 3] pp190–2)

22 A number of different English terms are introduced on the 'epistemological' side in order to try to capture the several possible shades of meaning of the original Greek words.

23 Euclid's codification of geometrical knowledge was still to come in Plato's day; but the general Greek method of geometrical proof was already well established.

24 This view has not always commanded assent. As we shall see (p154), John Stuart Mill regarded mathematical results as inductive generalisations from experience. Mill's view on this question has been resurrected by representatives of the sociology of knowledge school (see p348) such as David Bloor. (D Bloor *Knowledge and Social Imagery* Routledge & Kegan Paul London 1976)

25 The word 'dialectic' has several meanings. One of them is explained here; but it can also mean simply 'logic'. The term as used by Hegel and Marx refers to a historical process in which one condition or state naturally gives way to its opposite, and from the unification of these opposites a third state may be created — which will again generate its opposite ... and so on. In Hegel, the dialectical history of the so-called 'Absolute Idea' is supposedly traced out in human history. In Marx, it is means of production, and thence economic and social systems, that supposedly follow a 'materialist' dialectical evolutionary process.

26 Note that I am forced to used the word 'concept' here, even though I said above (p7) that this word did not give an adequate interpretation of the meaning of Plato's term 'Idea', which had a strong ontological as well as epistemological component.

27 The term is mine, not Plato's!

28 Of course, by the same token, Popper has not taken kindly to Plato, See K R Popper *The Open Society and its Enemies* 2 vols Routledge London 1945
29 See below, pp59–66.
30 Strictly, this description is not exact, as will become apparent in Chapter 2. One way of envisaging Bacon's methodology is as a series of 'mini-arches'.

Figure 5

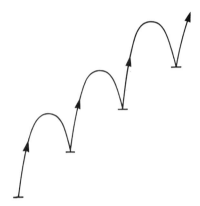

If this picture is correct, then the problem of getting up the 'left-hand side' remains (at the theoretical level, for the metascientist), even if the distance to be climbed at each step is very small. But for more on this see Chapters 2 and 4.

31 See pp114–16.
32 Aristotle *Topica* 105a (Aristotle *Posterior Analytics* by Hugh Tredennick... *Topica* by E S Forster... Harvard University Press Cambridge [Mass] and Heinemann London 1960 p303)
33 It is, for example, fallacious to argue from the statement that *some* princesses are beautiful to the conclusion that *all* princesses have beauty, even though the conclusion *might* be true as a matter of fact (that is, contingently so) at some particular point in time.
34 Or, at least, one of the most painfully irritating ones.
35 It may be objected here that we seem to be conflating a whole cluster of concepts. Deduction seems clear enough — though why it should be called *synthesis* is perhaps obscure. But hypothesis formulation, induction, and analysis seem to be very dissimilar. We are, indeed, forced to agree. But it is true, nonetheless, that they have frequently been conflated in the past. And there *is* some overlap, or warrant for an overlap. Perhaps the difficulty resides in over-indulgence in the 'arch' metaphor, in which the acquisition of knowledge is pictured as a two-way process: upward and downward. Finally, the role of 'dialectic' is ambiguous in Plato's system, just as it is in modern usage. On the one hand, it appears as the 'coping-stone' of the arch; and on the other it appears as the means whereby the arch may be climbed — the forms or Ideas discovered or guaranteed.
36 This identification is to be found, for example, in the writings of the Neo-Platonic commentator, Proclus (410?–485AD), who wrote: '[T]here are...certain most excellent methods delivered, one which reduces the thing sought, by resolution [that is, analysis] to its explored principle, which, as they say, Plato delivered [communicated], to Leodamus, and from which he is reported to have been the inventor of many things in geometry...'. ([T Taylor trans] *The Philosophical and Mathematical Commentaries of Proclus, on the First Book of Euclid's Elements...* 2 vols printed for the author London 1792 Vol 2 p25)
37 As we shall see in Chapter 2, they sometimes became confused or 'crossed'.
38 This was true, even for the sceptics of the ancient world. See note 109, below.
39 There are, of course, many forms of deduction besides that involved in the logic of classes that Aristotle studied. See, for example, note 89 below. In twentieth-century philosophy, particularly since they later work of Wittgenstein, there has been considerable debate as to whether deduction is a self-evident procedure, which holds universally, or whether it depends upon particular cultural situations, and social interests.
40 Plato offered an example of this in his dialogue, the Sophist. Somewhat playfully, he 'deduced' a definition for the term 'angling'. The method should be self-explanatory from the summary of the process shown in Figure 6.

Figure 6

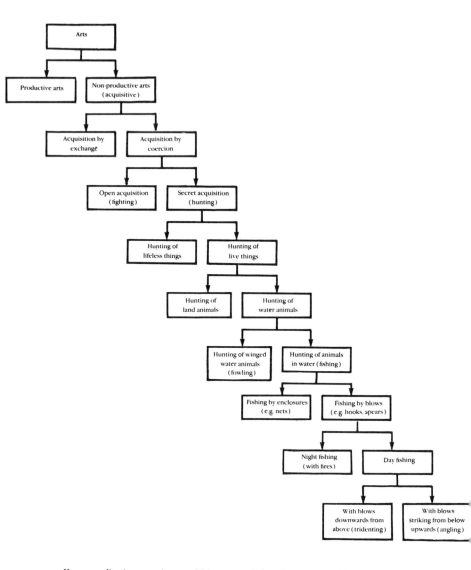

Hence, angling is a coercive, acquisitive art, carried out in secret, in which live animals living in water are hunted during the day by blows that strike upwards from below! Plato *Sophist* 219a–221a (*Plato in Twelve Volumes VII Theaetetus Sophist* with an English translation by Harold North Fowler Harvard University Press Cambridge [Mass] and Heinemann London 1967 pp273–81)

41   We may mention here that Plato seems to have envisaged a kind of hierarchy of forms, so that one would be, as it were, 'subordinate' to another within the realm of Ideas. In the *Republic* the 'highest' Idea was that of the 'good', which, if apprehended, would serve as a general guide for apprehending moral notions. In the *Parmenides*, Plato may have been seeking to reduce mathematics ultimately to the Idea of the 'one' as the leading notion in the hierarchy of mathematical concepts, (See F M Cornford 'Mathematics and Dialectics in the *Republic* VI–VII' *Mind* 1932 Vol 41 pp37–52 & 173–90 [at p179].) In Aristotle, as we shall see, there was supposedly a separate set of first principles for each science.

42 For introductory texts on Aristotle, see, for example: A E Taylor *Aristotle* Jack London & Edinburgh 1912; W D Ross *Aristotle* Methuen London 1923; J H Randall Jr *Aristotle* Columbia University Press New York 1960; G E R Lloyd *Aristotle: Structure of his Thought* Cambridge University Press Cambridge 1968

43 But some commentators have seen a persistent strain of Platonism, and even Neo-Platonism (see p31), in Aristotle's thought. See, for example, L Merlan *From Platonism to Neo-Platonism* Nijhoff The Hague 1953

44 Of special importance were the *Physics, On Coming-to-Be and Passing Away*, the *Generation of Animals*, the *Parts of Animals*, the *Movement and Progression of Animals*, the *History of Animals*, the *Meteorologica*, *On the Soul*, *On the Heavens*, and a number of minor works.

45 Greek *organon* = 'instrument' (of thought; that is, the acquisition and deployment of knowledge). By 'analytic', Aristotle meant (roughly) 'logic'.

46 Authoritative translations of these are: *Aristotle...The Categories On Interpretation* by Harold P Cooke.. *Prior Analytics* by Hugh Tredennick... Harvard University Press Cambridge (Mass) and Heinemann London 1973; *Aristotle Posterior Analytics* by Hugh Tredennick.. Harvard University Press Cambridge (Mass) and Heinemann London 1960; *Aristotle Sophistical Refutations Coming-to-Be and Passing-Away* by E S Forster... *On the Cosmos* by D J Furley... Harvard University Press Cambridge (Mass) and Heinemann London 1955.

47 *Aristotle The Physics* with an English translation by Philip H Wicksteed... and Francis M Cornford... 2 vols Harvard University Press Cambridge (Mass) and Heinemann London 1929–34; *Aristotle the Metaphysics* with an English translation by Hugh Tredennick...2 vols Heinemann London and Putnam New York 1933–35

48 See below, pp127-32.

49 See Cooke *op cit* (note 46) p2, quoting Theodor Comperz. But the doctrine of categories was already present in embryo in Plato's dialogue, *Theaetetus*.

50 *Categories* 1b–2a (Aristotle *op cit* [note 46] pp17 & 19) (emphases added to indicate the standard list of Aristotelian 'categories')

51 This practice was frequently employed by Aristotle. For example in the *Physics*, from the fact that linguistic predicates must have subjects to which they can meaningfully refer in a proposition, Aristotle supposed (by analogy) that there must be an underlying material substrate in the cosmos – the so-called *hyle* — which provides a subject for material (for example, chemical) change. Thus the philosophical activity of language analysis was used as a basis for physical hypotheses. (See E McMullen 'Matter as Principle' in E McMullen ed *The Concept of Matter in Greek and Medieval Philosophy* Notre Dame University Press Notre Dame 1963 pp173–212)

52 Strictly, for Aristotle, science here would mean a particular branch of philosophy or of knowledge, and would include, for example, politics as well as physics. But we would not, I think, find it strange to talk about the several sciences having their own separate first principles.

53 Greek *arche* (pl. *archai*) = first principle. The word has many derivatives in English; for example, archetype, architectonic, architecture, arch-bishop (or arch-rogue), and even just 'arch'. This last reminds us of the extended metaphor that we are finding continually useful in this first part of our text.

54 For a reliable text on introductory logic, see, for example, M R Cohen & E Nagel *An Introduction to Logic and Scientific Method* Routledge & Kegan Paul London 1934

55 Aristotle did not, however, deal with the logic of relations, which involves arguments such as: If *A* is greater than *B*, and *B* is greater than *C*, then *A* is greater than *C*.

56 Plato, we may recall, did not mention the syllogism, but only the so-called 'method of division', as described in note 40.

57 The *Topics* is a treatise, the greater part of which is concerned with showing the reader how to confute an opponent in a dialectical argument.

58 Actually, Aristotle's preferred simile seems to have been that of a race-course, rather than an arch. In his *Nichomachean Ethics* (1095a–1095b), he wrote: '[T]here is a difference between arguments from and those to the first principles. For Plato, too, was right in raising this question and asking, as he used to do, 'are we on the way from or to the first principles?' There is a difference, as there is in a race-course between the course from the judges to the turning-point and the way back. For while we must begin with what is known, things are objects of knowledge in two senses – some to us, some without qualification. Presumably, then, we must begin with things known to us.' (R McKeon ed *The Basic Works of Aristotle* Random House New York 1941 p937)

59 But, we may remind the reader there were supposedly ten kinds of predicates (categories).

60 *Topics* 102a & 103b (Aristotle *op cit* [note 46] pp283 & 293)

61 For example, if a thing is a man it is also a 'featherless biped'; and if it is a 'featherless biped' it is also a man. By contrast, if a thing is a man it is hairy; but if it is hairy it is not necessarily a man. (In deference to feminist susceptibilities, for 'man' in the preceding example please read 'man or woman'.)

62 *Topics* 101b (Aristotle *op cit* [note 46] p281)

63 Aristotle, needless to say, displayed no recognition of the fact that different languages are often based on quite different taxonomic (classificatory) boundaries, according to the specific needs and interests of the speakers of those languages. There are, unfortunately, no universal 'natural kinds'.

64   This term 'essentialism' is due to Karl Popper. It means adherence to and belief in the efficacy of, for philosophical analysis, the doctrine of essences.

65   How they arose is, needless to say, the topic of Darwin's *Origin of Species*. The biological kinds do in fact change — but slowly; so Aristotle gave no thought to the evolutionary process that Darwin envisaged. For Aristotle, kinds were fixed and distinct one from another. They seemed, therefore, to give a good warrant for his essentialist philosophy.

66   *Topics* 102a (Aristotle *op cit* [note 46] p283)

67   *Topics* 102b (*ibid* p285)

68   *Topics* 153a (*ibid* p657)

69   *Posterior Analytics* 97b (*ibid* p237)

70   See p29, in relation to the 'Tree of Porphyry'

71   These are modern, biological terms and do not appear in Aristotle's own text. (See note 75 below)

72   This history of biological taxonomy was, at least until the time of Darwin, deeply influenced by Aristotelian metascience. In philosophy, as we shall see, the controversy ran in the direction of considering whether or not taxonomic groupings were 'real' or 'nominal'. (See p32)

73   For example, in *Posterior Analytics* 71b (Aristotle *op cit* [note 46] p31)

74   The spatial imagery here is worthy of note.

75   Aristotle had no word, such as family or order, for the upper levels of a taxonomic hierarchy. Every taxon (classificatory group) was represented as a genus with respect to the level below and a species with respect to the level above. Here, since he is apparently talking about the very highest level, he naturally uses the word 'genus'.

76   *Posterior Analytics* 100a–100b (Aristotle *op cit* [note 46] pp257, 259 & 261)

77   They need not necessarily be biological species. One could envisage books, newspapers, diaries, pamphlets, etc., as separate species of written matter, which could readily be formed into higher groupings such as 'written works' or 'printed works'. But, strictly, what is the 'species-level', in such cases? Is 'book' at the 'ground-level'? Or 'biography', 'novel', 'French book', 'tome', 'incunnabulum'? Clearly the biological analogy can be most misleading when applied in non-biological contexts. And Aristotle's vision of the whole world of things being packaged into discrete, non-overlapping classes does little justice to the complexity of things or the ways in which different human groups may elect to classify and name things. In fact, one finds that items are ordered differently in different languages and different 'knowledge systems'.

78   In one passage, *Posterior Analytics* 93b (Aristotle *op cit* [note 46] p205), Aristotle does say that the essences become clear to us through syllogism and demonstration. That is, essences cannot be proved as the conclusions of a syllogism, but their use in a syllogism enables one to see the relationships between facts, or classes of things. In a sense the point is similar to that which we were making above (p13). When you know your way down a mountain, you have also found your way up.

79   *Posterior Analytics* 89b (Aristotle *op cit* [note 46] pp171 & 173)

80   *Posterior Analytics* (78a–78b (*ibid* pp85 & 87)

81   *Physics* 184a (Aristotle *op cit* [note 47] Vol 1 p11)

82   *Physics* 194b–195a (Aristotle *op cit* [note 47] Vol 1 pp129 & 131)

83   One may define the final cause rather clumsily as 'the end or purpose for the sake of which the process takes place'. It is, of course a teleological notion. (Greek *telos* = end)

84   *Aristotle Generation of Animals* with an English Translation by A L Peck Harvard University Press Cambridge (Mass) & Heinemann London 1943

85   The famous scholastic device for representing and remembering Aristotle's matter theory is shown in Figure 7.

Figure 7

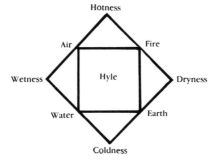

86   See note 6.
87   Plato specifically developed a theory of the onomatopoeic origin of language in his dialogue *Cratylus*
     (Plato in Twelve Volumes IV Cratylus Parmenides Greater Hippias Lesser Hippias Harvard
     University Press Cambridge [Maşs] and Heinemann London 1970), thereby seeking to establish a
     kind of one-to-one correspondence between language and reality. As may be anticipated, the scheme
     was not a success. Even so, the program of achieving an understanding of the nature of 'reality' by
     examination of the features of human language has been a persistent theme in Western philosophy.
88   On Aristotle's biological observations, see, for example, Ross *op cit* (note 42).
89   (a) If p implies q,
     and p is true,
     Then q is true. (The '*modus ponens*' argument)
     (b) If p implies q,
     and q is false,
     Then p is false. (The '*modus tollens*' argument)
     (c) If it is not the case that both p and q are true, and p is true
     Then q is false. (Or if q is true, p is false.)
     (d) If either p or q is true, but not both,
     and p is true
     Then q is false.
     (e) If either p or q is true, but not both, and p is false,
     Then q is true.
     We may also mention here an important logical fallacy (the fallacy of 'affirming the consequent'):
     If p implies q,
     and q is true,
     Then p is true.
     This is of great importance in philosophy of science, for recognition of this fallacy reminds us that
     evidence collected in favour of a scientific hypothesis does not prove that that hypothesis is true.
     (But as we shall note in Chapter 6, that which is to be regarded as logically valid or invalid is regarded
     by some commentators today as subject to social factors, and is not as it were laid up in some kind
     of Platonic heaven. In which case, some people might take issue with the unqualified assertion that
     the fallacy of affirming the consequent is a fallacy.)
90   J S Kieffer Galen's *Institutio Logica: English Translation, Introduction, and Commentary* Johns
     Hopkins Press Baltimore 1964
91   T L Heath *The Thirteen Books of Euclid's Elements [of Geometry] Translated from the Text of
     Heiberg with Introduction and Commentary* 3 vols Cambridge University Press Cambridge 1908
     Vol 1 p138. See also: Vol3 p442.
92   *Ibid*
93   The notion of logical *connection* (joining together) seems to have made the geometers suppose that
     'synthesis' was the appropriate word for this deductive process.
94   Quoted in Heath *op cit* (note 91) Vol 1 p138
95   *Ibid* Vol 3 pp442–3. For further discussion of the role of analysis in Greek geometry, see M S
     Mahoney 'Another Look at Greek Geometrical Analysis' *Archive for History of Exact Sciences* 1968–
     69, Vol 5 pp318–48. This article emphasises the role of geometrical analysis as as problem-solving
     device, rather than a general feature of scientific methodology as a whole. It should also be noted that
     the author questions the genuineness of the Pappus definition of analysis and synthesis quoted
     above, p27. For an even more detailed account, which has, however, attracted some fairly critical
     comment, see: J Hintikka & U Remes *The Method of Analysis: Its Geometrical Origin and its
     General Significance* Reidel Dordrecht & Boston 1974
96   Writing in 1833 in a review of an article by R D Hampden on Aristotle in the *Encyclopaedia
     Britannica*, William Hamilton said: 'In one respect, says Aristotle, the Genus is called a part of the
     Species; in another, the Species a part of the Genus. ... In like manner, the same method, viewed in
     different relations, may be styled either Analysis or Synthesis. This, however, has not been ack-
     nowledged; nor has it even attracted notice, that different logicians and philosophers, though
     severally applying the terms only in a single sense, are still at cross purposes with each other. One
     calls Synthesis what another calls Analysis; and this both in ancient and modern times.' (*Edinburgh
     Review* Vol 57 1833 p236)
97   For a discussion of the relationship between the rhetorical tradition and the development of
     scientific method, see R McKeon 'Philosophy and the Development of Scientific Methods' *Journal
     of the History of Ideas* 1966 Vol 27 pp3–22
98   The example was given by Porphyry in a work entitled *Isagoge* – an 'introduction' to Aristotle's
     Categories. Porphyry's work was in its turn the subject of a commentary by Boethius (480–524AD),
     and a pictorial representation of the 'Tree' may be found in this, whereas Porphyry himself merely
     spelled it out in words. See Boethius *In Porphyrium Dialogi a Victoriano Translati: Dialogus
     Primus* in J P Migne *Patrologiae Cursus Completus, sive Bibliotheca Universalis... Series (Latina)
     Prima* 221 vols Paris 1844–64 Vol 64 columns 9–48 (cols 41–2 for diagram). For an English
     translation of Porphyry's text, see E W Warren *Porphyry the Phoenician: Isagoge* Pontifical Institute
     of Mediaeval Studies Toronto 1975

99   The method was, as we have seen, known to Plato. See note 40 above.
100 *Infima Species* = the 'lowest species'. See also notes 75 and 77.
101 See p21. But the syllogisms into which the Aristotelian was to place his middle terms, found with the aid of understanding, were all deductive.
102 See note 95
103 *Galen's Method of Physick [Methodus Medendi]: Or, his Great Master-Peece; being the Very Marrow and Quintessence of all his Writings.... Translatour Peter English* George Suinton & James Glen Edinburgh 1656 p36
104 *Ibid*
105 *Ibid* pp28–9
106 *Ibid* p5
107 That is, the formulation of hypotheses, the working out of the empirical implications of these, and the testing of those by experiment and/or observation. But Galen did not lay emphasis on the need for hypotheses in this process of inquiry. The ancients always sought certainty in their knowledge, and tried to eschew hypotheses.
108 On Pyrrho and Pyrrhonism, see R H Popkin *The History of Scepticism from Erasmus to Descartes* Harper & Row New York 1968
109 See *Sextus Empiricus with an English translation by... R G Bury... In Three Volumes I Outlines of Pyrrhonism* Harvard University Press Cambridge (Mass) and Heinemann London 1939 pp257 and 259. Sextus also devoted considerable effort towards seeking to display the inadequacies of knowledge supposedly gained with the assistance of the 'hypothetical syllogism' (see note 89).
110 St Thomas Aquinas *Summa Theologica* Question 85 Article 6 (A C Pegi ed *Basic Writings of Saint Thomas Aquinas* 2 vols Random House New York 1945 Vol 1 p825)
111 *Ibid* p824
112 Plotinus *The Enneads* trans S MacKenna 2nd edn Faber & Faber London 1956
113 *Ibid* pp380–400 & *passim*
114 See *The Celestial and Ecclesiastical Hierarchy of Dionysius the Areopagite now first translated into English from the original Greek by the Rev John Parker M.A.* Parker London 1894
115 Proclus *op cit* (note 36) ppcxxvi-cxxx. Here Proclus's translator, the eighteenth-century Neo-Platonist Thomas Taylor, gives a translation of some excerpts from Proclus's writings in the fifteenth-century Neo-Platonist, Marsilio Ficino. The quasi-religious character is strongly evident. But Ficino himself developed Neo-Platonism into a system that was as much religious as philosophical.
116 Boethius was responsible for translating into Latin and commenting on the works in Aristotle's *Organon*. His role in transmitting and systematising the knowledge of the ancient world to mediaeval scholars was of profound importance, and by clarifying the contents of the Greek education system Boethius was instrumental in the subsequent establishment of the mediaeval education system of the *trivium* (grammar, rhetoric and logic) and the *quadrivium* (arithmetic, geometry, astronomy and music).
117 See note 98.
118 In his *Commentary on the Sentences*, Ockham accepts that some kind of reality attaches to universals; but their status is logical rather than existential. They appear in syllogisms, but this does not confer the status of reality upon them. It is believed that Roscelin of Compiègne (1050–c1125), few of whose writings have survived, took a stronger position, namely that universals are nothing but names — or even mere utterances. 'Common natures' are wholly subjective.
119 The realist/nominalist debate has been admirably treated by M H Carré in his volume *Realists and Nominalists* Oxford University Press London 1946
120 A C Crombie *Robert Grosseteste and the Origins of Experimental Science 1100–1700* Clarendon Press Oxford 1953.
121 R Grosseteste *Commentary on Aristotle's Posterior Analytics*, quoted in Crombie *op cit* (note 120) p63
122 *Ibid* pp64–5
123 *Ibid* p67
124 The reader may be left in doubt as to the specific anatomical device which affords camels their protection. But we are informed that an angry male camel is not a creature with which one should trifle.
125 Crombie *op cit* (note 120) p87
126 *Posterior Analytics* 98a (Aristotle *op cit* [note 46] p243)
127 A number of up-to-the-minute historians of science claim that it is 'bad form' to hunt for precursors of scientific or philosophical ideas. Nevertheless, most ideas, like people, do have predecessors. To note the appearance of an idea, perhaps in a simple form, in one writer and its subsequent development in later work, is not of course in itself a proof of historical indebtedness or connection between the earlier and the later writers, though such connections there may in fact be.
128 See below, pp150–53.
129 Duns Scotus *Philosophical Writings* trans A Wolter Nelson Edinburgh 1962 pp109–10 It is common practice in elementary philosophy classes to poke fun at the 'method of agreement'. For example there was the man who drank whisky, gin and water, got drunk; there was the man who drank vodka, brandy and water and got drunk; and there was the man who drank beer, cider and water and got

drunk — which showed that water was the cause (or aptitudinal union?) of inebriation. Such an example certainly demonstrates the inadequacy of the method to stand alone as a method of proof in science. But it does not show that the method has no place whatsoever in scientific investigation.

130 As a 'proposition reposing in his soul'

131 Duns Scotus here allowed the human will to be independent of the causal nexus.

132 Duns Scotus *op cit* (note 129) p109

133 William of Ockham *Sentences* i d 45 q I D, quoted in J R Weinberg *Abstraction, Relation and Induction* University of Wisconsin Press Madison 1965 p146. Weinberg gives a very lucid account of mediaeval writings on induction, and points out how many of the European ideas on this topic were derived from Aristotle via Arab scholars, notably Avicenna (Ibn Sina) (980–1037).

134 On this, see particularly J H Randall Jr. 'The Development of Scientific Method in the School of Padua' *Journal of the History of Ideas* 1940 Vol 1 pp177–206. See also, for example N W Gilbert *Renaissance Concepts of Method* Columbia University Press New York & London 1960. A number of Randall's interpretations have been revised in an important paper by Nicholas Jardine 'Galileo's Road to Truth and the Demonstrative Regress' *Studies in History and Philosophy of Science* 1976 Vol 7 pp277–318. They were also severely challenged by Gilbert in a review of Randall's revision and extension of his earlier work, J H Randall *The School of Padua and the Emergence of Modern Science* Padua 1961: N W Gilbert 'Galileo and the School of Padua' *Journal for the History of Philosophy* 1963 Vol 1 pp223–31.

135 Pietro d'Abano *Conciliator Differentiarum Philosophorum, et Praecipue Medicorum* 1310 (printed Venice 1496), quoted in Randall *op cit* (note 134) pp185–6

136 See above, p23.

137 *Jacobi de Forlivio super Tegni Galeni* Padua 1475, quoted in Randall *op cit* (note 134 1940) p189

138 Hugh of Sienna *Expositio Ugonis Senensis super libros Tegni Galeni* Venice 1498, quoted in Randall *op cit* (note 134 1940) p190

139 Paul of Venice *Summa Philosophiae Naturalis Magistri Pauli Veneti* Venice 1498, quoted in Randall *op cit* (note 134 1940) p191

140 A Nifo *Augustini Nifi Philosophi Suessani Expositio... de Physico Auditu* Venice 1552, quoted in Randall *op cit* (note 134 1940) p192

141 Jardine *op cit* (note 134) p291

142 P Pompanazzi *Quaestio de Regressu* 1503(?), quoted in Jardine *op cit* (note 134) p291

143 A Nifo *Aristotelis Libri Octo de Physico Auditu Interprete atque Expositore Magno Augustine Nipho* Venice 1543, quoted in Jardine *op cit* (note 134) p292

144 G Zabarella *Opera Logica* 1578, quoted in Jardine *op cit* (note 134) p296

145 See H Skulsky 'Paduan Epistemology and the Doctrine of the One Mind' *Journal for the History of Philosophy* 1968 Vol 6 pp341–61.

146 Quoted in Jardine *op cit* (note 134) p302

CHAPTER 2

PHILOSOPHY OF THE NEW SCIENCE

Some historians of science, such as Butterfield, Kuhn and many others, have seen the science of the seventeenth century as marking a radical departure from the science of all previous periods; and they speak confidently of 'The Scientific Revolution'. Others such as Duhem and Crombie, however, have been inclined to trace science back from the seventeenth century into the Renaissance and the Middle Ages, and in a sense, therefore, have to some degree played down the novelty of the seventeenth-century scientific movement, emphasising rather the continuities with earlier theories and practices, though not thereby underestimating the seventeenth-century achievement.

It would also be possible to emphasise historical continuity or discontinuity at the metascientific level as well as at the level of science itself. Undoubtedly, some seventeenth-century writers on metascience, notably Bacon and Descartes, emphasised the novelty of their methodological pronouncements. But as we shall see below, it is not difficult to observe the ways in which certain aspects of their thought clearly betrayed their intellectual ancestries. And the same may be said of Galileo's metascientific remarks, even though his own scientific work was so very distinctive, involving specifically a mathematisation of physics, with the selection and isolation of particular features of the observed phenomena for experimental examination and mathematical description.

The actual scientific achievements of the seventeenth century are too numerous to be described in any detail here; but just a few landmarks may be mentioned. Already in the sixteenth century, Nicholas Copernicus (1473–1534) had claimed — contrary to the usual ancient Greek beliefs — that the Sun was at the centre of the solar system, and that the Earth was a rotating object, moving round the Sun once a year. However, there was some doubt as to whether Copernicus's theory was intended to be thought of as a mere calculating device — which 'saved the appearances' and which made no claim to describe the real physical constitution of things — or whether it was intended as a factual description of the way things actually were in the physical world.[1] Then in 1609 Johannes Kepler (1571–1630) published his conclusion that the planets moved in elliptical, rather than circular, orbits, and later he successfully arrived at the relationship between the periods of the planets' motions in their orbits and the sizes of their orbits.[2] Subsequently, Galileo overthrew the ancient Aristotelian system of physics, with its doctrine of 'natural place' (such that motion supposedly occurred when objects were displaced from their natural situations in the cosmos), and he began to develop the modern concept of inertia. Galileo also provided observational evidence for the heliocentric system of Copernicus, with the aid of the newly invented telescope, and began to work out the general principles of kinematics, later to be subsumed by Newton's mechanics. Galileo also argued for the motion of the Earth, contrary to the apparent evidence of the senses. Newton's own achievements were so numerous and important as to defy any attempt at brief description. He developed a 'corpuscular' theory of light and colour with great success, and was able to establish a general physics that satisfactorily accounted for the observed motions of the heavenly bodies. He stated the general principle of inertia, namely that a body will remain at rest or move in a straight line with constant velocity unless acted upon by external forces. This was quite contrary to Aristotelian physics, based on common everyday notions, which supposed that bodies would only move if pushed, so to speak.[3] Newton also gave a general rule for the calculation of mechanical forces and envisaged a theory of universal gravitation, with the idea that gravitational forces acted throughout the universe, not merely at the surface of the Earth. Such forces could be calculated by means of the celebrated 'inverse square law'.

In the biological sciences, the work of William Harvey (1578–1657) on the circulation of the blood was perhaps the most noteworthy. There were less obvious achievements in chemistry, though a unified chemical theory — now thought to be false — was developed in terms of the so-called mechanical philosophy. That is, all chemical phenomena were accounted for in terms of the physical properties, motions and interactions of miniscule hypothetical corpuscles. (For example, acids were thought to be made up of spike-shaped corpuscles.) The theory of chemical affinity was established, and the foundations of the sciences of magnetism and electricity were laid at the beginning of the century by William Gilbert (1540–1603). A great deal of taxonomic work with plants and animals was carried out by, for example, John Ray (1627–1705). Some early crystallographical and geological studies were carried out by Nicholas Steno (1638–1686), and Marin Mersenne (1588–1648) made fundamental contributions to music theory and acoustics.

Such a listing could be extended with great ease but it is inappropriate to do so here. The point I wish to make is simply that nearly all the main branches of modern science were established in the seventeenth century, and a great deal of fundamental work from this period has survived, whereas by far the greater portion of earlier work has been quite superseded. Moreover, the seventeenth century also saw the foundation of the scientific movement as a social system, with a character that has, in some measure, lasted till the present day. For example, some of today's most prestigious scientific societies and journals were established in the seventeenth century.[4]

All these points are no doubt familiar enough to most readers; but they need to be mentioned in order to give the general background to the seventeenth-century metascientific writings that we shall shortly be discussing more fully, and with which they were intimately related. Seventeenth-century science was markedly more 'successful' (as it appears to us today) than had hitherto been the case. Precisely why all this happened is an extremely difficult historical question, on which historians of science still find it hard to reach agreement. But it is certainly the case that many old scientific ideas and theories were to a large degree superseded in the seventeenth century. Did the new science have also a new and distinctive metasciencc that distinguished it clearly from what had gone before, and which might account for its remarkable successes? Was a new methodology developed that was in some way vastly superior to previous ideas on this question? Or must we look elsewhere if we wish to account for the achievements of seventeenth-century science? Well, to have any hope of answering such questions we must examine the metascientific writings themselves and make some comparison between metascientific theory and scientific practice.

Galileo

It will be convenient to begin our discussion by considering the work of Galileo Galilei (1564–1642). His most famous work was entitled *Dialogue of the Two Chief World Systems*[5] in which he set forth his views on the new Copernican system, but did so in a manner that was supposed to accommodate the ecclesiastical powers by presenting the arguments in the form of a discussion between three speakers: Salviati (in effect representing Galileo's opinions); Sagredo (an intelligent layman, who, however, always ended up agreeing with Salviati); and Simplicio (who represented orthodox Aristotelian physics and the geocentric astronomy of Ptolemy [*c.* 100–170 AD]). By means of this literary device, Galileo no doubt hoped that he would be thought to be presenting the arguments in a suitably impartial manner. In fact, however, as might be expected, the device failed to placate the traditionalists and, as is well known, Galileo found himself at odds with the Inquisition. Nevertheless, we meet Salviati, Sagredo and Simplicio once again in the *Dialogues Concerning Two New Sciences*,[6] which work was concerned with the basic principles of mechanics, rather than astronomy. This masterpiece, written in Galileo's old age, was less controversial than the *Chief World Systems* and offered a lucid summation of his life-work in physical science. There were also a number of lesser works, such as *The Starry Messenger* and *The Assayer*.[7] All these texts contain interesting methodological pronouncements, but one must be wary of accepting them as indications of Galileo's actual scientific practice, since they were what we may call 'apologetic'[8] works, rather than being direct accounts of his own research procedures. That is to say, in these works, Galileo was presenting his ideas to the public, seeking to persuade those who might be sceptical of his revolutionary new ideas. The works should, therefore, be regarded as specimens of skilful rhetoric (and splendid pieces of literature, incidentally) as much as clear and distinct accounts of Galileo's own procedures and thought processes at the time when he was actually carrying out his scientific researches.

One of the hallmarks of Galileo's scientific methodology was his use of specially devised experiments to test particular scientific ideas; and it has been suggested by the Galileo scholar, Stillman Drake, that Galileo may have been initiated into the art of mechanical experimentation by his father, the musician Vincenzo Galilei,[9] who had performed useful empirical investigations on the problem of tuning stringed instruments with the aid of the monochord at the end of the sixteenth century. If Drake's argument is correct, and there seems no reason to doubt it, then we have an interesting connection between the 'modern' experimental science of Galileo and the ancient interests in Pythagorean harmonies and the like. This gives added weight to the supporters of the hypothesis of historical continuity between seventeenth-century science and the work of earlier centuries. Yet, in fact, it in no way detracts from the remarkable advances made by Galileo at the experimental, theoretical and metascientific levels.

In his empirical investigations, Galileo saw that it was necessary to abstract from the great mass of data that may be collected by the senses, and focus attention on just a few factors — specifically those that could be described in mathematical terms. Moreover, he sought to deal with relatively simple problems first, in a somewhat piecemeal fashion, rather than trying to work out a whole cosmology and philosophy all at once. His investigations of the law according to which bodies fall towards the Earth under the influence of gravity is a classic example, and we may use it as a basis for our discussions.

In the *Dialogues Concerning Two New Sciences* — as noted above, an apologetic text in which he was presenting his ideas to the world — Galileo started from definitions of uniform motion and uniformly accelerated motion:

> By steady or uniform motion, I mean one in which the distances traversed by the moving particle during any equal intervals of time, are themselves equal;[10]

and

> A motion is said to be uniformly accelerated, when starting from rest, it acquires, during equal time-intervals, equal increments of speed.[11]

But despite the fact that we are given definitions here, and it is clear that although he was using the popular literary device of the 'dialogue' Galileo intended to set out his argument in a quasi-formal way (*in more geometrico*), it is evident that there lurked only a little way below the surface an *assumption* about the way in which the motion of a uniformly accelerated body depends upon the *time* of motion (as opposed, say, to the distance of travel). We find this assumption similarly when Salviati speaks in the following words:

> We may picture to our mind a motion as uniformly and continuously accelerated when, during any equal intervals of time whatever, equal increments of speed are given to it. Thus if any equal intervals of time whatever have elapsed, counting from the time at which the moving body left its position of rest and began to descend, the amount of speed acquired during the first two time-intervals will be double that acquired during the first time-interval alone; so the amount added during three of these time-intervals will be treble; and that in four, quadruple that of the first time-interval.[12]

Galileo subsequently gives us his geometrical analysis of the situation, representing time intervals by increments on the vertical axis, and velocity increments on the horizontal axis. Distances travelled are identified with areas on the graph. Modifying his diagram suitably, we can readily understand his argument (see Figure 8).

Figure 8

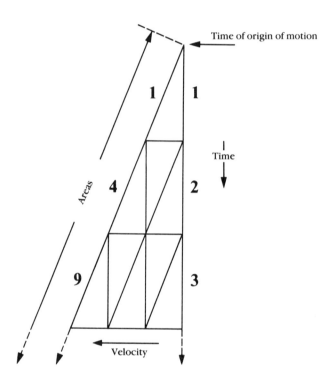

For:
in 1 unit of time 1 unit of distance is travelled;
in 2 units of time 4 units of distance are travelled;
in 3 units of time 9 units of distance are travelled; and so on.
Thus the distance travelled depends upon the *square* of the time.[13]

Having thus shown that the distance travelled is theoretically proportional to the square of the time, Galileo's discussion then went on to what purported to be an experimental investigation of the matter. Salviati speaks thus, in a particularly famous passage:

The request you [Simplicio] as a man of science, made, is a very reasonable one; for this is the custom — and properly so — in those sciences where mathematical demonstrations are applied to natural phenomena, as is seen in the case of perspective, astronomy, mechanics, music, and others, where the principles, once established by well-chosen experiments, become the foundations of the entire superstructure... So far as experiments go... I have attempted in the following manner to assure myself that the acceleration actually experienced by falling bodies is that above described [i.e. such that distance travelled is proportional to the square of time].

A piece of wooden moulding or scantling, about 12 cubits long, half a cubit wide, and three finger-breadths thick, was taken; on its edge was cut a channel a little more than one finger in breadth; having made this groove very straight, smooth, and polished, and having lined it with parchment, also as smooth and polished as possible, we rolled along it a hard, smooth, and very round bronze ball. Having placed this board in a sloping position, by lifting one end some one or two cubits above the other, we rolled the ball, as I was just saying, along the channel, noting, in a manner presently to be described, the time required to make the descent. We repeated this experiment more than once in order to measure the time with an accuracy such that the deviation between two observations never exceeded one-tenth of a pulse-beat. Having performed this operation and having assured ourselves of its reliability, we now rolled the ball only one-quarter the length of the channel; and having measured the time of its descent, we found it precisely one-half of the former. Next we tried other distances, comparing the time for the whole length with that for the half or with that for two-thirds, or three-fourths, or indeed for any fraction; in such experiments, repeated a full hundred times, we always found that the spaces traversed were to each other as the squares of the times, and this was true for all inclinations of the plane...

For the measurement of time, we employed a large vessel of water placed in an elevated position; to the bottom of this vessel was soldered a pipe of small diameter giving a thin jet of water, which we collected in a small glass during the time of each descent, whether for the whole length of the channel or for a part of its length. The water thus collected was weighed, after each descent, on a very accurate balance. The differences and ratios of these weights gave us the differences and ratios of the times, and this with such accuracy that although the operation was repeated many, many times, there was no appreciable discrepancy in the results.[14]

This passage seems to give a highly circumstantial account of a beautifully conceived empirical investigation, and one may feel some confidence that Galileo did carry out the experiments, just as they are described here. But then Simplicio makes a remark that has given rise to a good deal of discussion among historians of science. He says:

I would like to have been present at these experiments; but feeling confidence in the care with which you performed them, and in the fidelity with which you relate them, I am satisfied and accept them as true and valid.[15]

So one may be left wondering whether Galileo did actually perform the experiment which he described so carefully, or whether the whole description was merely a literary (or apologetic) device, designed to give the appearance of an experimental approach, whereas in fact what Galileo actually did was no more than a piece of mathematical reasoning, in which he arrived at his result with ruler, pencil and paper, rather than messing about with some kind of water clock. Perhaps all the talk of a 'hard, smooth, and very round bronze ball' was no more than a literary fiction!

Roughly speaking, this was the interpretation put forward by Alexandre Koyré, who did so much to establish modern research in the history of science, in his celebrated *Galilean Studies*.[16] In his subsequent book, *Metaphysics and Measurement*, Koyré represented Galileo as a Platonist, so that the revolution in science that he wrought involved a repudiation of Aristotelianism in favour of Platonism, turning towards the world of forms, where physics could be mathematical, where the distance travelled exactly equalled half the product of acceleration and the square of time, and so on — away from the generalisations of everyday experience that characterised the physics of Aristotle. To quote

Koyré's representation of Galileo's way of thinking:

> [N]ecesse determines esse. Good physics is made a priori. Theory precedes fact.
> Experience is useless because before any experience we are already in possession of the
> knowledge we are seeking for. Fundamental laws of motion (and of rest), laws that
> determine the spatio-temporal behaviour of material bodies, are laws of a mathematical
> nature. Of the same nature as those which govern relations and laws of figures and of
> numbers. We find and discover them not in Nature, but in ourselves, in our mind, in our
> memory, as Plato long ago has taught us.[17]

Unfortunately, however, Koyré's interpretation may well have been a
reflection of his own Platonism, as much as that of Galileo. For although
the interpretation of Galileo as a Platonist was sustained until well into
the 1960s, a number of more recent studies, based more particularly on
examinations of Galileo's personal laboratory notes, seem to show that
there is ample evidence that Galileo did carry out experimental work on
falling bodies, and many other physical phenomena besides.[18] In a
number of cases, his experiments have been repeated and excellent
agreement found with the original results.[19] We do not find, however,
that Galileo used a water-clock, as described in the Two New Sciences,
when he first discovered the distance/time law (though he may have used
such an apparatus subsequently). Rather, he employed a 'rolling-board'
fitted with frets, such as one might tie to a lute. Then the positions of the
frets were adjusted until the ball ran over them in equal time intervals.
This could be judged quite accurately, by listening for a regular sound of
the ball running over the frets[20]. The resulting distances would, of course,
be in the ratio 1 : 4 : 9 , etc., and could easily be measured.

It is true that, according to Stillman Drake's analysis of a document in
the Galileo archives, Galileo may have had his very first insight into the
relationship between speed and time of travel, during uniformly
accelerated motion, by a fortuitous 'play of numbers' on paper when he
was reflecting on the Mediaeval 'mean-speed' formula[21] for accelerated
motion.[22] However, this a priori insight (seemingly resulting from a
'pencil-and-paper doodling'!) was not followed up experimentally by
Galileo at the time, and was perhaps subsequently lost sight of. For not
long after, he was supposing that speed was proportional to distance of
travel (not time) and only later did he return to to the correct
relationship between speed and time. The 'paper-and-pencil' work might,
I suppose, be construed as evidence for some measure of Platonism on
Galileo's part; but the argument would not be very convincing. Any
physicist may do rough calculations or 'thought experiments', without
being in any way a Platonist. And in fact I do not think Koyré would have
placed much stress on the document that Drake discovered and analysed.
For Koyré, it was the apparant certainty and ideality of reasonings such as
that shown in relation to Figure 8 that counted, seemingly placing the
whole question of uniformly accelerated motion beyond the need for
experimental verification.

However, as indicated above, it seems clear that Galileo did perform
experiments. And as he saw it, though one might have an unshakable
mathematical representation of uniformly accelerated motion, it is an
empirical question as to whether falling bodies on Earth actually instance

uniform acceleration. Moreover, the experiment with the fretted board suggests that Galileo did approach the phenomena of accelerated motion by an empirical route (a little after his 'pencil-and-paper doodlings', referred to above).

It is, of course difficult to judge exactly what constituted scientific method for Galieo, on the basis of his actual work, for each piece of work was unique, and the historical record, though examined with great care by Drake and many other scholars, is necessarily incomplete. It will be interesting, therefore, to consider Galileo's public statements on 'scientific method', for though he may not have 'practised what he preached' exactly, it is really the history of public statements on method etc. that we are examining in this book. When Galileo 'went public', so to speak, he was in effect switching hats from that of a scientist to a metascientist; and it is his metascience that particularly interests us.

An important document, stating Galileo's 'official' views on methodology of science, is a letter written to one Pierre Calcavy in Paris in 1637, late in Galileo's life when he was under 'house arrest' at his villa at Arcetri. It was written in answer to an inquiry from the notable French mathematician, Pierre Fermat (1601–1665), concerning the *Two Chief World Systems*, which had been forwarded to Galileo by Calcavy. In Galileo's letter, one finds the following interesting statement:

> [A]s you and your friend can see from my book which is already in the press [i.e. *Two New Sciences*] ... I argue *ex suppositione*, imagining for myself a motion towards a point that departs from rest and goes on accelerating, increasing its velocity with the same ratio as the time increases, and from such a motion I demonstrate conclusively many properties. I add further that if experience should show that such properties were found to be verified in the motion of heavy bodies descending naturally, we could without error affirm that this is the same motion I defined and supposed; and even if not, my demonstrations, founded on my supposition, lose nothing of their force and conclusiveness... But in the case of the motion supposed by me it has happened that all the properties that I demonstrate are verified in the motion of heavy bodies falling naturally.[23]

How should this text be construed? On the face of it the quotation seems to offer no special difficulties, and may be related to the hyopothetico-deductivism of modern philosophy of science. That is to say, one may readily envisage Galileo meaning that one should argue from some assumed supposition or hypothesis (*ex suppositione*) to conclusions that flow deductively from the hypothesis. Then these might be tested experimentally; and if the experimental findings agreed with the data one might say that the hypothesis had received support — though it would not necessarily be *true*. Alternatively, one might find disagreement with the predictions, in which case the hypothesis would have to be rejected and a new one found to put in its place. This would be the easy construction to place on Galileo's text — thinking in twentieth-century terms. In the specific case we have been considering, the hypothesis would be that the velocity achieved by a falling body would be proportional to the time of fall, not the distance already travelled.

Unfortunately, however, this neat and tidy interpretation of Galileo's 'official' methodology is historically unsatisfactory, as W.A.Wallace has recently shown,[24] quite apart from the question of whether Galileo practised what he preached. To begin with, *ex suppositione* did not mean the same thing as *ex hypothesi*. Rather, in the writings of Thomas Aquinas, and more generally of the scholastic tradition, it meant 'reasoning backwards' from known effects to the causes which would give rise to those effects. (It was, I believe, analogous to what later writers were to call *abduction*.[25]) According to St Thomas,[26] such a procedure would then allow of a 'scientific' demonstration of effects from their causes. In which case, one would have something better than mere hypothetical knowledge; one would have true scientific knowledge of the causes on account of which (*propter quid*) effects occurred.

We may put the matter more succinctly with the help of the following formalism: If p then [if (if p then q) then q].[27] Here p is supposed to refer to experimental results or effects; while q refers to theoretical explanations or *causes*. The modern philosopher of science would deny that it is ever possible to deduce q from p; but it would seem that Aquinas was seeking to deploy a logical form whereby this might be accomplished. Doubtless this was so since, in seeking to forge a synthesis of Aristotelian thought and Christianity, there was a felt need for knowledge that was certain, in the realms of both natural philosophy and Christian theology.

Unfortunately, however, the device of reasoning *ex suppositione* does not achieve what is required of it. The form of argumentation can be boiled down to:

if p then q
and p
―――――――――
therefore q

(the '*modus ponens*' argument)

And in order to state the first premiss of this syllogism, one has to presume that experimental results can logically entail a theoretical explanation or a cause. But this is not so, and the dream of *deducing* (or educing with certainty) explanatory causes from phenomena was not brought any closer to realisation by the somewhat distracting logic of the *modus ponendo ponens*, or talk of reasoning *ex suppositione*.

Nevertheless, it seems very likely that Galileo had in mind something like the mode of reasoning that Aquinas tried to deploy when he (Galileo) spoke of reasoning *ex suppositione*; and that he made use of it to great advantage moreover. There are two reasons for saying this. First, an examination of Galileo's early notebooks has revealed his deep interest in and concern with Thomistic writings, with little reference to Plato or nominalists such as Ockham.[28] Second, some of Galileo's own experimental work seems to give weight to this interpretation. For example, in one series of experiments[29] he dropped a ball from different heights to an inclined surface at the edge of a table, whence it bounced through different parabolic arcs to different landing places, according to the prior vertical distances of fall. Galileo calculated the expected landing points for different distances of vertical fall and found good, but not perfect, agreement with the results of his calculations, errors being presumed to be due to friction and air resistance. He could readily show (*mathematically*) that his results were only compatible with the condition that velocity was proportional to time. In that sense, then, he was 'reasoning backwards' from the phenomena to a sort of theoretical explanation. He was reasoning *ex suppositione*, and in a way that was profitable for the understanding of physical phenomena.

In this case of falling bodies (as referred to in Figure 8, and by Galileo in his letter to Calcavy), we can perhaps reconstruct Galileo's reasoning as follows. He has a *supposition*, namely that for uniformly accelerated motion, as defined by him with speed increasing uniformly with time, the distance travelled is proportional to the square of the time. This connection between defined acceleration and the distance-time relationship is mathematical and certain, and not capable of being refuted by experiment. It is however, an empirical question as to whether bodies falling on the surface of the Earth do so according to the definition given by Galileo for accelerated motions. Galileo claimed that his experiments showed that distances travelled were indeed proportional to the times squared. Therefore, he concluded, a body falling on the surface of the Earth provides a physical example of uniformly accelerated motion, as mathematically defined.

This mode of reasoning was eminently satisfactory and secure, provided that *only* uniformly accelerated motion (as defined) will produce the effect of distance travelled being proportional to time squared. Also, the conclusion about the motions of real terrestrial bodies can only be made within the limits of error in the experimental set-up used — a point that Galileo did not dwell on. So in the case under consideration there was a component of the argument for which there could be certainty (in the mathematical reasoning), as, no doubt, Galileo desired. The *empirical* component was also present, however, and here there was an element of uncertainty as to what actually occurred in the real world. Also, Galileo was unable to offer any certain method for 'reasoning backwards' from empirical facts to explanatory causes. However, we don't doubt that he had the right mathematical expression for the motions of uniformly-accelerating bodies.

In general, in science, we may have several hypotheses with similar experimentally testable consequences. So the successful experimental testing of a hypothesis does not prove the truth of that hypothesis; and to assume it does involves one in the 'fallacy of affirming the consequent'. However, the success or failure of the experimental tests does not touch the question of the truth or otherwise of the mathematical component of the physics, as Galileo was evidently aware. Incidentally, as we shall see in Chapter 6, there is a fundamental philosophical problem as to the manner in which the mathematical and the empirical components of a physical enquiry may be supposed to 'lock together', so to speak. But this twentieth-century problem was obviously of no concern to Galileo.

It must be emphasised, then, that the 'reasoning backwards' of *ex suppositione* could only be performed satisfactorily by Galileo when the problem to be elucidated was capable of being expressed in *mathematical* terms. In this, it seems clear that he was strongly influenced by the works of Archimedes (*c.* 287–212 BC), which were then coming to be widely known in Europe.[30] But perhaps more interesting from our point of view is the way in which Galileo's methodology, although apparently so very distinct from what had gone before by reason of its mathematical analysis of phenomena, had strong intellectual affinities with the ancient tradition of Aristotelian enquiry. Yet also it looked forward to the mathematical physics of the future. Galileo was, it seems, still governed by the search for certainty in science — so that the 'arch of knowledge' could be ascended with just as much security as was thought to be possible in the deductive descent. And in a sense, one might say that he was successful. But while Galileo's method undoubtedly had its successes, reasoning *ex suppositione* does not enable one to deduce *efficient* causes — *why*, for example, in the case considered above, nature behaved in such a way that the speed of freely-falling bodies was proportional to their time of fall; or putting it another way, *why* freely-falling bodies undergo uniform acceleration. And it is efficient causes that are of prime concern in modern science. We should note, however, that Galileo effectively eliminated the Aristotelian doctrine of final causes (that is, teleology) from his physical science. In this, his thinking laid the foundations for the modern scientific movement in the physical sciences. However, *physics* involves more than going up and down (or down and up) a mathematical or geometrical ladder. Our arch has both logical-mathematical and empirical bricks.[31] So we can never be *certain* that even the strongest and best constructed scientific arch will stand forever! A mathematical structure may do so, so long as one retains the chosen assumptions. But the edifice of certainty offered by mathematics should not be confused with scientific knowledge.

Before we leave Galileo, let us say a little more about the general character of his work, and what distinguished it from that of his pre-decessors. Even though, as argued above, there were Aristotelian (or Thomistic) elements in Galileo's methodology, this does not alter the fact that he was reinterpreting the whole notion of motion. In Aristotle's physics, motion was regarded as a quality: a body would be moving or stationary according to whether or not it possessed the quality of motion. But it was not possible to treat such an Aristotelian property satisfactorily in terms of numbers, and hence examine the phenomena of motion in more than a general manner. By contrast, Galileo's falling bodies were, so to speak, 'mathematical' entities, moving in 'mathematical' space. All their attributes (such as colour, smell, weight, etc) were disregarded (for the purpose of the experiment on falling bodies), and attention was focussed solely on position and time. So in a sense Galileo was no longer dealing with real bodies moving in real space, but with 'mathematical fictions'. Consequently, by means of an abstract mathematical analysis of the problem he could indeed envisaged the book of nature as having been written in mathematical terms; and in this sense one can understand why Koyré described Galileo as a Platonist. But the historical records do not allow us to suppose that Galileo's results were obtained independently of experiment. Really, what he was doing was constructing a new method of experimental physics and hence a new kind of 'natural philosophy', not a 'neo-neo Platonism'! And as I say, Galileo did seek to connect his mathematical models to the world by means of experiment. For a Platonist that would not have been necessary.

In the work entitled *The Assayer*, in which he was debating the question of the nature of comets (and getting the wrong answer, as it happened), Galileo gave an important statement in relation to the question of mathematical abstraction, in which we can recognise what has come to be known as the doctrine of primary and secondary qualities:

> I say that whenever I conceive any material or corporeal substance, I immediately feel the need to think of it as bounded, and as having this or that shape; as being large or small in relation to other things, and in some specific place at any given time, as being in motion or at rest; as touching or not touching some other body; and as being one in number, or few or many. From these conditions I cannot separate such a substance by any stretch of my imagination. But that it must be white or red, bitter or sweet, noisy or silent, and of sweet or foul odor, my mind does not feel compelled to bring in as necessary accompaniments. Without the senses as our guides, reason or imagination unaided would probably never arrive at qualities like these. Hence, I think that tastes, odors, colours, and so on are no more than mere names so far as the object in which we place them is concerned, and that they reside only in the consciousness. Hence if the living creature were removed, all these qualities would be wiped away and annihilated.[32]

What Galileo seems to have been saying here was that a body, to be a body at all, must have at least the qualities of shape, position, motion or rest, contiguity (or spatial relation to other bodies), and number. Such so-called *primary* qualities really inhere in bodies themselves; others such as taste and feel, which came to be called secondary qualities, only inhere in the mind of the observer.[33] This is an epistemological/ontological doctrine that was to engender a considerable amount of philosophical discussion in the seventeenth and eighteenth centuries, and we shall refer to it again on a number of occasions. The point to be noted here is that Galileo seems to have distinguished those qualities that were susceptible to investigation by means of mathematical analysis from those that were not. Here, then, we have one of the chief sources of the doctrine of primary and secondary qualities, which, as we shall see, was to interact in philosophically interesting ways with the seventeenth-century doctrines of matter — either corpuscularian or atomistic.

Bacon

Besides the significant seventeenth-century change of scientific style that was brought about by Galileo (and others whom we have not space to consider here), we may examine the methodological revolution brought about by the English philosopher–politician Francis Bacon, Lord Verulam (1561–1626), directing our attention particularly to his major text on philosophy of science, the *Novum Organum*, or *New Organon*, first published in 1620.[34] Unlike Galileo, Bacon made no significant contributions to science itself, but he exerted a profound influence through his writings *about* science. (He was a metascientist *par excellence*.) Like Galileo, Bacon represented himself as a revolutionary, seeking to overthrow all the old methods associated with Aristotelian philosophy and scholasticism. But as will become apparent, there were important elements of Aristotelianism within Bacon's system, despite all his anti-Aristotelian rhetoric. And the overall 'shape' of his methodology was quite traditional. Whereas Galileo's 'new science' had the distinctive feature of examining natural phenomena with the help of mathematical analysis, the science that Bacon envisaged was qualitative in character and based on taxonomic principles.

The *New Organon* was written in a series of aphorisms, which in Bacon's day meant concise statements or principles of a science, rather than maxims or pithy sayings, as we would understand the word today. The text is divided into two Books. The first Aphorism of Book I marks Bacon as an empiricist, and indicates that he believed that only through observation could one come to understand nature. However, as we shall see, there were also significant rationalist elements in his thinking. The second Aphorism tells us that the investigator is not limited to the use of his bare senses alone, but is to employ such instruments as he can construct in order to assist his researches. In fact, man is to try to 'command' nature, and wrest her secrets from her, thereby producing fruits for the benefit of mankind. It is this injuction that Western man has followed so successfully yet also with such potentially disastrous results. Bacon, it seems, believed that his 'new instrument' would indeed provide the method whereby man might enrich himself through knowledge. He clearly recognised the association of knowledge and power, and in the first Aphorism of Book II represented them as conjoined, thus giving expression to an important social truth.

The third Aphorism of Book I tells us that 'where the cause is not known the effect cannot be produced'; and this informs us that Bacon, like Aristotle, believed that the search for causes was the proper role for the scientific investigator. Nevertheless, Bacon's constant refrain was the inadequacy of the Aristotelian methodology. In particular, he pointed out (Aphorisms 8, 11–14) that the syllogism could not, in itself, furnish a way of gaining new information; it provided no route to the first principles. (As noted above, p 30, this point had been made long before by Sextus Empiricus.)

As we have seen, Aristotle had supposed that 'induction' provided the route leading to new knowledge and to the first principles of the several sciences. But Bacon was well aware (Aphorism 46) that 'induction by simple enumeration'[35] provided no satisfactory and secure way to the acquisition of knowledge, even though, as we shall see, he relied on something very like the 'arch simile' as a picture of the structure of the processes for the acquisition and deployment of knowledge. He objected that the Aristotelians (or those who employed the 'way...now in fashion') gave an utterly inadequate account of the inductive ascent, flying directly from 'senses and particulars' to 'the most general axioms' (Aphorism 19), instead of climbing steadily and securely. Bacon's new method, however, was intended to provide a very different way of ascent. What this method actually was I shall now endeavour to explain, following the example that Bacon himself gave in Book II of the *New Organon*, in illustration of his intended procedures.

The case considered was the determination of the cause of heat. To this end, Bacon suggested, one should collect a wide range of information respecting things that displayed the 'form' of heat — or were hot. These should be listed in a table. One should also draw up a list of analogous objects or phenomena which did not display the form of heat. Finally, one tabulated similar objects or phenomena that showed the form of heat in varying degrees.

Thus in the table of hot bodies or situations (the 'Table of Essence and Presence'), we find items such as the Sun, quicklime acting with water, summer, flames, and so on. Then in the table of cold things or circumstances (the 'Table of Deviation, or of Absence in Proximity') we have such things as the Moon and the stars, ashes mixed with water, winter, and St Elmo's fire. Finally, in the third table ('Table of Degrees or Comparison in Heat') we have such things as the planets (which Bacon tells us are, by tradition[!], of different temperatures), dung (which might be hot or cold), daily variations in temperature, flames of different strengths, and so on. Clearly, Bacon was attempting a systematic collection of information respecting the phenomena of heat. Or, in the language of his day, he was attempting to determine the 'history' of heat.

The next step was to eliminate certain possibilities, so far as the cause of heat was concerned. From his tables, Bacon could say something about what was *not* the cause of heat, even if he was unable at this stage to say what was the cause. Thus, for example, examination of the table would show that 'lightness and brightness' were *not* the cause of heat, for the Moon was light and bright, yet (supposedly) cold. Other possible causes could also quickly be eliminated in a similar manner. However, such a process of 'Exclusion or Rejection' could only be carried so far (fourteen possibilities were explicitly mentioned by Bacon), so that one was not — unfortunately — left with a positive statement of the cause of heat as a result of the process of elimination.

In fact, therefore, Bacon's method had broken down at this point, if it was supposed to be providing an automatic method that would allow anyone to ascend directly to the first principles of a science, or even a short step on the way. Bacon, however, was well aware of the problem, so at this point in the description of his method, he made a plea for an 'Indulgence of the Understanding' on the part of the reader, so that he might draw the 'First Vintage' of the form of heat (Book II, Aphorism 20). Or, in modern terminology, Bacon made a guess (or more grandly, he formulated an hypothesis) as to the form of heat; and then he *tested the consequences of this guess against new data.*

As it happens, we believe that Bacon guessed rightly about the nature of heat, saying 'Heat itself, its essence and quiddity ["whatness"], is Motion and nothing else' (Book II, Aphorism 20). So one might test other cases — over and above those that appeared in the original tables — to see whether heat and motion were always associated, one with another. By this means (I suppose), one would be able to draw second, third, fourth ... vintages, and thereby approach with ever greater certainty towards a knowledge of the form of heat. This was the nub of the Baconian method of induction; and of course it depended for its success upon the collection of a wide base of empirical information, on which the intellectual construction might rest. We should not, however, be seduced into believing that the method really worked, even if in the case of heat Bacon obtained a result to which modern science might give its assent. All he did was adopt one of the various theories of heat that were being canvassed in his day, and used it in illustration of his method. He might well have been less fortunate and produced an answer that the modern would find quite unacceptable.

I think it is reasonable to say that the general account of the acquisition of scientific knowledge given by Bacon was something like the hypothetico-deductivism of modern authors, and one might be inclined to see the resultant structure as analogous to the 'arch of knowledge' of Plato and others, with a new suggestion added as to how one might mount the ascending limb — which was to be constructed with a very wide base. But if we look more closely at Bacon's text we find that he was not contemplating a single arch and one set of causal principles, but a whole hierarchy of principles. So we may represent the total structure somewhat as in Figure 9.[36]

Figure 9

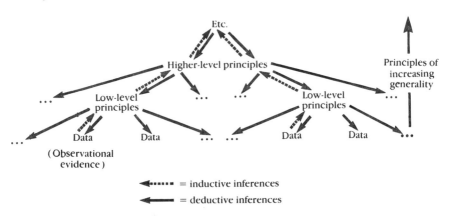

The upward ascent, then, was to be made with the help of the intellect, formulating the principles in terms of hypotheses, which should be rendered more certain, in the manner that has been described. However, Bacon did not actually use the term 'principle'. Rather, he employed words like 'law', 'axiom', 'nature' and 'form'; so we must try to understand the meaning of these terms within his system. The word 'nature' (or 'simple nature') presents no particular difficulty. It simply meant 'phenomenon', 'property' or 'quality'. Examples might be 'heat', 'yellowness', 'motion', 'the tides', or 'the Corporeal Substance of the Moon'. It was the object of Bacon's science to explain the causes or 'forms' of the 'natures'. (In modern jargon, the 'natures' were the '*explananda*', or those things that were to be explained, as opposed to the '*explanantia*' [sing. '*explanans*'], which were to do the job of explaining.)

In Aphorism 2 of Book II, Bacon referred to Aristotle's doctrine of the four causes, and he seems to have accepted the traditional four-fold classification. However, he stated that 'the discovery of the formal [cause] is despaired of'. Nevertheless, he made significant use of the ancient word 'form', and in fact we can probably interpret him as meaning that a 'form' was a law of nature, for he said: '[I]t is this law, with its causes, that I mean when I speak of *Forms*; a name which I the rather adopt because it has grown into use and become familiar'. So, for example, the 'law of heat' was one and the same as the 'form of heat'. (One may usefully reflect on the fact that even today we say that 'heat is a form of motion'.) Perhaps, therefore, we may suggest that within Bacon's system a 'form' was a necessary and sufficient condition for the appearance of a 'nature'; it was its causal explanation. This interpretation may be confirmed by reference to Book II, Aphorism 4:

> For the Form of a nature is such, that given the Form the nature infallibly follows. Therefore it is always present when the nature is present, and universally implies it, and is constantly inherent in it. Again, the Form is such, that if it be taken away the nature infallibly vanishes. Therefore it is always absent when the nature is absent, and implies its absence, and inheres in nothing else.

The object of Bacon's science, then, seems to have been the mastery of nature by the discovery of forms. Suppose, for example, one wished to make gold. The separate 'forms' of yellowness, weight, ductility, malleability and so on, should be found, using the technique of Baconian induction (with tables, etc.) for each. Then each form would have to be 'superinduced' on to any chosen substance, and thereby it might be transmuted into gold (Book II, Aphorism 5). Clearly, Bacon was thinking here along lines suggested by the alchemists of his day, and the whole imaginary process has a strong Aristotelian flavour, despite the *New Organon*'s constant animadversions against the Aristotelian methodology.[37]

There are other aspects of the method that may remind us of Aristotle. Motion was claimed to be the 'form' of a simple nature — heat. But motion was itself a simple nature, and some motions were apparently associated with hotness, while others were not. On the other hand, all heat was (seemingly) associated with motion. This suggests that the problem of method was linked in Bacon's mind with some kind of classificatory procedure — which we might well anticipate if we are looking for Aristotelian aspects of the Baconian science.

To take this point further, we may note the following passage from Book II, Aphorism 4:

> [T]he true form is such that it deduces the given nature from some source of being which is inherent in more natures, and which is better known in the natural order of things than the Form itself. For a true and perfect axiom of knowledge then, the direction and precept will be, *that another nature be discovered which is convertible with the given nature, and yet is a limitation of a more general nature, as of a true and real genus*.

Here, unfortunately, Bacon's meaning is far from clear, but from what he says elsewhere about the relationship between heat and motion it appears that a taxonomy representing the phenomena of heat might be expressed as in Figure 10.

Figure 10

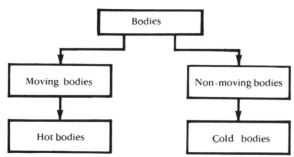

However, this does not seem to mesh exactly with the passage quoted above. Perhaps the matter would have been clarified for us if Bacon had completed the part of *The Great Instauration* that was to have been known as *The Ladder of the Intellect*, but that is little comfort to the modern would-be interpreter. Fortunately for us, the matter has been carefully examined by Professor Mary Hesse,[38] who has found further exemplifications of Bacon's methodology in an early work entitled *Valerius Terminus*.[39] Here Bacon considered the problem of finding the form of whiteness, of blackness, of transparency and of colour.[40] It seems that he used his method to arrive at the hypothesis that different colours are due to the different size ratios of the constituent parts (atoms?) of bodies. A small portion of Bacon's whole scheme of nature might thus be represented as in Figure 11.

Figure 11

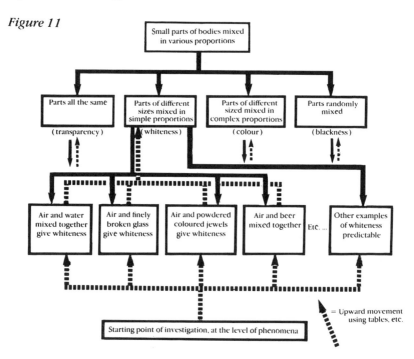

It may be noted that the lowest-level 'natures' (for example, whiteness) were subsets of a form of a more general nature; and this seems to accord with the italicised portion of the quotation from Book II, Aphorism 4. Each part of the 'taxonomy' was presumably supposed to be established by the help of the Baconian method, using tables, rejections and exclusions, etc. But it is hard to see how the method might really be effective in any large-scale way; and we may note that in this case Bacon did not get the 'right answer'. The detail is interesting, however, for the way in which it reveals Bacon's continuing attachment to the Aristotelian program of acquiring scientific knowledge through some kind of classificatory procedure.

A few further points may be made. There has been considerable discussion over the years as to whether Bacon believed that his method, if used properly, would allow a more or less automatic generation of scientific knowledge. Book I, Aphorism 61, supports this interpretation. We read:

> [T]he course I propose for the discovery of sciences is such as leaves but little to the acuteness and strength of wits, but places all wits and understandings nearly on a level. For as in the drawing of a straight line or a perfect circle, much depends on the steadiness and practice of the hand, if it be done by aim of hand only, but if with the aid of rule or compass, little or nothing; so is it exactly with my plan.

However, if one examines Bacon's curious unfinished novel, *The New Atlantis*,[41] which described an imaginary society whose members were devoting their several energies to various aspects of the Baconian method, then we find that some of them were required to undertake tasks that necessarily called for 'wit and understanding': for example, those whose job it was to 'raise the...discoveries by experiments into greater observations, axioms and aphorisms' — the Interpreters of Nature;[42] or those whose appointed task was to 'direct new experiments, of a higher light, more penetrating into nature than the former' — the Lamps.[43] (Others had more mundane tasks, like the Depradators who had to 'collect the experiments which are in all books'.[44])

Besides, in the *New Organon* itself there are strong indications that more than mere industry was required for the successful prosecution of the Baconian method. Consider, for example, the celebrated Aphorism 95 of Book I:

> Those who have handled sciences have been either men of experiment or men of dogmas.[45] The men of experiment are like the ant; they only collect and use; the reasoners resemble spiders, who make cobwebs out of their own substance. But the bee takes a middle course; it gathers its material from the flowers of the garden and the field, but transforms and digests it by a power of its own. Not unlike this is the true business of philosophy; for it neither relies solely or chiefly on the powers of the mind, nor does it take the matter which it gathers from natural history and mechanical experiments and lay it up in the memory whole, as it finds it; but lays it up in the understanding altered and digested. Therefore from a closer and purer league between these two faculties, the experimental and the rational, (such as has never yet been made) much may be hoped.

Here we see that Bacon envisaged a significant rational (arachnid!) component to the scientific enterprise, so that all wits would not be at the same level; the Baconian methodology would be much more than a mere handle-turning process. In particular, the processes of hypothesis formulation and testing would call for skills well above the usual run of things.

Hooke

I have hinted above that Bacon's methodology might appropriately be construed as a version of the ancient 'arch of knowledge'. It must be confessed, however, that in the *New Organon* nearly all attention was focussed on the upward ascending branch. Nevertheless, there is one passage (Book I, Aphorism 103) where Bacon referred to the road of inquiry rising and falling, and this feature of the intellectual terrain is well displayed in what is, I suggest, the best example of an attempt at direct application of Bacon's methodology that is to be found in the literature. I refer here to some remarks to be found in the *Posthumous Works* of Robert Hooke (1635–1702). Hooke was one of the first secretaries of the Royal Society of London, and some of the early members of that organisation specifically set out to perform scientific investigations in the manner recommended by Bacon.[46]

Hooke was interested in the cause of earthquakes. So he set out descriptions of where earthquakes occurred and where they did not, and he tried to carry out 'rejections and exclusions' in the approved Baconian manner. But having got thus far he found he could make no further progress — as one might expect. Consequently, he formulated four hypotheses, showed that three of them were unsatisfactory, and then proposed an ingenious method for testing the fourth — which involved the idea of pole-wandering as being the cause of earthquakes and other geological phenomena. Actually, the testing process was never carried out successfully, since the suggested rate of movement was too slow for Hooke to measure — or he did not carry out the observations over a long enough period to have any hope of success. (Indeed, we cannot be sure whether he even tried the experimental work at all.) The point, however, is that one can distil from Hooke's account of his procedures a picture of his general methodology of science. It looked like Figure 12.[47]

We still see there some variant of the 'arch of knowledge', although it was, so to speak, constructed of a series of subsidiary arches. Hooke seemed to appreciate — probably more clearly than Bacon — the necessity for hypothetical jumps (albeit as small as possible) in the process of ascent towards scientific knowledge. Even so, in this particular example, his endeavour was not notably successful; yet it is probably the best one can offer as a conscious effort to put the Baconian methodology into practice. Much of Hooke's other work, for example in optics, microscopy, mechanics and acoustics, where he did not tie himself so carefully to a specific methodological program, was rather more successful. The truth is that good science and good metascience are often quite distinct from one another.

One further interesting point is worth considering before we leave Hooke. The reader will notice that Hooke identified the 'upward' movement — that is, from effects to causes — as synthesis, whereas from

Figure 12

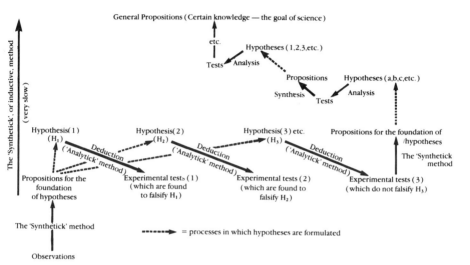

what has been said previously[48] one might have expected him to have called it analysis. We have, however, already noted the terminological confusion surrounding the terms analysis and synthesis, and some reasons for the difficulty have been suggested. Hooke's language will be specially pleasing to the connoisseur of these matters, since, as we shall see, his wording was the exact opposite of his contemporary, Newton. The conclusion seems fairly evident, namely that the terminological tradition that flowed from the remark of Pappus in his commentary on Euclid was highly confused and complex, as was perhaps the whole literature of metascience! More will be said on this matter in the discussions of Descartes and Newton that follow.

Descartes

As we have seen, Bacon thought that there should be both rational and empirical elements in any sound methodology of science. But the *Novum Organum* lay far the greater stress on the upward inductive movement from a wide empirical base, the very width of which indeed was somehow to give efficacy to the method. In France, however, Bacon's approximate contemporary, René Descartes (1596–1650), tended (at least, in his public methodological utterances) to emphasise the downward, deductive, 'rational' aspects of scientific enquiry. We shall, therefore, now turn our attention to this equally influential philosopher, mathematician, scientist and metascientist, and make some effort to unravel the very complex secondary literature that has grown up round this remarkable man and his writings. Descartes wrote two major works dealing with methodological matters. The first (which was never completed) was entitled *Rules for the Direction of the Mind*. Most of this work was probably composed in the years 1626–28, though some parts are now believed to have been written as early as 1619. The work, today usually known by its Latin title of *Regulae*..., remained unpublished until 1701.[49] The second important methodological work was the celebrated *Discourse on Method* of 1637.[50] The *Discourse* contained a résumé of Descartes' general philosophical position, but this was given its most detailed exposition in the *Meditations* (1641), published along with the *Objections and Replies*[51] that had been made to this work when it had previously been circulated privately in manuscript form. Descartes also published several chiefly scientific and mathematical works including the *Geometry* of 1637[52] in which he successfully applied algebraic reasoning to geometrical problems, thus laying the foundation of modern analytical geometry. Also most important was the *Principles of Philosophy* (1644)[53] in which the whole Cartesian system was set out in detail but in a straightforward and systematic manner, so that starting from certain philosophical principles the reader was gently carried to an under-standing of the whole structure of the cosmos, and to specific details of, for example, geological, meteorological, and chemical phenomena.

Descartes has traditionally been represented as an almost paradigmatic rationalist philosopher — perhaps one of the best examples of a Baconian philosophical spider. Thus, he has been supposed to have found a cunning device for ascending immediately to the apex of the 'arch of knowledge', descending thereafter from the grand metaphysical principles established at the apex to concrete facts of the physical world. This may seem a caricature of science, but it appears to have been one that Descartes himself desired to promote as representative of his own scientific practice in his polemical metascientific writings, particularly the *Discourse on Method*. We shall shortly be turning our attention to this most important metascientific text; but first let us consider the earlier *Regulae*, whose significance within the corpus of Descartes' writings has received much critical discussion,[54] and has recently been clarified considerably (for me) by the work of John Schuster.[55]

In 1618, when a young man residing in the Low Countries, Descartes met Isaac Beeckman (1588–1637) a Dutch physician, schoolmaster and engineer. The two scholars envisaged the notion of developing a new 'mechanical philosophy', which involved a combination of classical atomism (or corpuscularianism) and a 'general mathematics'. That is, they hoped they would be able to work out a general system of mathematics (capable, for example, of handling algebraic equations of different degree, but treated geometrically), which could find direct application to the presumed corpuscles of the 'micro-world'. This ambitious scheme eventually proved to be a chimera, but its mark is very evident in the youthful *Regulae*, much of which is written in a tone of high enthusiasm, as if all the problems of natural philosophy were to be dissipated forthwith by the application of Descartes' methodological directives.

Mostly, these directives consisted of attending carefully to problems and breaking complex steps of reasoning into their simpler constituent parts. But this unremarkable procedure could supposedly be integrated with the 'arch of knowledge' — with the traditional dual methodological pathway of analysis and synthesis. Thus we read in Rule 5:

> Method consists entirely in the order and disposition of the objects towards which our mental vision must be directed if we would find out any truth. We shall comply with it exactly if we reduce involved and obscure propositions step by step to those that are simpler, and then starting with the intuitive apprehension of all those that are absolutely simple, attempt to ascend to the knowledge of all others by precisely similar steps.[56]

Yet despite the seeming traditional character of the procedure, Descartes did have something new in mind. He supposed (in Rule 12) that the mind could form mental images of geometrical forms by the direct impression of ideas in the brain; and the *vis cognoscens* (cognitive awareness) could be aware of these 'ideas' — either directly in direct sensation or subsequently in acts of memory. This epistemological thesis was supposed to provide the underpinning of the methodological directives of the *Regulae*. For the mind could supposedly manipulate geometrical lines and figures in the brain (or the imagination), thereby solving certain algebraic problems geometrically.

However, towards the end of the *Regulae* it would seem that Descartes came to realise that his project was untenable. One could not do general algebra (for example, the solution of quadratic and higher-order equations) by the simple mental manipulation of lines and rectangles. So the programme came unstuck, and the *Regulae* were left incomplete, only being published posthumously. I would add, however, that even if Descartes had been able to formulate a satisfactory epistemological basis for his sought-for 'universal mathematics', it is hard to see how it might have found satisfactory application in a 'physico-mathematics' for the mechanical (corpuscularian) philosophy.

Be this as it may, when Descartes subsequently tried to give an account of his methodological principles (in the *Discourse on Method*), he sought to provide a metaphysical, rather than epistemological, basis for his scientific first principles and for the attendant scientific procedures. It is this argument of the *Discourse on Method*, rather than that of the *Regulae*, which is usually discussed in commentaries on Descartes' work. So we shall attend to it here in a little detail; but it should be emphasised that the metaphysical argument, which made up such an important component of Descartes' metascience, was in fact only a kind of attempted justification of his scientific procedures and results, rather than a statement of his actual scientific practice or method. So with this *caveat*, let us examine Descartes' 'official' metascience, as presented in the *Discourse*.

First, we should note that Descartes was interested in certainty, and knowledge that could be guaranteed to be true. Yet the tradition of philosophical scepticism to which he was heir[57] had led many to question whether there could be any knowledge that was totally immune to philosophical doubt. Undeterred, Descartes supposed that he could handle the sceptics' problem by arguing in the following manner.

Any person knows that he has thoughts, even though what he might be thinking on any given occasion could be false: one's sense impressions might lead one astray; one might be dreaming — or even be deceived by some evil demon. But the actual fact that one was thinking could not be gainsaid. Further, Descartes claimed, there could not be thoughts without there being some thinking subject having those thoughts. Hence he felt empowered to say: 'I think therefore I am' (or *'Cogito, ergo sum'*). Next, on considering this proposition, Descartes felt that its truth was utterly unassailable. There was, so to speak, no fuzziness round the edges. The proposition was (as is so often said) 'clear and distinct' to Descartes' mind; and also it was, he maintained, *true.*

So now Descartes had an indubitable, certain truth that had this characteristic quality of clearness and distinctness. Using this proposition, *'Cogito, ergo sum'*, as an exemplar, he could then hope to gauge the truth of *other* propositions, according to whether they did or did not have the quality of clearness and distinctness. For example, one could embark on a proof of the existence of God and examine each step in the reasoning process to ensure that it was clear and distinct.

The particular argument that Descartes deployed in favour of the existence of God was based on the notion of 'perfection'.[58] It went as follows. Descartes knew full well that he did not know everthing with certainty: he doubted many things and certainly wasn't omniscient. Therefore, he acknowledged himself to be imperfect, and he had the concept of imperfection in his mind. But for him to have the idea of imperfection, he also (he maintained) had to have the idea of perfection, with which it might be compared. Yet the idea of perfection could not have been in his mind unless there were a perfect being which could have given rise to it. So there must exist a perfect being — God.[59]

All this seemed to Descartes to be just as clear and distinct as the argument of the *'cogito'*. Therefore, by the criterion already established, the argument for the existence of God, by consideration of perfection, must be valid. So God's existence was (supposedly) proved by a 'clear and distinct' route. God was a perfect being; so he would not allow deception. In other words, God would act as a kind of guarantor for ideas that had the quality of clearness and distinctness. Such ideas must be true, just like the argument of the *'cogito'*.

At this stage, the reader will very likely wish to object that Descartes was arguing in a circle. The 'clear-and-distinct' criterion was used as an aid on the route towards the establishment of God's existence. But then God was used as a guarantor of the truth of clear and distinct arguments or propositions. The claim that the argument was circular was first raised by Antoine Arnauld (1612–1694) in the *Fourth Objections* to the *Meditations*,[60] and it has engendered an enormous philosophical literature since that time.[61] Fortunately, however, we need not interest ourselves in this issue; the point to be noted is that Descartes had seemingly 'established' (by a rapid ascent, using the magic wand of the *'cogito'*) the metaphysical coping stone of the 'arch of knowledge'. Thereafter, with the knowledge that there was a perfect God, who would act as a guarantor of Descartes' clear and distinct ideas, Descartes could then set forth these ideas in a logical and deductive (or quasi-deductive) manner, to give certain and scientific knowledge of physical things, thus moving from the realm of metaphysics (God) to physics (tangible things).[62]

God exists; and He created the cosmos. Further, Descartes thought that we have a clear and distinct idea of the notion of substance,[63] and also that we know clearly and distinctly that matter and space are *one and the same* for the notion of an extended void was, for Descartes, self-contradictory. So the more volume of space one considers, the more matter there is — exactly and equivalently so. However, space is infinitely divisible, and so also, as a consequence, is matter. Furthermore, there can be no vacuum — no space without matter, for space (or extension) and matter are one and the same. Also, since there is no (mathematical) limitation to extension, the cosmos will also be infinitely extended.

Given the equation of matter and space, Descartes considered also the problem of motion, the existence of which he seems to have taken for granted, even though one would hardly think it could be known *a priori* — by a clear and distinct introspection. If there was to be motion, therefore, and no vacuities, the only way this could happen would be by matter moving in a series of circles — or vortices.[64] Also, after the original division of space into its initial particles (presumably done by God, but Descartes doesn't specifically say so), and after God imparting the initial vortical motions to the system (which Descartes does mention), matter must have become divided into small fragments or corpuscles, the very smallest particles being flexible, or of variable shape, so that all space might be fully occupied with matter at all times. Further, as the vortical motions proceeded — after the Creation — the corpuscles gradually 'evolved', so that three main kinds came into being: (1) small spherical corpuscles, formed by the rubbing of the original larger pieces; (2) the very fine 'rubbings' which filled the interstices between the spherical corpuscles; (3) larger lumps of matter formed either by accretion of the very small particles at times when they happened to be stationary, or perhaps residual pieces left over from the original division of matter. These three types corresponded approximately with the air, fire and earth of earlier chemical theories, though Descartes thought the small spherical corpuscles constituted the matter by which light was tramsmitted.

Descartes then went on to give an account of the gradual evolution of the cosmos and the Earth. He suggested how a great many other kinds of corpuscles came into being, and how the 'macro' properties of bodies might be interpreted in terms of the 'micro' corpuscles: for example, sharp-pointed corpuscles for acids; long 'spaghetti-like' ones for oils; large round ones for mercury; and so on. The whole cosmos, therefore, was accounted for in mechanical terms. And in other works, such as the *Treatise on Man*,[65] Descartes attempted to give purely mechanical accounts of the structure and functioning of living organisms. In his view, animals were mere mechanical automata. Given this mode of explanation of phenomena, we customarily speak of Descartes as a leading exponent of the seventeenth-century mechanical philosophy.[66]

Descartes also (according to traditional interpretations) attempted to deduce some major laws of physics from his high-level metaphysical principles. For example, he argued that God, who imparted motion to the cosmos, was changeless in his essence. Hence, it seemed that the quantity of motion in the cosmos would be unchanging. Consequently, Descartes seemingly had a kind of rationalistic metaphysical basis for his laws of motion and impact. But he was not successful in this, being mistaken in his understanding of the laws of impact. In fact, while it has been a traditional interpretation of Descartes that he tried to 'deduce his physics from his metaphysics' one can dispute that this was so. D.M.Clarke, for example, has maintained recently that Descartes' laws of motion and impact were not thought of by Descartes as being formally deducible from his metaphysical principles. They were, rather, hypothetical in character, and could perhaps be in error. On Clarke's view, Descartes was more firmly wedded to his metaphysics than his physics, though even the metaphysics might require reconsideration if the Church so decreed.[67] This takes us a considerable distance from the traditional interpretation of Descartes.

Even so, from what has been said above, it might appear that Descartes' method and science were fundamentally rationalistic, aprioristic, or deductivist, with negligible empirical imput. But this would certainly present quite a false picture of his *actual* scientific practice, which involved a good deal of empirical work (for example, he engaged in dissection), geometrical analysis of optical problems, the use of explanatory models and analogies, hypotheses of various kinds, and indeed all the intellectual hand-holds that we might expect a scientist to try to employ.[68] It would be quite misleading to suggest that Descartes simply sat down in his study and, starting from the 'cogito', deduced his way right through to a description of the movements of the heavens and the constitution of oil or quicksilver. And as has been said, the asserted deductive link between the metaphysics and the first principles of physics is regarded as dubious today.

Indeed to imagine that this was Descartes' actual procedure would be to fall for his rhetorical metascientific pronouncements. These had their purpose, as we have seen. They sought to show that Cartesian science was firmly attached — by long chains of deductive reasoning — to first principles that were metaphysically well-grounded. With God as the non-deceiving guarantor, holding up the apex of Descartes' 'arch of knowledge', his system should have been secured against the attacks of the most sceptical of sceptics. However, it might be premature to dismiss all the arguments of the *Discourse* as mere philosophical rhetoric. Consider, for example, the following well-known passage:

> I have first tried to discover generally the principles or first causes of everything that is or that can be in the world, without considering anything that might accomplish this end but God Himself who has created the world, or deriving them from any source excepting from certain germs of truths which are naturally existent in our souls [that is, Descartes' well-known 'innate ideas']. After that I considered which were the primary and most ordinary effects which might be deduced from these causes, and it seems to me that in this way I discovered the heavens, the stars, an earth, and on the earth, water, air, fire, the minerals and some other such things, which are the most common and simple of any that exist, and consequently the easiest to know. Then, when I wished to

descend to those which were more particular, so many objects of various kinds presented themselves to me, that I did not think it was possible for the human mind to distinguish the forms or species of bodies which are on the earth from an infinitude of others which might have been so if it had been the will of God to place them there, or consequently to apply them to our use, if it were not that we arrive at the causes by the effects, and avail ourselves of many particular experiments. In subsequently passing over in my mind all the objects which have ever been presented to my senses, I can truly venture to say that I have not there observed anything which I could not easily explain by the principles which I had discovered. But I must confess that the power of nature is so ample and vast, and these principles are so simple and general, that I observed hardly any particular effect as to which I could not at once recognise that it might be deduced from the principles in many different ways; and my greatest difficulty is usually to discover in which of these ways the effect does depend upon them. As to that I do not know any other plan but again to try to find experiments of such nature that their result is not the same if it has to be explained by one of the methods, as it would be if explained by the other.[69]

A helpful interpretation of this interesting passage has been offered fairly recently by Daniel Garber.[70] He supposes that Descartes' plan was, as the text suggests, to start with certain general metaphysical principles, and from them deduce the fundamental laws of physics. (This interpretation does not mesh exactly with that of D.M.Clarke.) Descartes also had direct access to empirical information, from observation and experiment. He could then — or he thought he could — give all the possible causes of the effects that would both explain the phenomena and be consistent with the most general physical and metaphysical principles. Then he could supposedly carry out some 'crucial' experiments which would allow him to eliminate all the possible explanatory causes except one; and this would remain as the true explanation. By this means, Descartes could hopefully sustain certainty in science, use deductions from first principles, and invoke certain experimental procedures. And he would not be offering a 'mere' hypothetico-deductive methodology of science. However, although this may have been Descartes' original 'plan' (or, I could suggest, how he wished to present his system to the world), when it came to the crunch — that is, when he attempted to write down his whole system in the *Principles of Philosophy* (clearly and distinctly!) — he found that he was unable to make his text look anything like what his metascientific rhetoric might have led one to expect. And so he had to resort to a wide range of hypotheses, once he was a short distance from the alleged first principles. So the 'official' Cartesian methodological program was a failure, even though in point of fact Descartes made some extremely significant contributions to science, such as the discovery of the sine law of refraction. He could 'do' science well enough even though his actual practice did not mesh with his doctrines as to how it 'ought' to be done.

Leaving this question of the mesh of Descartes' theory and practice aside for the present, we may usefully refer once again to the 'shape' of his metascience. It should be clear, I think, that it was still structured according to the venerable 'arch'; but unlike Bacon's system, the emphasis was chiefly on the 'downward', deductive limb. The 'upward' ascent was made with remarkable expedition. Moreover, Descartes seems to have been trying to make the upward ascent (usually inductive) deductive, with his putative proof of the existence of God — possessed of certain characteristics useful to the well-being of a metascientist with a penchant for deductive chains of reasoning! But it must be emphasised that it was Descartes' metascience that was arch-shaped. I don't know that his actual scientific and mathematical investigations could be said to be of any particular architectural form.

The way in which Descartes' metascience was beholden to ancient traditions is a matter of considerable interest, revealing as it does the ambiguity existing within those traditions. One of the most interesting passages appears in the second set of replies in the *Objections and Replies*, where Descartes referred to the work of the Greek geometers, which he saw as being deductive (synthetic) from first principles, but with no account given to the means by which the first principles were discovered. Somewhat amusingly, Descartes suggested that the Greek geometers *did* have a method of discovery (analysis) but chose to keep it secret![71] Likewise, in the *Regulae* he spoke of the ancients suppressing their discoveries 'fearing, it would seem, lest the simplicity of their explanation should make us respect their discovery less, or because they begrudge us an open vision of the truth'.[72] Yet, it has recently been argued by Hintikka,[73] Descartes' own methodology (*in mathematics*) bore a considerable similarity to what must have been the ancient geometers' method of discovery. I think it may be useful to say something about what this 'analytic' method in Descartes' geometrical work may have been, for here perhaps — more than in the work in the empirical sciences — there was a closer relationship between theory and practice.

One might think that Descartes' 'analysis', if anything, was the move to the first principles — to God, for example — for this would have been an attempted movement from effects to causes. (It would, however, be deductive, so far as Descartes saw the matter.) This seems a reasonable interpretation, even though there are passages[74] which suggest that Descartes, like many others, was uncertain as to the 'direction' of analysis and synthesis, the problem having arisen because of the conflation of analysis and synthesis in the *methodological* sense with that in the geometrical sense, as originally proposed by Pappus in the fourth century A.D.[75] But in Descartes' own *geometrical* analysis — which gave rise to the whole branch of mathematics known as analytical geometry — the usage was, Hintikka has argued, remarkably similar to that first intended by Pappus.

In Descartes' *geometrical* analysis, what was being 'analysed' was a geometrical figure. The points and lines of the figure were expressed algebraically, so equations could be set up which would then be solved by ordinary algebraic methods. So, says Hintikka, Descartes' analysis should be seen as an analysis of geometrical configurations, not of stages of proof, or somehow finding a pathway whereby one could move from experience to explanatory principles. But when one abstracts certain variables (for example, position, in Galileo's work on motion which we considered earlier) from the complex of observable phenomena, then one is doing something analogous to geometrical analysis. At this point, then, the methodological and mathematical traditions of analysis appear to fuse, as we shall shortly see stated explicitly by Newton. The 'new' analysis, then, involved the search for functional relationships expressible in algebraic terms, not merely a breaking-down of the stages of a deductive argument into steps of an ever-smaller — and presumably more self-evident — kind. Nor was the 'new' analysis just another name for induction, though that meaning still tended to persist for a time.[76] All this had rather little to do with the arguments of the *'cogito'*, presented in the *Discourse* and the *Meditations*, and the grand cosmic vision of the *Principles of Philosophy*. So again we find the connection between science and metascience rather tenuous.

The Port-Royal logicians

As a transition from Descartes and Cartesianism to the empiricist philosophies of the British school (that is, the writings of Locke, Berkeley and Hume, each influenced in one way or another by the work of Isaac Newton), it may be helpful to say a few words about what was one of the most influential texts on logic, epistemology and method of the seventeenth century — the so-called *Port-Royal Logic*.[77] This was the (anonymous) work of two Jansenist[78] theologians, Antoine Arnauld (1612–1694)[79] and Pierre Nicole (1625–1695), written as a text-book of logic and method for the use of students at the Port-Royal College in Paris, and elsewhere. In their introduction, the authors stated that they had composed their work in response to a challenge to write down all that was of any use in logic, in four or five days. It was initially supposed that it could all be done in a day; but eventually 'four or five' were needed! The task, we are led to believe, took longer than expected because 'so many reflections presented themselves that it became necessary to write them down, in order to proceed'.[80] But this account is scarcely credible, for the text (at least as we know it today) was a carefully wrought piece of didactic prose, giving an admirable epitome of the philosophical thought of the period; and as such it became widely used and highly influential.

The *Port-Royal Logic* is particularly interesting to us as a blend of Aristotelianism and Cartesianism, the chief concern of its authors — to show the falsity of scepticism — being of lesser concern to us today. Aristotle's *Organon* had, of course, originally been intended as an aid to clear thinking, but over the centuries it was chiefly the logic of the syllogism that had exercised scholars' attention, and the purpose of logic — straight thinking and command of debate — had rather been lost sight of. Now, in this new and popular text, the original role of logic was resuscitated. However, the logic of the syllogism had not itself been changed, and neither had the leading features of the Aristotelian approach to scientific knowledge; so we still find Aristotle's doctrine of the five predicables and the ten categories given prominent treatment in the *Port-Royal*. Much of the text was, therefore, still basically Aristotelian, and its method of inquiry could be boiled down to the search for the essential definitions of things.

However, there were some distinctive seventeenth-century emendations of the ancient doctrine. It has been a well-known maxim of mediaeval Aristotelians that *nihil est in intellectu quod non prius fuerit in sensu* — nothing is in the mind but what is first carried in by the senses. Arnauld and Nicole, following Descartes, disagreed. The *cogito* was cited as an acceptable example of an innate idea.[81] And Descartes' doctrine of 'clear and distinct' ideas was given prominence.[82] The authors also offered a discussion of 'modes', 'substances' and 'relations' that certainly seems to have had some role in shaping the subsequent discussion of these terms in Locke's *Essay*.[83] A substance was presented as a self-subsistent thing. It could, however, be *mod*ified in some way — or have some particular quality. For example, 'body' was a substance; but 'round' was a mode. You could have a body without its being round, but you could not have roundness subsisting by itself, without body. So, said the authors, it was impossible to conceive of the mode 'roundness', without conceiving its relation to the substance of which it was a mode.[84]

Arnauld and Nicole also offered a simple linguistic theory.[85] After discussing the relationship between signs and the things that they signify, they stated that 'words are by institution the signs of thought'[86] — just as respiration may be taken as a sign of life. This was to provide a basis for much of the overall philosophical program of the French Enlightenment: if you wanted to get your thoughts straight, you needed a well-constructed language appropriate to the task.[87] So, working on the assumption that words were signs or indications of thoughts, the authors devoted the second part of their book to a study of the structure of language — to general grammar.

The third part of the *Port-Royal Logic* was devoted to an exposition of the standard rules of Aristotelian logic, which we shall not attempt to discuss here. The fourth section, which was regarded in its day as particularly innovative within a didactic text was devoted to 'method'. However, in point of fact what was included here was by no means so very innovative. As the reader may very likely anticipate, the consideration of 'method' was to a large degree a discussion of our old friends 'analysis' and 'synthesis'. And again the geometrical and methodological aspects of the two were conflated.

The actual definition of analysis and synthesis was orthodox:

> [T]here are two kinds of method, one for *discovering truth*, which is called *analysis*, or the *method of resolution*, and which may also be termed the *method of invention*; and the other for explaining it to others when we have found it, which is called *synthesis*, or the *method of composition*, and which may be also called the *method of doctrine*.[88]

The reader was also informed[89] that one might proceed *a priori* from causes to effects, or *a posteriori* from effects to causes.

As an actual example of analytical method, the authors described how the vexed question of the immortality (or otherwise) of the soul might be settled. The soul thinks. (Here the *cogito* argument was mentioned.) The idea of substance extended, or body, is not to be found within the concept of thinking. So thinking is not a mode of substance. So thinking substance and extended substance are two different kinds of substance. Therefore, destruction of extended substance does not involve destruction of thinking substance. Also, the soul, being neither divisible nor composed of parts (not being substance extended), cannot perish and is therefore immortal.[90]

One might well wonder what such a strange argument was doing in a book of *logic*. And the position is made yet more curious for us by the authors' claim that 'this is the analysis of the geometers'.[91] Which said, they then offered their version of a statement of Pappus's traditional account of geometrical analysis and synthesis.

One may also be puzzled today to understand how Arnauld and Nicole could see any similarity between their example of reasoning by analysis and the ancient method of geometrical analysis. The situation is probably best explained in sociological terms. The fact that the *Port-Royal* authors begged the question in their example of analytical reasoning was, I suggest, simply because the social group to which they belonged predisposed them to produce a particular result that was socially acceptable, by an argument having the appearance of sophistication and logical rigour. Besides, we should not criticise Arnauld and Nicole for failing to do the impossible — for failing to make an inductive argument deductive (or for failing to prove the existence of immortal souls). They did emphasise the importance of using both analysis and synthesis as complementary methods, for the two

> differ only as the road by which we ascend from a valley to a mountain does from that by which we descend from the mountain into the valley, which is no difference of road, but only a difference in the going.[92]

But the actual example of analysis that they gave seems to indicate that they had nothing to offer in the way of useful advice as to how to conduct enquiries so as to yield new information, or discover principles, by the 'analytic method'.

It is further worth noticing the conflation of logic and scientific method. The two have continued to be treated together in text-books until well into the twentieth century[93] yet they are uneasy bedfellows and really came together somewhat by historical chance, some aspects of which history we are exploring in this book. In the seventeenth century, however, the union would not have seemed in any way unnatural. The authors were concerned to make clear the manner in which men reason. Here the canons of logic, and some account of the way in which conclusions are reached, by analysis or synthesis, by induction, deduction, or whatever, were all relevant. But this does not mean that logicians had anything really useful to say about how to conduct investigations in order to obtain new theoretical knowledge. And it does not mean that it is correct to suppose that scientific inquiry is an essentially logical procedure, though it has long been the tradition that it is so. As we shall see, the divergence of studies of logic and scientific method is a feature of certain important trends within twentieth-century philosophy of science. Whether this is to be deplored or applauded is a matter that it is best not to pursue at this juncture. In any case, there is still a strong continuing tradition among some modern metascientists to concern themselves with the examination of the claimed logical structure of science.

Newton

The great rival to the Cartesian philosophy and cosmology in the seventeenth and eighteenth centuries was the system propounded by Sir Isaac Newton (1642–1727), the most consummately successful of all seventeenth-century scientists and mathematicians, and generally regarded as one of the greatest intellects in human history. Newton wrote two major books, the *Principia* of 1687[94] and the *Opticks* of 1704.[95] The former was concerned with the foundations of mechanics and their application to astronomy; the latter was concerned with optical phenomena and Newton's theory that white light was a mixture of all the different coloured lights of the spectrum. Newton wrote little directly on philosophy of science,[96] but what he had to say has been the subject of the most searching examination. And clearly any methodological pronouncements from a scientist of Newton's calibre must be of considerable interest to the present inquiry. So we shall seek to explicate Newton's ideas on scientific methodology, seen in the context of his chief scientific writings.

The *Principia* began with the three celebrated 'laws of motion' and certain definitions, which served as starting points for a deductive (synthetic) system carried out, *in more geometrico*, to a host of 'theorems' in mechanics which could find application to specific physical problems.[97] Further, when suitable empirical 'boundary conditions' were introduced, predictions of planetary motions and the like could be made with remarkable success. Involved in all this was Newton's postulate that gravity was a ubiquitous force in nature, which acted universally according to the inverse square law, its effects not being confined to the region of the Earth alone. Unlike Descartes, Newton was more than willing to entertain the notion of a vacuum; indeed the fact that the motions of the heavenly bodies could persist indefinitely required that there be no friction to impede their motion. Once a body was in motion it would, according to Newton, continue to move without acceleration or deceleration unless acted on by some external force. This was his principle of 'inertia', as stated in his first law of motion.

There was debate until well into the eighteenth century as to the relative merits of the Newtonian and Cartesian (vortex) cosmologies, with a good deal of English and French nationalistic fervour influencing the issue. Eventually, the matter was considered settled in favour of Newton. The two theories made different predictions as to the figure of the Earth, and when the shape was determined by direct survey it turned out that it was that which the Newtonian theory required. But this matter need not detain us. In any case, the Cartesians could never explain Kepler's third law[98] by means of the theory of vortices, thus failing to meet the challenge thrown out to them by Newton.[99].

We shall return in a moment to consider the status of the *Principia*'s three laws of motion, but first let us say something about the *Opticks*, the literary style of which was remarkably different from the awesome *Principia*. On the opening page of the *Opticks* we read:

My design in this book is not to explain the properties of light by hypotheses, but to propose and prove them by reason and experiments: in order to which I shall premise the following definitions and axioms.[100]

Following this, we find that rays of light, 'refrangibility' and 'reflexibility', angles of incidence, reflection and refraction, 'homogeneal' light, and colours of homogeneal and heterogeneal light, were carefully *defined*. It should be noted, however, that in the *definition* of homogeneal light Newton seems to have slipped in some element of theory as to white light being a mixture of lights of different colours — perhaps the major theoretical claim of the *Opticks*.

After the definitions came a set of eight axioms, and then a long series of propositions and theorems, with supporting argument for each. It is evident, however, that to a large extent Newton did not actually *deduce* the theorems directly from the definitions and axioms alone, for he frequently found it necessary to have recourse to *experiments* in order to carry through the arguments successfully. And there were few strictly mathematical deductions. (For that matter, there wasn't so very much mathematics in the book as a whole, which is no doubt the main reason why it has always been much more widely read than the *Principia*.) So we find that Newton constantly had to refer to the results of experiments that he had himself performed, in order to construct the arguments from the definitions and axioms to the various theorems.[101]

For example, the issue that particularly exercised Newton's attention was the cause of the colour of the spectrum. Descartes had proposed that colour was *produced* by the prism when a light ray struck the surface obliquely, different colours being due to the different spins received by the light corpuscles on their glancing contact with the glass.[102] Newton, however, thought the prism merely sorted out the colours already present in the white light. To distinguish between the two theories, he showed that that a second prism did *not* alter the colours. As shown in Figure 13,[103] it merely deviated them further, and a particular angle of deviation was associated with each colour. Also, the colours could be recombined to give white light by inverting the second prism.

Figure 13

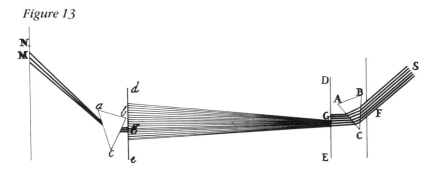

Significantly, Newton maintained that his conclusion in this matter could be reached without reference to any hypothetical commitments as to the nature of light — for example, whether it was corpuscular or wave-like in nature. Thus, he was wont to claim that his scientific work was in no way hypothetical in character. ('I frame no hypotheses'[104] he wrote in the *Principia*.) The optical and mechanical first principles could be inferred directly from experiments — or so Newton would have his readers believe.

With these introductory remarks, we are now in a position to consider what was perhaps Newton's major statement of his methodology of science. It occurs in the famous '31st Query'[105] to the *Opticks*, where we may read:

> As in mathematicks so in natural philosophy, the investigation of difficult things by the method of analysis, ought ever to precede the method of composition [i.e. synthesis]. This analysis consists in making experiments and observations, and in drawing general conclusions from them by induction, and admitting of no objections against the conclusions, but such as are taken from experiments, or other certain truths. For hypotheses are not to be regarded in experimental philosophy. And although the arguing from experiments and observations by induction be no demonstration of general conclusions; yet it is the best way of arguing which the nature of things admits of, and may be looked upon as so much the stronger, by how much the induction is more general. And if no exception occur from phaenomena, the conclusion may be pronounced generally. But if at any time afterwards any exception shall occur from experiments, it may then begin to be pronounced with such exceptions as occur. By this way of analysis we may proceed from compounds to ingredients, and from motions to the forces producing them; and in general, from effects to their causes, and from particular causes to more general ones, till the argument end in the most general. This is the method of analysis: and the synthesis consists in assuming the causes discover'd, and establish'd as principles, and by them explaining the phaenomena proceeding from them, and proving the explanations.
>
> In the two first books of these Opticks, I proceeded by this analysis to discover and prove the original differences of the rays of light in respect of refrangibility, reflexibility, and colour... And these discoveries being proved, may be assumed in the method of composition for explaining the phaenomena arising from them:...[106]

This very interesting passage is well worth several readings. Doubtless, it will quickly be noticed that we have here, once again, our 'arch of knowledge'; indeed we find it in a state of almost pristine purity. So let us give it a pictorial representation (Figure 14) using Newton's particular terms.

Figure 14

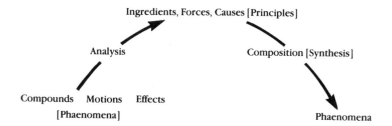

Ingredients, Forces, Causes [Principles]

Analysis   Composition [Synthesis]

Compounds   Motions   Effects   Phaenomena
[Phaenomena]

We will also further note that Newton's terminology was the exact opposite of Hooke's.[107] But since, as has been said, the whole question of the right 'direction' of analysis had become utterly confused, this need scarcely cause surprise. Note that Newton conflated the mathematical and the methodological traditions of analysis and synthesis, which conflation had been, I believe, partly responsible for the whole terminological confusion. It may be that Newton meant that analysis, as in Descartes' geometry, should refer to the determination of functional relationships between variables. But I am inclined to the view that this was not the case, or at most only partially so. Newton seems to have been concerned with analysis in a kind of 'chemical' sense as much as anything else — that is, from 'compounds' to 'ingredients'. His conception of the process of analysis was evidently much broader than that to be found in the mathematical tradition alone.

As we have seen, Newton was particularly anxious to avoid any accusation that his reasoning was in the least degree hypothetical.[108] His method had to be something much better than the testing of the implications of hypotheses. He was no mere hypothetico-deductivist in his program of metascience; some kind of certainty was to be sought. Newton was fairly specific about this. General conclusions were to be drawn inductively from experiments and observations, and be subject only to experimental refutation — not mere metaphysical quibbling. In so far as the principles were drawn directly from experiments and observations, then Newton's science should (he claimed) be certain and secure.

Readers will probably be wishing to raise the objection that general statements cannot be inferred from specific items of empirical information: this is the notorious 'problem of induction'. But although induction is certainly invalid from a logician's point of view, one may well ask whether science *has* to be deductive in all its aspects. Besides, the 'problem of induction' was really only raised to the status of a major philosophical issue by Hume in the eighteenth century, well after Newton's death.[109] Moreover, the seventeenth-century meanings of induction and deduction were not so strictly delineated as they are today. For example, Newton himself in one remarkable passage stated that he deduced general propositions from phenomena by induction![110] It would appear that Newton — like many modern scientists — possessed a general belief in the uniformity of Nature. And where experimental 'proof' was properly forthcoming, general principles could supposedly be 'deduced' (or inferred) from observations and experiments. It was speculations that ran beyond the limits of experimental inquiry that Newton rejected as hypothetical and unacceptable.[111]

Yet I venture to suggest that Newton erred in all this. For example, his first law of motion referred to the behaviour of objects that had never actually been observed — bodies moving uniformly in straight lines totally free from the influence of any external forces. It could hardly be said, therefore, that it was an unambiguous induction from phenomena. Or the law of reflection in optics employed the notion of a light ray, which was and is very much a theoretical construct, as opposed to a real object.[112] Again, as we shall shortly see, Newton believed in an atomic theory of matter, with very little direct experimental evidence in favour of what was, frankly, an hypothesis (albeit a most excellent one). Finally, as has been emphasised by I.B.Cohen, there is not much evidence that Newton's actual scientific practice had much to do with the public schema of analysis and synthesis that appeared in the *Opticks*.[113]

A little more light (but perhaps not so very much) is thrown on Newton's metascientific views by the four 'Rules of Reasoning in Philosophy', which were printed in the third edition of the *Principia*. These were:

1. We are to admit no more causes of natural things than such as are both true and sufficient to explain their appearances.

2. Therefore to the same natural effects we must, as far as possible, assign the same causes.

3. The qualities of bodies, which admit neither intensification nor remission of degrees[114] and which are found to belong to all bodies within the reach of our experiments, are to be esteemed the universal qualities of all bodies whatsoever.

4. In experimental philosophy we are to look upon propositions inferred by general induction from phenomena as accurately or very nearly true, notwithstanding any contrary hypotheses that may be imagined, till such time as other phenomena occur, by which they may either be made more accurate, or liable to exceptions.[115]

The first rule can be construed as a version of the famous 'Ockham's razor',[116] or principle of simplicity or parsimony in science. It says that other things being equal, and given a choice of theories, explanations, hypotheses, laws or whatever, the simplest one is to be preferred. The second rule gave expression to Newton's faith in the uniformity of nature and hinted at the doctrine of primary and secondary qualities. As an example of the rule in use, Newton could have excluded the Aristotelian claim that the laws that obtained in the superlunary and the sublunary realms were fundamentally different. Specifically, Newton himself used the rule to justify the supposition that the laws of the 'macro' world of physical objects held good in the 'micro' world of atoms. The fourth rule said very much the same as the methodological statement from the *Opticks*' 31st Query that has already been discussed.

We have seen that Newton set his face against the use of hypotheses in science: he was unsympathetic to speculative excesses or ideas that outran their empirical bases. Yet some very interesting suppositions lurked only just below the surface of the *Principia*. For example, Newton believed in the doctrines of absolute space and absolute time;[117] that is, he believed that there was a frame of reference against which all motions and times could be determined, located in God's 'sensorium'.[118] The notions of absolute space and time brought Newton a good deal of subsequent criticism, some of which we shall attend to later. But it is very interesting to note that even within Newton's own system he was forced to use some relative measure of space, rather than God's sensorium! Thus, in the first edition of the *Principia* (1687) there were, in fact, no less than nine acknowledged hypotheses. But by the time the third edition was published (1726) some of these had become transmogrified into the 'Rules of Reasoning' that we have just been discussing; some were renamed 'Phaenomena'; and one hypothesis remained unchanged, and with its same label, through the three editions. This hypothesis — which incidentally was tucked away deep in the text, and given no special prominence — simply said: 'The centre of the system of the world is immovable'.[119] But this was a remarkable statement, for it meant that in practice Newton's absolute space was a redundant scientific concept. He had to choose some physical object as a point of reference for the positions and motions of other bodies, and the Sun served this purpose. Newton did not actually regard the Sun as a fixed object with respect to absolute space, but the use of the Sun (or, it could have been, some other body) as a reference point was inescapable. He had no practical access to absolute space and time, and these were for practical purposes quite useless to Newton's science. They were, to be sure, of paramount importance in his metaphysics and system of natural theology, but if he had followed his own 'First rule of reasoning', he should perhaps have excised them.

Just a few words should be added here about Newton's ideas on matter and his version of the doctrine of primary and secondary qualities. The chief text is again in the 31st Query, where the following well-known passage may be found.

> [I]t seems probable to me, that God in the beginning formed matter in solid, massy, hard, impenetrable, movable particles, of such sizes and figures, and with such other properties, and in such proportion to space, as most conduced to the end for which He form'd them; and that these primitive particles, being solids, are incomparably harder than any porous bodies compounded of them; even so very hard, as never to wear or break in pieces; no ordinary power being able to divide what God Himself made one in the first creation.[120]

It will be seen that by virtue of the fact that Newton's atoms were supposed never to wear out he had a kind of 'chemical' basis for the doctrine of the uniformity of nature. The Newtonian cosmos, therefore, came to be envisaged in the eighteenth century as being like a great machine, which once created and set in motion by God would continue to run for evermore without further attention, and consequently generally to be admired by the rational component of the Creation: man.

We should note further that in saying that the original atoms of the universe were *solid*, *massy*, *hard*, *impenetrable* and *movable* Newton was specifying what he considered to be the primary qualities of bodies (though he did not himself employ that term). In order to be a body at all, a body had to be possessed of these five qualities.[121] Secondary qualities of matter, such as colour or taste, arose from the manner in which the atoms, endowed only with primary qualities, were arranged together, or acted upon one another. It does not appear, however, that Newton selected his primary qualities on the basis that they were amenable to mathematical description and manipulation. But they were supposedly the *real* qualities of bodies, not (like beauty) merely residing in the eye of the beholder.

Before we leave Newton, I should like to raise a question of fundamental importance in philosophy of science, although Newton dealt with it only by giving his own answer to the problem. He did not discuss it in any general way or as a general philosophical issue.

The problem is this. The mechanics of the *Principia* was set out deductively, in a 'theoretical language'. It had the formal appearance of a mathematical deduction system, analogous to the axiom system of Euclid's *Geometry*. Yet Newton's system *worked* — beautifully so. It seemed to mesh exactly with the real world. Why should this be, given that his three laws of motion can be construed as definitions, as much as anything else?[122] Why should an axiom system have anything to do with the real world? (One could, of course, ask the same question of Euclidean geometry.) What was the nature of the 'hook' linking Newton's theoretical science with the real world?

Newton would have answered this question, in part, by saying that his laws of motion were true general statements, based on induction from experience. It was by induction that he knew them. They could be 'induced/deduced' directly from phenomena. So that is where the main hook would have been. But, as we have seen, this claim that Newton's laws were direct inductions from phenomena seems fundamentally implausible today, though until almost the end of the nineteenth century the truth of Newton's mechanics and physics seemed secure, so that people might well have been swayed into accepting Newton's metascientific assertions. I shall not attempt to resolve this difficulty here and now; but notice is hereby given that the 'hook problem', if so it may be called, has constantly recurred in philosophy of science, right up to the present. And Newtonian physics may be seen as presenting a paradigm case of the problem. But Newton himself did not, I think, see the matter as a major difficulty. As previously mentioned, the 'problem of induction' was not a burning issue in Newton's day. And he had his own answer at the ready in his 'Fourth Rule of Reasoning', or regulative principle, in which every good Newtonian might take comfort. Subsequent metascientists have not found his a satisfactory means of escape from the difficulty. But again Newton would very likely not have been much perturbed. The section of the *Principia* entitled 'System of the World' showed how the general equations established in the theoretical part of the text applied with remarkable exactness to our own solar system when suitable observational data were introduced for planetary distances, lengths of planetary years, and so on, and accurate predictions could be and were made for the future positions of celestial objects. So there was further 'hooking' here, as well as that which was claimed at the level of 'first principles'. It was, no doubt, this 'lower-level hooking' that carried the greatest degree of conviction, rather than Newton's metascientific pronouncements.

Locke

Newton was not, of course, a philosopher in the modern sense (though he certainly regarded himself as a philosopher of nature), and he did not give much attention (publicly) to metascientific questions and the philosophical underpinnings of his whole system. This task was undertaken, perhaps somewhat gratuitously, by his contemporary, John Locke (1632–1704), in his very influential *Essay Concerning Human Understanding* (1690).[123] Locke had acquaintance with the methods of the new science through his membership of the Royal Society, and he collaborated in some of the investigations of the chemist, Robert Boyle (1627–1691). Locke was also quite a distinguished physician and an accomplished naturalist. There is a legend that he found the geometry of Newton's *Principia* too difficult to struggle through, and he asked the Dutch physicist Huygens to reassure him of its soundness; then he simply took Newton's results and used them as a framework for the development of his own philosophy.[124] This story is, however, probably only a half-truth, and Locke was able to write a worthy review of the *Principia* not long after its publication.[125]

A link between Newton's science and Locke's philosophy is immediately suggested by the well-known 'Epistle to the Reader' that one encounters at the beginning of the *Essay* and which contains the following celebrated passage:

> The commonwealth of learning is not at this time without master-builders, whose mighty designs in advancing the sciences will leave lasting monuments to the admiration of posterity; but every one must not hope to be a Boyle or a Sydenham, and in an age that produces such masters as the great Huygenius, and the incomparable Mr.Newton, with some other of that strain, it is ambition enough to be employed as an under-labourer in clearing the ground a little, and removing some of the rubbish that lies in the way to knowledge...[126]

It may appear, then, that it was to be Locke's self-appointed task to serve as an under-labourer in the Newtonian garden. He was to try to set straight the philosophical underpinnings of the Newtonian science, basing them on empiricist principles as Newton himself would have approved. But this cannot be the exact truth of the matter, for Locke's *Essay* was virtually complete by the time he first saw the *Principia*, the 'Epistle to the Reader' being a late addition. Rather, it appears that Newton and Locke shared similar views on the philosophy of the new science[127] and Locke was delighted to see its splendid vindication in the *Principia*. Both were antipathetic to the Cartesian system, Newton chiefly because of its alleged hypothetical character, and Locke because of its supposed dependence on the doctrine of 'innate ideas'. But even if Locke's *Essay* should not be seen as an attempt at a direct philosophical defence of the *Principia*, it may certainly be seen to have that role in relation to the new science and the mechanical philosophy as a whole.[128] It is possible that the 'under-labourer' passage marked the philosopher's acceptance of a subordinate role in the face of the technical impenetrability of science, already becoming manifest in Newton's formidable work.

Anyway, at the outset of the *Essay*, Locke, wearing his philosophical hat, presented a hard-hitting critique of the view that the human mind can be endowed with innate ideas. He envisaged the mind as being somewhat like a piece of blank white paper at birth,[129] which gradually had ideas marked upon it during the course of a life-time by sensations

and impressions (which, however, Locke on some occasions unhelpfully
called ideas). So sensations formed the prime source of ideas; this was the
empiricist doctrine:

> Our senses, conversant about particular sensible objects, do convey into the mind
> several distinct perceptions of things according to those various ways wherein those
> objects do affect them; and thus we come by those *ideas* we have of yellow, white, heat,
> cold, soft, hard, bitter, sweet, and all those which we call sensible qualities; which when
> I say the senses convey into the mind, I mean, they from external objects convey into
> the mind what produces there those perceptions.[130]

The suggestion, then, was that these simple ideas entered the mind
through the sense organs, in single file,[131] so to speak, following which
the mind could build up the simple ideas into complex ideas,[132] rather as
pieces of meccano can be fitted together to form a model. Also, the mind
supposedly had the capacity to think about or reflect upon the ideas with
which it was stocked by courtesy of the sense organs.[133]

Having thus stated the first essentials of his empiricist psychology, or
theory of the mind, Locke proceeded to give an attempted taxonomy of
the different sorts of ideas that might be found in the mind. It can be represented in summary form as follows:[134]

Simple Ideas (for example, coldness, whiteness, smell of rose)
(a) that enter the mind by one sense only
(b) that enter the mind by more than one sense
(c) that arise from reflection
(d) that are suggested 'by all the ways of sensation and reflection'
Complex Ideas
1. Modes[135]
(a) Simple (repetitions of simple ideas of one kind)
  (i) space
  (ii) duration
  (iii) number
  (iv) infinity
  (v) modes of motion, sounds taste, etc.
  (vi) pleasure and pain
  (vii) power [?simple]
(b) Mixed (consisting of combinations of simple ideas of different
     kinds, not concerned with particular things or substances); for
     example, obligation, drunkenness, a lie
2. Substances
(a) Single substances; for example, lead, man, sheep
(b) Collective ideas of several substances; for example, a flock of
     sheep
3. Relations, produced by the comparison of ideas; for example, cause
     and effect, identity and diversity, whiter, good and evil.

One may or may not think that this taxonomy represents an accurate
classification of the sorts of ideas (or concepts) that are typically found in
the mind. But this is not the point at issue. The significant thing is, I
suggest, the *kind* of philosophical program in which Locke was engaged.
It is true that Locke's system was not one of unalloyed empiricism, for in

fact the mind had to do quite a lot more than merely receive impressions like a piece of blank, white blotting-paper. It also had to be able to make comparisons between sense impressions, and form complex ideas out of simple ones. Nevertheless, Locke's empiricism was certainly one of the 'more pure' variety, though it may not have been, historically speaking, the very purest. Be that as it may, it was his *plan* to show how the attributes of the human mind, in all its intricacy, could be accounted for almost entirely by the build-up of sensory information during the course of a life-time. And if the performance of this plan may seem inadequate, it should be realised that even twentieth-century writers such as Russell or Carnap have also found it virtually impossible to develop a wholly empiricist system in a plausible and coherent manner.[136]

Locke intended his empiricist philosophy to be used as an underpinning for, or philosophical justification of, certain major metaphysical assumptions of the science of his day, being chiefly concerned with giving a warrant for the mechanical philosophy and the principle of causality — that like causes are always followed by like effects, every effect having a cause, there being no uncaused (physical) events.[137] The mechanical philosophy in the seventeenth century took on a great variety of forms, but in general it assumed that there was a world of atoms or corpuscles[138] below the level of the visible world. And the coalitions and interactions between these small particles, and perhaps their intrinsic attributes, might be held responsible for, or explain, the phenomena of the world of everyday experience. In addition, the mechanical philosophers supposed that all physical, chemical and biological phenomena, and to some extent mental phenomena as well, were to be accounted for in mechanical terms. So in many seventeenth- and eighteenth-century scientific writings we find a heavy reliance on mechanical models and analogies. Both Descartes and Newton were, in their different ways, mechanical philosophers. Boyle and Hobbes were philosophers of this kind *par excellence*.

How, then, was Locke's empiricist theory of knowledge to serve in the cause of the mechanical philosophy, employed by the natural philosophers of his day? We can see the connection in relation to the doctrine of primary and secondary qualities. It has sometimes been asserted, and there are passages in the *Essay* that support this view, that Locke supposed that for *some* perceived qualities there was a one-to-one relationship between the ideas and the actual qualities of the objects giving rise to those ideas; and these were the primary qualities: solidity, extension, figure, motion (or rest) and number. But for the secondary qualities such as taste, colour, smell and so on, there was no such one-to-one correspondence. Locke's words were:

> [T]he ideas of primary qualities of bodies are resemblances of them, and their patterns do really exist in the bodies themselves; but the ideas produced in us by these secondary qualities have no resemblance of them at all. There is nothing like our ideas existing in the bodies themselves. They are, in the bodies we denominate from them, only a power to produce those sensations in us: and what is sweet, blue, or warm in idea, is but the certain bulk, figure and motion of the insensible parts in the bodies themselves, which we call so.[139]

But if we take a 'one-to-one' correspondence interpretation for the primary qualities, it is not easy to see how in itself it might have helped Locke greatly in his campaign on behalf of the mechanical philosophy. So an alternative approach should perhaps be tried. One such is that offered by M.Mandelbaum in his book *Philosophy, Science and Sense Perception*.[140] This author suggests that for Locke a primary quality was one that *produced* ideas in the mind, rather than being one that in some manner *resembled* them. In fact, this meshes quite well with Locke's definition of 'idea' and 'quality':

> Whatsoever the mind perceives in itself, or is the immediate object of perception, thoughts or understanding, that I shall call an *idea*: and the power to produce any idea in our mind I call quality of the subject wherein that power is.[141]

Let us see where this approach in interpretation leads us.

According to Locke, the primary qualities were *always* found in bodies, whereas one might often find bodies without colour, odour, and so on: 'Sense constantly finds [them] in every particle of matter which has bulk enough to be perceived'.[142] Moreover, by a shift ostensibly legitimised by Newton's 'Third rule of reasoning', Locke could say that the bodies of the 'micro-world' of atoms or corpuscles would also be endowed with the primary qualities, as were the tangible objects of the 'macro-world'.[143] The secondary qualities, therefore, arose from the 'powers' of the combinations of the primaries.[144] For example, the metal gold is yellow, dense, shiny, etc. But this did not mean that the atoms or corpuscles of gold had themselves these qualities. It meant that the particles were of such shapes, configurations etc, that they gave rise to yellowness, denseness and shiningness in tangible specimens of gold. Looking back to the quotation cited above in favour of the 'one-to-one correspondence' interpretation we may see that it can in fact bear the interpretation that is here being advanced; namely that for Locke the primary qualities were the ones that *produced* ideas in our minds — of secondary qualities as well as primary.

But, it may be inquired, how did the primary/secondary quality distinction, seen in the proposed way, help Locke's campaign for the mechanical philosophy? The true answer is, I fear: not so very much. Nevertheless, one can see the direction of his line of thought. He would, for example, have liked to have been able to show that the micro corpuscles were solid, just like macro objects. So we find the following passage concerning solidity:

> [Solidity]... seems the idea most intimately connected with and essential to body, so as nowhere else to be found or imagined, but only in matter; and though our senses take no notice of it, but in masses of matter, of a bulk sufficient to cause a sensation is us; yet the mind, having once got this idea from such grosser sensible bodies, traces it farther, and considers it, as well as figure, in the minutest particles of matter that can exist, and finds it inseparably inherent in body, wherever or however modified.[145]

Sadly, however, this is no adequate proof of the solidity of the components of the micro-world. We could just as readily erect the hypothesis that the solidity of objects arose from the interactions of point atoms exerting force fields, or something of this kind, for all that Locke's arguments could show to the contrary. In fact, it must be acknowledged that there was really an intolerable tension between Locke's empiricist epistemology and his wish to give philosophical support to the tenets of the mechanical philosophy, which invoked the notion of *invisible*, yet existent, particles at the micro-level.

A somewhat similar situation is to be found with respect to Locke's discussion of 'substance', which, it will be recalled, he had designated a 'complex idea'. Locke thought that it was not possible for there to be a collection of qualities occurring together in objects without there being some 'support' for them: the qualities of yellowness and shiningness of gold must inhere in '*something*'. The 'something' Locke called 'substance':

> The idea...we have, to which we give the general name substance, being nothing but the supposed, but unknown support of those qualities we find existing, which we imagine cannot exist *sine re substante*, without something to support them, we call that support *substantia*; which...is in plain English, standing under, or upholding.[146]

This 'substance', Locke supposed, was something over and above — or rather, under and beneath — the primary qualities. Unfortunately he was forced to admit that 'we know not what it is'.[147] Nevertheless, the idea of substance was allowed into the class of complex ideas, even though it was *not* arrived at through the senses. Again, we find a conflict in Locke's general line of argument. To be consistent, he would have had to have retreated into a kind of phenomenalism, and abandon the mechanical philosophy, as in fact Berkeley did not long afterwards.[148]

In his discussion of substances, Locke stated that when we think of corporeal things we always combine together three kinds of idea: primary qualities, secondary qualities and power.[149] And of the notion of power he wrote:

> He that will examine his complex idea of gold, will find several of its ideas that make it up to be only powers: as the power of being melted, but of not spending itself in the fire, of being dissolved in *aqua regia*, are ideas as necessary to make up our complex idea of gold, as its colour and weight: which, if duly considered, are also nothing but different powers. For...yellowness is not actually in gold, but is a power in gold to produce that idea in us by our eyes when placed in a due light: and the heat which we cannot leave out of our idea of the sun, is no more really in the sun than the white colour it introduces into wax.[150]

This notion of power played an important role in Locke's philosophy, even though it was illegitimate within the confines of a strict empiricist philosophy:[151] a man is directly aware of phenomena, not causal powers.[152] I suggest that in fact when Locke was talking of 'power' he was really concerned with the philosophical problem of *causality*.[153] He could see and feel the quality of heat apparently emanating from the Sun. He could see and feel this happen many, many times. He could see the same antecedent (cause); he could see the consequent (effect). But he could not detect, as such the 'causal power', attributed to the Sun, with his senses. So strictly Locke should have retreated — as did Hume[154] — to the doctrine of 'constant conjunctions' with respect to causality. And in a few passages he did just this:

> [W]hen a countryman says the cold freezes water, though the word freezing seems to
> import some action, yet truly it signifies nothing but the effect; viz., that water, that was
> before fluid, is become hard and consistent; without containing any idea of the action
> whereby it is done.[155]

However, he seems not to have been able to sustain this position consistently. And in his section on 'Cause and effect' he almost went so far as to imply that causal power could be observed directly:

> In the notice that our senses take of the constant vicissitude of things, we cannot but
> observe that several particular[s], both qualities and substances[,] begin to exist; and
> that they receive this their existence from the due application and operation of some
> other being. From this observation we get our ideas of *cause* and *effect*.[156]

So we see that a substantial philosophical problem was beginning to emerge, arising out of the empiricist theory of knowledge. If science does, in fact, make extensive use of the cause/effect relation, it is difficult to give a satisfactory account of it in purely empiricist terms. As we shall see, this problem was to come to a head in the writings of David Hume.

However, if Locke was largely unsuccessful in his attempt to give a philosophical justification of the science of his day — and particularly its 'mechanical philosophy' — on the basis of an empiricist epistemology, there was nevertheless one aspect of his enterprise that can be counted a substantial success, and which had a very significant impact on science in a number of ways. I refer here to Locke's 'nominalism'.

We have already encountered this in connection with the work of the mediaeval scholar, William of Ockham.[157] We have also seen that the Aristotelian program of scientific inquiry involved the search for appropriate definitions of the essences of things, from which the properties could supposedly be deduced.[158] This philosophy, involving a belief in essences, may, following the terminology of Karl Popper,[159] be called essentialism. It presupposed that there are actually existent entities or things — such as Platonic or Aristotelian forms — corresponding to classes of objects. Essentialism, so to speak, reifies classes.

The opposing view, already presented by some mediaeval scholastics, was that only particular things can actually *exist*.[160] This position is called nominalism, reflecting the fact that essentialists and nominalists differ about the status of words. Suppose, for example, that we are Aristotelians and we successfully find the essence of gold. We might perhaps agree that the essential definition of gold is that it is a yellow, dense, malleable, fusible, fixed (that is, non-volatile) substance. Now if there were some such essence of gold actually existing — perhaps in the physical 'form' of gold, in some Platonic heaven, or as some hidden chemical constituent within the metal — then the essentialist doctrine would be true; and perhaps the further properties of gold could somehow be logically deduced or otherwise inferred from knowledge of this essence.

But all this seems extremely implausible, especially when put in this bald way; and Locke argued against the doctrine with considerable cogency. He said that a defining formula (such as that suggested for gold above) did not express the real essence of a thing; it expressed only the *nominal* essence.[161] (It might, perhaps, have been better if Locke had kept the word 'essence' out of the picture altogether; but we cannot alter his language now.) It said how the word 'gold' was used in common parlance — what its meaning was. It told us something about how humans had classified the world. They made a mental sorting of the objects in the world, and those that were yellow, dense, malleable, fusible and fixed they decided to call gold. That is, they created a sign (ie a word) to designate this particular class of objects. And of course it was quite impossible to deduce other properties of gold from its nominal definition. To do this, one had to have recourse to observations or scientific experiments. In saying all this, it is very clear that Locke had made a final repudiation of the Aristotelian methodology of science.

Nevertheless, it did not mean that Locke wanted to say that the notion of essence was absurd in its entirety. It was only the idea that essences could reside in definitions that he found objectionable; and rightly he pointed out that definitions were 'nominal essences', not the 'real essences'. But there *could* be 'real essences' of things. Thus he wrote in respect to gold:

> [T]he real essence is the constitution of the insensible parts of that body on which those qualities and all the other properties of the gold depend.[162]

So it was in the inward chemical constitution of gold that the essence of gold resided. Perhaps the modern chemist might be inclined to agree. He would not think it absurd to say, for example, that the essential feature of hydrogen lies in the fact that its atoms are made up of one proton and one electron. However, Locke supposed that the 'real essence' of a body was strictly unknowable. He did not imagine that science would ever find a way of showing the shapes of molecules, or anything of that kind. On the other hand he was prepared to entertain the idea of real essences, causally responsible for the phenomena of the macro-world. In this respect, his doctrine of 'real essences' was somewhat analogous to the doctrines of 'substance' and causal power.

The nominalism of Locke, and the wider move towards empirical methods in science with which it was associated, reflected the increasing interest in the seventeenth century in taxonomies and natural history, and tabular representations of knowledge. It also may be linked with attempts at language reform and the establishment of 'scientific languages', which were based upon the categorisation of concepts in taxonomic hierarchies.[163] More specifically, one can show, for example, how certain naturalists such as the influential John Ray (1627–1705) altered their classificatory systems during the course of their lifetimes, under the impact of Locke's nominalism.[164] Elsewhere, I have argued for a beneficial shift from essentialism to nominalism in eighteenth-century chemistry,[165] and D.L.Hull has argued that such a shift played a significant role in the establishment of evolutionary biology.[166] The important point, of course, is that the new philosophy claimed that new knowledge was to be obtained by experimentation, not by analysis of language or by establishing the correct definitions of things. If you wanted to know more about the properties of gold than anyone had ever known before you would need a chemical laboratory, not a dictionary! Thus, in this respect at least Locke's campaign on behalf of the new science bore some worthwhile fruit.

NOTES

1   Copernicus's *De Revolutionibus* was published on his behalf by one Andreas Osiander (1498–1552) and included a preface (by Osiander) which presented the theory merely as one intended to 'save the appearances'. The probability is that this was not Copernicus's own intention, but since the publication occurred in the last year of his life he was unable either to 'own' or 'disown' the preface. Some commentators have supposed that the preface was merely a discreet smoke-screen to distract attention from the supposedly theologically objectional treatise. We may note, however, that the idea of 'saving the appearances' was quite a commonplace in Greek thought and in mediaeval writings on metascience. (See O Barfield *Saving The Appearances: A Study in Idolatry* Faber & Faber London 1957 Chapter 7.) Also the doctrine has much in common with the nineteenth-century philosophy of instrumentalism, which we shall discuss in Chapter 5.

2   The broad outline of what Kepler achieved is admirably portrayed by Arthur Koestler in his well-known book *The Sleepwalkers* (Hutchinson London 1959; Penguin Harmondsworth 1964). For a more detailed reconstruction of Kepler's thought, see C Wilson 'How did Kepler Discover his First Two Laws?' *Scientific American* 1972 Vol 226 93–106. Kepler's thinking was enormously complex and rested in part on certain ancient Pythagorean and neo-Platonic notions of cosmic harmony. His work offered a remarkable blend of ancient and modern attitudes and procedures.

3   More precisely, Aristotle supposed that each of the four 'elements' had its own 'natural place' in the geocentric cosmos. If displaced from their 'natural' positions, objects would tend to revert to their natural places. Otherwise they would maintain their natural places indefinitely. In addition, motion of the heavenly bodies in their courses was supposedly sustained by what Aristotle called the 'Unmoved Mover' — in effect, his God.

4   The social character of seventeenth-century science was, however, naturally very different from that of today. Until the nineteenth century, science was largely an amateur activity. But the advent of professionalisation has drastically altered the whole character of the scientific movement, and its present incestuous relationship with the military-industrial complex is a far cry from the situation that obtained three hundred years ago. Even so, Francis Bacon had some vision of what social changes might attend the rise of science. And there has always been an intimate connection between technology and warfare.

5   G Galilei *Dialogo...dove nei Congressi di Quattro Giornate si Discorre sopra i Due Massimi Sistemi del Mondo Tolemaico e Copernicano: Proponendo Indeterminatamente le Ragioni Filosofiche e Naturali tanto per l'una quanto per l'altra parte* Florence 1632; *Dialogue Concerning the Two Chief World Systems — Ptolemaic & Copernican* trans Stillman Drake University of California Press Berkeley & Los Angeles 1953

6   G Galilei *Discorsi e Dimonstrazioni Matematiche Intorna a Due Nuove Scienze Attenenti alla Mecanica et i Movementi Locali... Con una Appendice del Centro di Gravita d'Alcuni Solidi* Leyden 1638; *Dialogues Concerning Two New Sciences* trans H Crew & A de Salvio Macmillan New York 1914 and McGraw-Hill New York 1963

7   G Galilei *Siderius Nuncius, Magna Longeque admirabilia...Spectacula Pandens, Suspiciendaque Proponeus Unicuique, Praesertim Vero Philosophis atque Astronomis, quae a G.Galileo... Perspicilli nuper a se Reperti Beneficio sunt Observata in Lunae Facie, Fixis innumeris, Lacteo Circulo, Stellis Nebulosis, Apprime Vero in Quaturo Planetis circa Jovis Stellam Disparibus Intervallis atque Periodis, Celerate Mirabili Circumvolutis; Quos Nemini in Hanc Usque Diem Cognitos, Novissime Auctor Depraehendit Primus, atque Medicea Sidera Nuncupandos Decrevit* Venice 1610; G Galilei *Il Saggiatore: Nel Quale con...Bilancia Squisita... si Ponderano le Cose Contenute nella Libra Astronomica e Filisofica di L Sarsi Sigensano, Scritto in Forma di Lettera all' Illmo...V.Cesarini* Rome 1623; *Discoveries and Opinions of Galileo: Including The Starry Messenger (1610) Letter to the Grand Duchess Christina (1615) and Excerpts from Letters on Sunspots (1613) The Assayer 1623)* trans Stillman Drake Doubleday New York 1957

8   The term, 'apologetic' works, now seems to be in quite common currency today but my own usage of it here derives from my colleague, Dr G A Freeland.

9   See S Drake 'Renaissance Music and Experimental Science' *Journal of the History of Ideas* 1970 Vol 31 pp 484–500

10   G Galilei *op cit* (note 6 1963) p 148

11   *Ibid* p 155

12   *Ibid*

13   Today we would do the calculation by means of an algebraic analysis:
Distance (s) = average velocity (v) × time (t)
(since the object is assumed to start from rest)
$= v/2 \times t$.
But v is proportional to t (*ex hypothesi*)
or v = acceleration (a) × t
so $s = \frac{1}{2}at^2$.

14   G Galilei *op cit* (note 6 1963) pp 171–2

15   *Ibid* p 172

16   A Koyré *Études Galiléennes* Hermann Paris 1939; trans A Mepham Harvester Press Hassocks 1978

17   A Koyré *Metaphysics and Measurement: Essays in Scientific Revolution* Chapman & Hall London

1968 p 13; the essay here is republished from an earlier article in *The Journal of the History of Ideas* 1943.

18   S Drake 'The Role of Music in Galileo's Experiments' *Scientific American* 1975 Vol 232 pp 98–104; P Ariotti 'From the Top to the Foot of the Mast of a Moving Ship' *Annals of Science* 1972 Vol 28 pp 191–203; T R Girill 'Galileo and Platonistic Methodology' *Journal of the History of Ideas* 1970 Vol 31 pp 501–30; S Drake *Galileo at Work: His Scientific Biography* University of Chicago Press Chicago 1978

19   J MacLachlan 'A Test of an "Imaginary" Experiment of Galileo' Isis 1973 Vol 64 pp 374–9

20   See Drake *op cit* (note 18 1975)

21   According to this formula, the distance travelled in accelerated motion was calculated by multiplying the time of travel by the speed at the middle instant of notion.

22   S Drake 'Galileo's Discovery of the Law of Free Fall' *Scientific American* 1973 Vol 228 pp 84–92

23   G Galilei *Le Opere di Galileo Galilei, Edizione Nazionale* eds A Favaro & I del Lungo 20 vols Florence 1890–1909 Vol 17 p 90 (The translation given here is from W A Wallace *Prelude to Galileo: Essays on Medieval and Sixteenth-Century Sources of Galileo's Thought* Reidel Dordrecht & Boston pp 79–104.)

24   W A Wallace 'Galileo and Reasoning *Ex Suppositione*' *op cit* (note 23) pp 124–59

25   See below, p 185

26   *S. Thomae Aquinatis… In Aristotelis Libros Peri Hermeneias at Posteriorum Analyticcarum Expositio cum textu et recensione leonina* eds P Fr Raymundi & M Spiazzi Marietti Italy 1955 pp 349–50 (commentary on Book 2 Chapter 8 Section 343 of Aristotle's *Posterior Analytics*)

27   This form of reasoning was traditionally called '*modus ponendo ponens*'.

28   Wallace *op cit* (note 24) p 136

29   *Ibid* pp 146 & 150–6

30   Some commentators have been inclined to place great emphasis on the influence of Archimedes' writings as a major factor in the rise of the modern scientific movement. See, for example, H F Kearney *Science and Change* Weidenfeld & Nicolson London 1971

31   For further discussion of Galileo's reasoning *ex suppositione*, see, for example: D W Mertz 'The Concept of Structure in Galileo: Its Role in the Methods of Proportionality and *Ex Suppositione* as Applied to the Tides' *Studies in History and Philosophy of Science* 1982 Vol 13 pp 111–31. For a critique of Mertz's paper, see: W L Wisan 'On Argument *Ex Suppositione Falsa*' *Studies in History and Philosophy of Science* 1984 Vol 15 pp 227–36.

32   G Galilei *op cit* (note 7 1957) p 274

33   One can, if one is so inclined, postulate tertiary qualities, such as beauty or ugliness, which lie even more definitely within the subjective domain.

34   F Bacon … *Instauratio Magna. Distributio Operis. Eius Constituuntur Partes Sex, Prima; Partitiones Scientiarum. Secunda; Novum Organum siue Indicia de Interpretatione Naturae. Tertia; Phaenomenon Universi, siue Historia Naturalis & Experimentali ad Condendam Philosophiam. Quarta; Scala Intellectus. Quinta; Prodromi, siue Anticipationes Philosophiae Secundae. Sexta; Philosophia Secunda, Siue Scientia Actiua* London 1620. *The New Organon* was, as its title suggests, intended as a replacement for Aristotle's ancient 'instrument' of knowledge. It was not intended to stand alone, but was to form the *second* part of Bacon's *Great Instauration*, or great reformation of learning. As indicated in the Latin title above, this was to be in six main parts: (1) *The Division of the Sciences*; (2) *The New Organon: or Directions Concerning the Interpretation of Nature*; (3) *The Phenomena of the Universe; or a Natural and Experimental History for the Foundation of Philosophy*; (4) *The Ladder of the Intellect*; (5) *The Forerunners; or Anticipations of the New Philosophy*; (6) *The New Philosophy; or Active Science*. Only a fraction of this grand project was ever completed, and even the *New Organon*, which was one of the more finished parts of the whole, remained incomplete. I have used the version of the *New Organon to be found in The Philosophical Works of Francis Bacon… Reprinted from the Texts and Translations, with the Notes and Prefaces, of Ellis and Spedding* ed J M Robertson Routledge London and Dutton New York 1905 pp 256–387

35   That is, a simple induction in which one collects information about the members of a class of objects and then makes a generalised statement about all the members of the class on the basis of the limited range of information that has been assembled.

36   A representation such as that in Figure 9 is supported by a remark in another of Bacon's works, *The Advancement of Learning* (1605): 'For knowledges are as pyramides, whereof history is the basis.' (Bacon *op cit* [note 34 1905] p 95)

37   On this, see R E Larson 'The Aristotelianism of Bacon's *Novum Organum*' *Journal of the History of Ideas* 1962 Vol 23 pp 435–50

38   M Hesse 'Francis Bacon's Philosophy of Science' in B Vickers ed *Essential Articles on Francis Bacon* Shoe String Press Hamden 1968 pp 114–39

39   F Bacon *Valerius Terminus or the Interpretation of Nature…* (c 1603), in Bacon *op cit* (note 34 1905) pp 187–205. *Valerius Terminus*, which exists only as a fragment, seems to have been a preliminary version of *The Advancement of Learning* and the *New Organon*.

40   *Ibid* pp 196–97

41   Bacon *op cit* (note 34 1905) pp 712–32. The work was probably written about 1624, and was first published in 1627.

42 *Ibid* p 732

43 *Ibid*

44 *Ibid* p 731

45 That is (roughly), empirics or rationalists.

46 Further details of Hooke's 'Baconian' approach to a geological problem are to be found in my paper 'Robert Hooke's Methodology of Science as Exemplified in his *Discourse of Earthquakes*' *British Journal for the History of Science* 1972 Vol 6 pp 109–30

47 *Ibid* p 119

48 See above, p 26

49 R Descartes *Regulae ad Directionen Ingenii* Amsterdam 1701; English translation in *Descartes' Philosophical Writings* selected & trans Norman Kemp Smith Macmillan London 1952

50 The *Discourse* was composed as an introduction to Descartes' treatises on optics, meteorology and geometry: *Discours de la Méthode Pour bien Conduire sa Raison, & Chercher la Verité dans les Sciences, Plus La Dioptrique. Les Metéores. Et La Géometrie. Qui sont des Essais de cete* [*sic*] *Méthode* Jan Maire Leyden 1637. There are several modern English translations of the *Discourse*, including one in the Kemp Smith edition (note 49), an Everyman edition and a Penguin edition.

51 *Renati Des-Cartes, Meditationes de Prima Philosophia in Qua Dei Existentia et Anima Immortalitats Demonstratur Soly* Paris 1641. A convenient English translation is to be found in E Anscombe & P T Geach eds *Descartes: Philosophical Writings*, Nelson Edinburgh 1954. An English version of the *Objections and Replies* will be found in E S Haldane & G R T Ross eds *The Philosophical Works of Descartes* 2 vols Cambridge University Press Cambridge 1911 Vol 2

52 For an English translation, see *Discourse on Method, Optics, Geometry, and Meteorology René Descartes* trans P T Olscamp Bobbs-Merrill Indianapolis 1965

53 *Renati Des-Cartes Principia Philosophiae* Elzevier Amsterdam 1644. For English translations of portions of this, see *Descartes A Discourse on Method* trans J Veitch introduction A D Lindsay Dent London and Dutton New York 1912. For the recent full translation, see *Principles of Philosophy* trans R P Miller & V R Miller Reidel Dordrecht 1983

54 See, for example; L J Beck *The Method of Descartes: A Study of the Regulae* Clarendon Press Oxford 1952; N Kemp Smith *New Studies in the Philosophy of Descartes: Descartes as Pioneer* Macmillan London 1963 Chapter 2; G Buchdahl *Metaphysics and the Philosophy of Science: The Classical Origins Descartes to Kant* Blackwell Oxford 1969 Chapter 3

55 J A Schuster 'Descartes' *Mathesis Universalis* 1619-28' in S Gaukroger ed *Descartes: Philosophy, Mathematics and Physics* Harvester Sussex 1980 pp 41–96. I am also indebted to Dr Schuster in these matters in the comments which he made on a draft of the present volume.

56 Haldane & Ross eds *op cit* (note 51) Vol 1 p 14. It should be noted, perhaps, that the spatial metaphor used here by Descartes is the converse of that arch-like structure to which we have grown accustomed in this book. Changing for the moment to a geological simile, since I cannot think of the precise 'opposite' of an arch, it would seem that Descartes can be thought of here as referring to a 'synclinal' methodological structure, whereas most methodologists favoured analogies with anticlines!

57 On this see, for example, R H Popkin *A History of Scepticism from Erasmus to Descartes* Harper New York 1968

58 Discussion of the idea of 'perfection' formed an important component of mediaeval theology. In basing his argument on this notion, Descartes displayed some of his intellectual ancestry, revealing its traditional rather than its revolutionary character.

59 Another way of carrying out the 'proof' (the so-called 'ontological' argument) is as follows. God is, by definition, the most perfect being that one might conceive. It is impossible to envisage a being more perfect than God. But a being will be more perfect it it exists than if it is imaginary or unreal. Therefore, God exists. This argument has greatly exercised philosophers' attention over the centuries and still has a few admirers. Of course, the word 'perfection', which was widely deployed in the scholastic literature that formed the educational staple of Descartes' day, had a great deal more connotations of a theological nature then than now.

60 Haldane & Ross eds *op cit* (note 51) Vol 2 p 92

61 Despite the apparently clear and distinct(!) circular character of the argument, as presented above, some authors contend that it can be reconstructed so as to avoid the charge of circularity. See, for example, T Tlumak 'Certainty and Cartesian Method' in M Hooker ed *Descartes: Critical and Interpretive Essays* Johns Hopkins Baltimore 1978 pp 40–73

62 In his Preface to the French translation of the *Principia*, Descartes wrote: 'All philosophy is like a tree, of which metaphysics is the root, physics the trunk, and all the other sciences the branches that grow out of this trunk...' (Descartes *op cit* [note 53 1912] p 156)

63 Other philosophers such as Berkeley have thought exactly the opposite on this issue. Evidently one man's philosophical meat is another man's poison; and one may well be forgiven for being sceptical of the 'clear and distinct' criterion for truth.

64 For a full account of the vortex theory, see J F Scott *The Scientific Work of René Descartes* Taylor & Francis London 1952; or E J Aiton *The Vortex Theory of Planetary Motions* History of Science Library London 1972

65 R Descartes *Traité de l'Homme et de la Formation du Foetus*... Guillaume le Jeune Amsterdam 1680
66 The sketch of Descartes' system that has just been given is drawn from his *Principles of Philosophy*, the most mature and lucid statement of his cosmology, in relation to his metaphysical first principles. For my more detailed description of Descartes' geological and mineralogical theories, see my paper 'Mechanical Mineralogy' *Ambix* 1974 Vol 21 pp 157–78
67 D M Clarke *Descartes' Philosophy of Science* Manchester University Press Manchester 1982 Chapter 4
68 See, for example, G Buchdahl 'The Relevance of Descartes' Philosophy for Modern Philosophy of Science' *British Journal for the History of Science* 1963 Vol 1 pp 227–49
69 Haldane & Ross eds *op cit* (note 51) Vol 2 pp 48–9
70 D Garber 'Science and certainty in Descartes' in M Hooker ed *op cit* (note 61) pp 114–51
71 Haldane & Ross eds *op cit* (note 51)
72 *Ibid* Vol 1 p 6
73 J Hintikka 'A discourse on Descartes's method' in Hooker ed *op cit* (note 61) pp 74–88
74 Haldane & Ross eds *op cit* (note 51) Vol 2 pp 48 & 49
75 This conflation was by no means confined to Descartes. It was, for example, equally evident in the writings of his British counterpart, the rationalist Thomas Hobbes (1588–1679): W Molesworth ed *The English Works of Thomas Hobbes* 11 vols John Bohn London 1839 Vol 1 pp 309–12. For a discussion of this, and of Hobbes's methodology in general, see W Sacksteder 'Hobbes: The Art of the Geometricians' *Journal of the History of Philosophy* 1980 Vol 18 pp 131–46
76 See, for example, the quotation from Newton's *Opticks* on p 80.
77 [A Arnauld & P Nicole] *La Logique ou l'Art de Penser; Contenant Outre les Régles Communes, Plusieurs Observations Nouvelles, Propres á Former de Jugement* Paris 1662. I have used the English translation of T S Baynes *The Port-Royal Logic translated from the French with Introduction, Notes and Appendix*, 2nd edn Sutherland & Knox Ediburgh 1851. There is also a modern edition edited by J Dickoff & P James, published by Bobbs-Merrill, Indianapolis & New York 1964
78 Jansenism (which followed the teachings of Cornelius Jansen) was a branch of the Roman Catholic Church which held that the human will is by nature perverse and incapable of good.
79 We have already noticed above (p 70) that Arnauld was one of the contributors to the *Objections and Replies* to Descartes' *Meditations*.
80 [Arnauld & Nicole] *op cit* (note 77 1851) p xlv
81 *Ibid* p 33
82 *Ibid* pp 61–3
83 See p 86
84 [Arnauld & Nicole] *op cit* (note 77 1851) p 37
85 This was drawn in part from a logical work of Pierre Gassendi (1592–1655), published post-humously in 1654.
86 [Arnauld & Nicole] *op cit* (note 77 1851) p 44
87 This notion provided, for example, a stimulus to the reconstruction of the language of chemistry in the work of Lavoisier and his colleagues at the end of the eighteenth century.
88 [Arnauld & Nicole] *op cit* (note 77 1851) p 309
89 *Ibid* p 308
90 *Ibid* pp 313–14
91 *Ibid* p 315
92 *Ibid*. Readers may note with interest here the retention of the spatial metaphor of ascent and descent.
93 See, for example, M R Cohen & E Nagel *An Introduction to Logic and Scientific* Method Routledge & Kegan Paul London 1934
94 I Newton *Philosophiae Naturalis Principia Mathematica* Royal Society London 1687 (facsimile edition Dawson London no date). There are a number of modern editions in English translation. The one most commonly used is F Cajori ed *Sir Isaac Newton's Mathematical Principles of Natural Philosophy and his System of the World Translated into English by Andrew Motte in 1729*... University of California Press Berkeley & Los Angeles 1934 (with several subsequent reprintings)
95 I Newton *Opticks, Or, A Treatise of the Reflexions, Refractions, Inflections and Colours of Light. Also Two Treatises of the Species and Magnitude of Curvilinear Figures* London 1704; 4th ed 1730 republished Bell London 1931
96 Most of what he wrote that will be of interest to students is to be found in: *Newton's Philosophy of Nature: Selections from His Writings: Selections From His Writings* ed H S Thayer Hafner New York 1953
97 It has been emphasised by Truesdell that it was not until the middle of the eighteenth century, chiefly due to the work of Euler, that applied mathematics was sufficiently developed to allow the representation of equations of motion for compound systems by means of differential equations. It was only then that one might say that 'Newton's equations' were truly employed in the modern sense (C Truesdell *The Rational Mechanics of Flexible or Elastic Bodies 1638–1788: Introduction to Leonhardi Euleri Opera Omnia Vol. x et xi Seriei Secundae* Orell Fussli Zurich 1960 pp 424–5). It

may be noted that in his Preface to the *Principia* Newton defined rational mechanics as 'the science of motions resulting from any forces whatsoever and of the forces required to produce any motions, accurately proposed and demonstrated'.

98   Kepler discovered that the periodic times of the planets' motions were proportional to the 3/2th powers of their distances from the Sun.

99   For the Newtonian attack against the Cartesian theory of vortices, see particularly the Preface written for the Second Edition of the *Principia* by Roger Cotes (H S Thayer *op cit* [note 96] pp 116–134)

100   Newton *op cit* (note 95 1931) p 1

101   Because of its overt dependence on experimental results, so that it obviously offers anything but a purely deductive system, structured in more geometrico, the *Opticks* is sometimes called an 'analytic', as opposed to a 'synthetic', text.

102   For details of Descartes' theory, see A I Sabra *Theories of Light from Descartes to Newton* Oldbourne London 1967 p 65

103   Newton *op cit* (note 95 1931) p 47. Newton's account of this experiment was first published in the 80th Number of the *Philosophical Transactions of the Royal Society*, London 1672 pp 3075–87

104   Newton *op cit* (note 94 1934) p 547. Some commentators used to translate '*Hypotheses non fingo*' as 'I feign no hypotheses', but this is not the preferred interpretation today. It may be that this repudiation of hypotheses in the *Principia* should only be thought of as referring to speculations about the cause of gravity, rather than the use of all hypotheses in scientific inquiry.

105   Having taken an 'official' stand against hypotheses, Newton had to try to exclude them as far as possible from his 'apologetic' texts — as opposed to his laboratory notebooks, for example. However, like any other mortal sceintist, he could not avoid making conjectures or speculations on matters which could not be 'proved' to his satisfaction. Therefore, we find that a number of fascinating 'Queries' were appended to the *Opticks*, where Newton's imaginative genius was given free rein. These 'Queries' provided topics for investigation by other natural philosophers right through the eighteenth century. The last Query in particular — the 31st — also offered Newton's readers some insight into his general philosophy of nature and his theological opinions.

106   Newton *op cit* (note 95 1931) pp 404–5

107   Cf p 67 above

108   We have, in fact, much evidence that he did use hypotheses and not just in the 'Queries'. See I B Cohen *Franklin and Newton: An Inquiry into the Speculative Newtonian Experimental Science and Franklin's Work in Electricity as an Example Thereof* American Philosophical Society Philadelphia 1956 pp 575–89; and Newton *op cit* (note 94 1934 pp 671–5). Indeed, it has been said of Newton that he considered his hypotheses quite respectable, though they should not be referred to by that unseemly word. It was (for Newton) Descartes' speculations that were to be deplored because of their hypothetical character.

109   See below, p 114

110   J Edleston ed *Correspondence of Sir Isaac Newton and Professor Cotes, Including Letters of Other Eminent Men, Now First Published from the Originals in the Library of Trinity College, Cambridge; together with an Appendix, Containing other Unpublished Letters and Papers by Newton; with Notes, Synoptical View of the Philosopher's Life, and a Variety of Details Illustrative of his History* J W Parker London 1850 p 156. Here Newton was writing to Cotes in 1713, in response to some concern expressed about where the source of attraction lay in two mutually attracted massive bodies. But Newton's previous letter to Cotes set out the matter a little more fully and with rather less incongruity. He wrote: '[A]s in geometry the word hypothesis is not taken in so large a sense as to include the axiomes & postulates, so in experimental philosophy it is not to be taken in so large a sense as to include the first principles or axiomes w[h]ich I call the laws of motion. These principles are deduced from phaenomena & made general by induction... And the word hypothesis is... used by me to signify only such a proposition as is not a phaenomenon nor deduced from any phaenomena but assumed or supposed w[i]thout any experimental proof.' (pp 154–5)

111   But see again note 105.

112   See S E Toulmin *The Philosophy of Science: An Introduction* Hutchinson London 1953 pp 29–30

113   I B Cohen *The Newtonian Revolution: With Illustrations of the Transformation of Scientific Ideas* Cambridge University Press Cambridge 1980 p 16. Cohen's discussion is, however, directed towards the *Principia* rather than the *Opticks*.

114   A quality like temperature, could, for a given body, be more or less 'intense'. But this was not possible for a quality like mass (in Newtonian mechanics). If the mass alters, the body itself is no longer 'the same'.

115   Newton *op cit* (note 94 1934) pp 398 & 400

116   The term is generally attributed to William of Ockham, whom we mentioned in the previous chapter. But the attribution is today regarded as incorrect.

117   Newton *op cit* (note 94 1934) pp 6–12

118   Newton *op cit* (note 95 1931) p 403. It is not easy to comprehend exactly what Newton meant by 'God's sensorium', but it seems to have been something like God's 'sensory field' — a sensorium being the 'seat of sensation' of the mind, which allows a person to hold a kind of picture of the external world in his head, and with the help of which the external world is 'mapped'. (It may be

helpful, by way of analogy, to imagine oneself equipped with a kind of 'cinema screen' in one's head, on to which are projected 'pictures' of the external world. This 'cinema screen' would be a person's 'sensorium'.) Newton reckoned that *God* would be aware of the motions of an object (detectable in His sensorium), even if there were no other objects in the universe by means of which a relative motion might be determined or compared. So absolute as well as relative motion was a meaningful concept.

119   Newton *op cit* (note 94 1934) p 419
120   Newton *op cit* (note 95 1931) p 400
121   In the *Principia*, Newton wrote: 'The extension, hardness, impenetrability, mobility, and inertia of the whole, result from the extension, hardness, impenetrability, mobility, and inertia of the parts; and hence we conclude the least particles of all bodies to be also extended, and hard and impenetrable, and movable, and endowed with their proper inertia' (*op cit* note 94 1934 p 399). One can recognise the following pairs *vis-à-vis* the *Opticks* text: extension/solid; hardness/hard; impenetrability/impenetrable; mobility/movable; inertia/mass. Only the first of these pairs doesn't quite 'fit'.
122   This is the view associated chiefly with the French philosopher/mathematician, Henri Poincaré (see p 190). It certainly doesn't command universal assent, but is mentioned here as a means of raising a problem as to how one should envisage scientific theories.
123   J Locke *An Essay Concerning Humane Understanding* Thomas Basset London 1690. There are numerous modern editions. I have used *An Essay Concerning Human Understanding by John Locke* abridged and edited by A S Pringle-Pattison Clarendon Press Oxford 1924
124   This story is thought to have its origin in a passage in J T Desaguliers *A Course of Experimental Philosophy* 2 vols Senex Innys Manby Osborn & Longman London 1734 Vol 1 preface (pages unnumbered).
125   See J L Axtell 'Locke's Review of the *Principia*' *Notes and Records of the Royal Society of London* 1965 Vol 20 pp 152–61
126   Locke *op cit* (note 123 1924) pp 6–7
127   On the relation between Newton and Locke, see G A J Rogers 'Locke's *Essay* and Newton's *Principia*' *Journal of the History of Ideas* 1978 Vol 39 pp 217–32
128   In a sense, by defending the mechanical philosophy Locke was defending Descartes as well as the science of the British school. But Descartes' route to 'mechanism' was very different from that of his British counterparts.
129   Locke *op cit* (note 123 1924) p 42
130   *Ibid* p 43
131   *Ibid* p 43
131   The 'single-file' doctrine is seen even more vividly in an ingenious 'mechanical' psychology proposed by Locke's contemporary Robert Hooke. See D R Oldroyd 'Some "Philosophicall Scribbles" attributed to Robert Hooke' *Notes and Records of the Royal Society of London* 1980 Vol 35 pp 17–32
132   Locke *op cit* (note 123 1924) pp 92–3
133   *Ibid* pp 43–4. Locke likened this to an 'internal sense'.
134   *Ibid* pp 93–5
135   The modern meaning of 'mode' is 'way', 'manner' or 'fashion'. But Locke's usage was more specifically technical. He defined a 'mode' as referring to 'such complex ideas which, however compounded, contain not in them the supposition of subsisting by themselves, but are considered as dependences on, or affections of substances; such as are the ideas signified by the words triangle, gratitude, murder, &c' (*Ibid* p 94). This usage was essentially the same as that of Arnauld and Nicole in their *Port-Royal Logic*, discussed above. But Locke's usage was beginning to wander from that prescribed by his definition. It is difficult to see, for example, how 'space' can be thought of as some 'affection of substance'. We shall notice the further development of the meaning of the term in our discussions of the work of Hume.
136   For an account of Carnap's attempt, for example, see below, p 235
137   We need not embark here upon a disquisition concerning the freedom of the will!
138   On the mechanical philosophy, see for example R S Westfall *The Construction of Modern Science, Mechanisms and Mechanics* Wiley New York 1971; or Kearney *op cit* (note 30)
139   Locke *op cit* (note 123 1934) p 69
140   M Mandelbaum *Philosophy, Science and Sense Perception* Johns Hopkins University Press Baltimore 1964
141   Locke *op cit* (note 123 1924) p 66
142   *Ibid* p 67
143   It is convenient to speak here of the macro-and the micro-worlds; but they are modern terms, not Locke's.
144   Locke *op cit* (note 123 1924) p 67
145   *Ibid* p 57
146   *Ibid* p 156. It may be noted that arguments of this kind were by no means new. In his *Physics*, Aristotle had argued for the existence of a material substrate to the world by considering the grammar of the Greek language: every predicate must be predicated of some subject; likewise every quality must inhere in something.

147 *Ibid*
148 See below, p 108
149 Locke *op cit* (note 123 1924) p 159
150 *Ibid* p 160
151 It is not, of course, our prerogative to insist that Locke be a hidebound empiricist. But one can point out the difficulties within the empiricist's position. We would like him to have been consistent in his asrguments through the *Essay*; but he was not.
152 The doctrine of 'causal powers' has recently undergone an interesting revival in modern philosophy. See R Harré & E H Madden *Causal Powers: A Theory of Natural Necessity* Blackwell Oxford 1975
153 This connection can also be seen in Locke's own text: Locke *op cit* (note 123 1924) p 153
154 See below, p 113
155 Locke *op cit* (note 123 1924) p 153
156 *Ibid* p 180
157 For the relationship between Locke and the mediaeval nominalists, see J R Milton 'John Locke and the nominalist tradition' in R Brandt ed *John Locke Symposium Wolfenbuttel 1974* Walter de Gruyter Berlin & New York 1981 pp 128–45
158 See above, p 19
159 For example, K R Popper 'Three Views Concerning Human Knowledge' in his *Conjectures and Refutations* 3rd ed Routledge & Kegan Paul London 1969 pp 97–119
160 On this, see D M Armstrong *Nominalism and Realism* Cambridge University Press Cambridge 1978 p 12
161 Locke *op cit* (note 123 1924) pp 242–3
162 *Ibid* p 243
163 The best-known example of such a work in Britain, sponsored by the Royal Society, was J Wilkins *An Essay Towards A Real Character and a Philosophical Language* London 1668. For a discussion of theories of artificial languages in the seventeenth and eighteenth centuries, see J Knowlson *Universal Language Schemes in England and France, 1600–1800* University of Toronto Press Toronto 1975
164 On this, see P R Sloan 'John Locke, John Ray, and the Problem of the Natural System' *Journal of the History of Biology* 1972 Vol 5 pp 1–53
165 D R Oldroyd 'An Examination of G.E.Stahl's *Philosophical Principles of Universal Chemistry*' *Ambix* 1973 Vol 20 pp 36–52
166 D L Hull 'The Metaphysics of Evolution' *The British Journal for the History of Science* 1967 Vol 3 pp 309–37

CHAPTER 3

PROBLEMS FOR EMPIRICISM AND FOR THE PHILOSOPHY OF THE NEW SCIENCE

I n the previous chapter we have given a brief indication of the work of some of the leading seventeenth-century writers on the philosophy and methodology of science, trying to display some of the chief metascientific positions that were adopted as the modern scientific movement was establishing itself. We noted the new mathematical physics of Galileo, the empiricism of Bacon, the attempted rationalist enterprise of Descartes, the outstanding success of Newton's actual science (but the somewhat doubtful metaphysical and methodological principles on which it was based), and finally the attempt by Locke to establish a general empiricist philosophy of science that would satisfy the needs of the seventeenth-century mechanical philosophy.

In the present chapter, we shall look at some of the eighteenth-century criticisms of seventeenth-century science and metascience, and the new philosophical ideas that were generated as a result of the examination of earlier writings, particularly those of Locke. We shall look first at the notable German philosopher–librarian–historian–lawyer–..., Gottfried Wilhelm Leibniz (1646–1716). We shall then look at the ideas of Berkeley and his criticisms of the work of Locke; and at the empiricism of Hume and the problems that he uncovered in relation to causality and the uniformity of nature. This will lead us to a consideration of the difficult but all-important philosophy of Kant.

Leibniz

It is beyond the scope of this book to attempt to give more than the briefest indication of Leibniz's contributions to philosophy of science, let alone the totality of his philosophical system. We shall, however, say something about his criticisms of Newton's ideas on space and time, and of the atomic theory of supposed hard, massy, impenetrable bodies at the micro-level. For these criticisms, although grounded in what one might be inclined to regard as somewhat curious metaphysics, were certainly most cogent, and can, in a way, be regarded as providing an early presentiment of the modern theory of relativity. In Leibniz, therefore — though he was in one sense the seventeenth/eighteenth-century rationalist *par excellence* – we have ideas that have come to form important components of more recent science and philosophy of science.

The personal relations between Leibniz and Newton were severely strained over many years, chiefly in connection with an acrimonious priority dispute as to who was the first to invent/discover the calculus.[1] Newton himself, however, did not engage directly in the debate to any great extent but directed matters from afar through the mouthpieces of various disciples or epigoni, who were anxious to see the Newtonian science, and its associated metaphysics, vanquish all possible rivals — particularly those emanating from Continental sources. Chief amongst these Newtonian spokesmen was the Reverend Samuel Clarke (1675–1729), and it is through the published correspondence between Leibniz and Clarke[2] that one may most readily come to an understanding of the fundamental differences between the positions of the two intellectual giants.

It was Leibniz's general program, like that of Descartes, to seek to deduce the metaphysical basis of the whole cosmos from a few supposedly self-evident first principles that could be known *a priori*. The two most important principles (for mathematics and natural philosophy, respectively) were those of contradiction and sufficient reason:

> Our reasonings are grounded upon *two great principles, that of contradiction*, in virtue of which we judge *false* that which involves a contradiction, and *true* that which is opposed or contradictory to the false;
> *And that of sufficient reason*, in virtue of which we hold that there can be no fact real or existing, no statement true, unless there be a sufficient reason, why it should be so and not otherwise, although these reasons usually cannot be known by us.[3]

Or more briefly, one might say that nothing exists or occurs without there being a 'sufficient reason' why it should do so. That is, there must be a reason or cause adequate to account for the existence of every thing, or the occurrence of every event. If there isn't a reason sufficient or adequate for this task, then the thing won't exist, or the event won't occur.

Leibniz saw God as providing the ultimate 'reason' of things; and he deployed the so-called cosmological argument to prove (as he thought) the existence of God. There could, Leibniz argued, be no sufficient reason — within themselves — for the existence of material bodies. Therefore, their reason for existence must reside in some mental entity: God.[4] Monotheism could be inferred with the help of the principle of sufficient reason — for one God would be enough.[5] Also, everything that God created would be as perfect as possible,[6] though not everything was necessarily perfect. (If there were more than one absolutely perfect being, there would be more than one God.) So God chose, out of the 'logically possible' worlds[7] that He might have created, the one that was possessed of the highest degree of perfection. Further, the chosen world would necessarily be one of universal harmony, even if not of total perfection; for God, the creator of the world, was perfect. And 'existence' being regarded as a perfection, the universe would be a plenum, without any void or vacuities, which (in Leibniz's view) would make it less than maximally perfect.[8] Finally (for our present purposes), we may mention here the important general principle of the 'identity of indiscernibles', which could supposedly be inferred from the principle of sufficient reason. Leibniz denied that there could be two *different* things that were identical in all respects: 'for surely it must be possible to give a reason why they are different'.[9]

It was with these principles[10] as weapons that Leibniz ventured forth to assail the citadel of Newtonian physics, with the redoubtable Samuel Clarke as commandant of the defending forces. Newton, it will be remembered, had thought of space as a kind of 'receptacle', in which bodies could freely move. The 'points' of (absolute) space would be known to God, 'in His sensorium', even in the total absence of matter. The same considerations applied to time: God would know of the flow of (absolute) time, even in the absence of all motion in the universe, or indeed if there were no matter at all — that is, if the 'receptacle' of space were empty.

But Leibniz found the whole Newtonian scheme metaphysically objectionable, and he used the principles of sufficient reason and identity of indiscernibles to attack the Newton–Clarke doctrines of absolute space and time. The different parts of absolute space would be indiscernible, he claimed, God's sensorium notwithstanding; they would not differ from each other in any way whatsoever. Consequently, it would have been possible for the world to have been created at any region in absolute space. But there was no good reason for God to have preferred one region to another, for all were identical and indistinguishable. Therefore, the whole idea of absolute space failed to mesh with the Leibnizian principle of sufficient reason. Consequently, Leibniz rejected the idea as unintelligible.[11] He also made an analogous argument with respect to time.[12]

Leibniz's own view of the matter[13] was that space and time were relational, not absolute; and they were *ideal* rather than real.[14] That is to say, insofar as they existed they did so in the mind; they were mental constructs.

It may require a little elucidation to make clear the idea of space as relational. Suppose we have an object, X, located somewhere in relation to other objects, assumed to be fixed with respect to one another. (For example we could say that X is one kilometre south of the ticket office of the Sydney Opera House.) If another object, Y, moves so that it comes to occupy the same relation to the reference point (the ticket office) as X had previously, then we would say that it occupied the same place as X.[15] And space is simply the collection of all places![16] Time was perhaps more obviously ideal and relational. The idea of time as an existent entity is scarcely intelligible. Certainly there are time *intervals*, measurable with instruments in various ways; but to say this immediately invokes the relational nature of time. There would be no passage of time if there were no change in the universe.

Newton and Clarke were not, of course, unable to see the arguments in favour of a relational theory of space and time; but Newton in the *Principia*[17] and Clarke in his fourth letter to Leibniz (para.13)[18] adduced arguments in favour of absolute motion, maintaining that inertial effects would manifest themselves in physically detectable ways for accelerated motion — even with respect to 'absolute space'.[19] Discussion of these issues has been going on for centuries now, and is by no means exhausted, but it would distract us too much from our present task to pursue these interesting matters here. What can be said is that most modern commentators have come to accept a relational view of space and time, which is thought to be implicit in Einstein's theory of special relativity. Admittedly, Einstein's subsequent work in general relativity was compatible with an absolutist approach to 'space–time', but it would not be appropriate to pursue this issue here. What is interesting for our present purpose is that the remarkable metaphysical superstructure that Leibniz erected enabled him, as early as the beginning of the eighteenth century, to throw grave doubt on the metaphysical underpinnings of what seemed to be the supreme intellectual achievement of seventeenth-century science.

We may also mention here that Leibniz rejected the idea of the possibility of empty space, or the void, on the basis of his principle of the identity of indiscernibles; for two parts of totally empty space would be indistinguishable.[20] Further, he maintained (as we have already mentioned) that 'matter is more perfect than a vacuum';[21] so God, in making the world as perfect as possible, would necessarily create a plenum rather than atoms and empty space. Besides, Leibniz could see no reason for there to be a limit to the divisibility of matter. Like space, which is mathematically divisible to infinity, matter should also be infinitely divisible.[22] He also argued against atoms (not just empty space) through the principle of identity of indiscernibles,[23] and more interestingly he claimed that the very idea of Newtonian atoms cohering together to form solid bodies was unintelligible. What could hold them together? If it were some kind of 'glue', of what would the 'glue' be made? If there were hard 'hooks and eyes' on the atoms, how could the hardness of *these* be explained?[24] A Newtonian atom supposedly possessed volume; but how did the separate regions of the atom hold together?

I suggest that Leibniz was here putting his finger on an important point, namely that nothing can be explained in terms of itself. In modern matter theory, we find that there is no ultimate explanation of the nature of things, and we certainly make no attempt to explain the phenomena of the macro-world in terms of a micro-world that is qualitatively the same. In fact, the ultimate level of explanation at any given epoch in the history of science — for example, the quanta of modern physical theory — must remain essentially mysterious, for the lowest ontic level cannot be explained in terms of some yet more fundamental entities. So, as is well known, physical explanation runs off into mathematics, and we are unable to visualise the ultimate level of matter satisfactorily by means of any coherent model drawn from the macro-world.

Leibniz's own ontology, expounded in his *Monadology*, was indeed one that had some of the basic features of a modern matter theory such as I have just mentioned. This was so, even though his scheme was arrived at in an aprioristic manner by reasoning from first principles, rather than by working from experimental data — as did Niels Bohr, for example, in his first formulation of the quantum model of the hydrogen atom. Leibniz started from the proposition that since there are 'compounds' in the world there must necessarily be 'simples'; and these he called *monads*.[25] It would take us beyond the scope of this book to attempt to give a satisfactory exposition of the theory of *monads*.[26] Suffice it to say that Leibniz envisaged a micro-world of point entities — something like a huge network of centres of active force, but with no invervening spaces between the centres — which were all in a condition of pre-arranged harmony in their relations one to another. Each monad had its own pre-arranged 'life', so that it would go through a predetermined sequence of actions, rather like an infinite piece of recording tape that has already been recorded so that it will necessarily play a predetermined pattern of musical sounds, or whatever. The 'events' for all the monads would always be 'in harmony' one with another. For example, no two monads would ever be pre-programmed to behave in such a way that their actions would be mutually incompatible or contradictory. However, the monads would have no mutual causal interactions. Or, using Leibniz's metaphor, they were 'windowless'.[27] Also, since the monads were supposedly without volume, the difficulty of understanding how the 'parts' of a Newtonian atom might 'hold together' simply did not arise.

All this may seem somewhat strange, to say the least, and actually it may be nearer to Leibniz's original intention to think of monads as numerical solutions to some grand cosmic equation rather than some strange kind of volumeless particles or centre of force. However, the point I wish to emphasise here is simply that Leibniz the rationalist metaphysician, in criticising the Newtonian theory, put forward ideas in relation to the doctrines of space and time, and to matter theory, that have some fundamental analogy to modern theories, no matter how strange they may appear in points of specific detail. Thus, from the beginning of the eighteenth century at least, some minds could recognise fundamental difficulties in the edifice of the Newtonian–Lockian scientific– philosophical system, its great empirical success notwithstanding.

It may be asked whether there is any discussion of analysis and synthesis, and the arch of knowledge, within the corpus of Leibniz's writings. Indeed there is, in many places, notably in a paper (thought to have been written in 1679) entitled 'On universal synthesis and analysis, or the art of discovery and judgment'.[28] And roughly speaking, the two movements of analysis and synthesis corresponded for Leibniz to induction and deduction respectively. For 'truths of reason', the analysis and synthesis (operating in harness) would supposedly provide certain truth. But for 'truths of fact' the results of synthesis would need to be tested by experience, so that something like a hypothetico–deductive methodology was envisaged. For God, of course, all knowledge would be deductive; but man could never be in that happy position. So empirical methods and the deployment of hypotheses were required. Actually, Leibniz did have in mind the possibility (never realised) of an almost God-like form of scientific knowledge — using a sort of 'universal characteristic' as the basis of a 'philosophical algebra'. He envisaged the construction of a general science of symbols which could be applied to experience. The scheme has been described by Loemker in the following words:

> [Leibniz's general science] would contain the sets of axioms applicable to any particular science and, derived from them and from the definitions of the symbols, the appropriate

rules for transforming the symbolic formulas which constitute the methodology of the science. Every science... [was] thus thought of as capable of mathematical organization through a general theory represented in a language by means of an appropriate set of symbols and developed by fitting operations.[29]

Such a project — evincing the seventeenth-century optimism of the new scientific philosophy — remained unfulfilled in Leibniz's lifetime and remains so to this day. Perhaps Leibniz realised he was over-optimistic, and in one passage at least[30] he proposed a methodology that was as much a sample of Baconian empiricism as of the rationalism that was envisaged for his general science. So although it may often seem appropriate to regard Leibniz as a perfect example of one of Bacon's spiders, in fact he was in certain respects a Baconian bee — but not, I think, a Baconian ant.

Berkeley

We may now turn our attention to another important eighteenth-century critic of Newtonian science and Lockian epistemology: George Berkeley (1685–1753), Bishop of Cloyne (in Ireland), the second of the celebrated trio of British empiricists, Locke, Berkeley and Hume. Berkeley's philosophy emerged as a result of his examination of Locke's *Essay,* and it paved the way for the subsequent criticisms of Hume, which in their turn were to bring about a complete restructuring of philosophy in the work of Kant.

Berkeley's ideas were formulated early in his life, and as a young man he formed the firm conviction that the very notion of substance, as a metaphysical entity, was unintelligible. He wrote first a *New Theory of Vision* (1709),[31] and then *A Treatise Concerning the Principles of Human Knowledge. Wherein the Chief Causes of Error and Difficulty in the Sciences, with the Grounds of Scepticism, Atheism, and Irreligion, are Inquir'd Into* (1710), which contained his leading exposition of his theory of knowledge. The reception accorded to this work was so hostile, however, that he brought out a more 'popular' version in the form of *Three Dialogues between Hylas and Philonous* (1713). Unfortunately the critics didn't find this much to their liking either. Later works were: *De Motu* (1721), a critique of Newtonian mechanics, which, however, repeated arguments given in the *Principles*; *The Analyst* (1734), a critique of the doctrine of infinitesimals in Newton's and Leibniz's calculus; *Alciphron, or the Minute Philosopher* (1732), a theological work which, however, concerned itself with some of the ideas of the mechanical philosophy and contained some modification of Berkeley's earlier views; and *Siris* (1744), a medical/chemical text which advocated the use of tar water as a useful general medicament.[32]

Many pejorative adjectives have been used to describe Berkeley's philosophy: puzzling, ambiguous, sceptical, quixotic, and so on. And certainly on first acquaintance his position does seem to have been a very strange one. But as we shall see shortly, Berkeley's position was a direct outcome of his thoroughgoing and uncompromising empiricism — which in itself does not appear to be a specially peculiar philosophical doctrine.

Berkeley began his *Principles* with the claim that the 'objects of human knowledge' are ideas, occurring either as a direct result of sensation or 'by attending to the passions and operations of the mind',[33] or by the action of memory and imagination. Locke would have had no disagreement with this. But Berkeley rejected the view that there was any 1:1 correspondence between the sensed objects and the sensations (or ideas) to which they gave rise. Indeed, he went further to question whether objects did give rise to ideas at all, for, unlike Locke, he repudiated the idea of there being matter or substance endowed with any kind of causal power.

Locke, of course, had accepted the category of substance.[34] It was the bearer of the various qualities that we actually *perceive* with our senses. But Berkeley was more consistent in his empiricism, and pointed out that one never could perceive substance, only qualities. Therefore, he claimed, one should eliminate the notion of substance; reality should only be accorded to minds and the ideas of which minds had direct experience. This meant that for Berkeley the physical world only had existence in so far as it was perceived:

> For what is said of the absolute existence of unthinking things without any relation to their being perceived, that seems perfectly unintelligible. Their *esse* is *percipi*, nor is it possible they should have any existence, out of the minds of thinking things that perceive them.[35]

In summary: '*To be* is *to be perceived*.'

The immediate reaction of the reader might be to scoff, or respond like Dr Johnson, who kicked a large stone with his foot in order to demonstrate the real existence of the external world.[36] One might well have sympathy with Dr Johnson's response; but really it does nothing to combat Berkeley's thesis. All Johnson succeeded in doing was to generate some new sensations in his leg. This did but little to undermine Berkeley's position.

A somewhat more sophisticated response would be to point out that some objects might not be being perceived, at any given time, by any sentient being. So would they spring into or out of existence as and when they were observed or ignored, if 'to be' was 'to be perceived'? Such a situation would have been an absurdity, even for Berkeley, and so he proposed that all things were perceived at all times by the mind of God, thereby being sustained in a steady condition of being.[37] The doctrine has been expressed with clarity and wit in the famous limerick of Ronald Knox:

> There was a young man who said, God
> Must think it exceedingly odd
>   If he find that this tree
>   Continues to be
> When there's no one about in the Quad[rangle].
>
> Dear Sir: your astonishment's odd:
> I am *always* about in the Quad.

And that's why the tree
Will continue to be
Since observed by
Yours faithfully,
God.

But in a sense to invoke God in this way was an unnecessary admission. If one were really strong-minded about the doctrine that 'being' and 'being perceived' were absolutely one and the same, it would not, I suppose, matter greatly whether things sprung into and out of being at the turn of a head!

Berkeley, then, seemed to find the whole idea of inert matter (substance) acting as some kind of active agent metaphysically quite repugnant — or aesthetically so, one should perhaps say. For him, only minds or spirits could be active causal agents or powers. But the argument against Locke wasn't presented simply in terms of an aesthetic repugnance towards matter. The weapons of philosophical debate were brought into action. Berkeley objected to the possibility of the mind being acquainted with general, abstract ideas or universals. For example, he maintained that if you try to conjure in your mind the idea of a triangle, the idea of some *specific* triangle always and necessarily arises — not some general abstract notion.[38] It happens that we come to believe in abstract ideas because of the use we make of words; but, said Berkeley: 'it seems that a word becomes general by being made the sign, not of an abstract general idea but, of several particular ideas, any one of which it [the word] indifferently suggests to the mind'.[39] So, although Berkeley could give an account of why we believe that there are such entities as abstract ideas, this did not mean that he admitted that they had any philosophical legitimacy. Rather, he preferred to suppose that in the formation of general ideas, a particular idea, marked by a word, served as a sign, representing to us all the other things of the same kind. The Lockian (or Aristotelian) notion of 'substance' would (for proponents of such a notion) be a 'universal'; but there was no sense impression of a 'particular' which could serve as a sign for the universal notion of substance. Therefore, for Berkeley, it was inadmissible.

Berkeley also objected to Locke's distinction between primary and secondary qualities. He claimed that there could be no body devoid of secondary qualities: a body existing without colour, taste, smell, temperature, and so on, was simply unimaginable.[40] This would hold for the bodies of both the 'macro' and the 'micro' world — according to Berkeley. So in his view Locke's micro-world, giving rise to the perceived qualities of the macro-world, was unintelligible and therefore philosophically unacceptable. The mechanical philosophy, therefore, was incompatible with the tenets of a hard-line empiricist theory of knowledge. This point we have already made in a general way in our discussion of Locke's philosophy. Here, we may emphasise that an analogous comment was made by Berkeley a long time ago now. But because the overall position from which he mounted his attack was so idiosyncratic he achieved few followers in his rejection of Locke, despite the fact that Locke's position was really most unsatisfactory.

If we wish to pigeon-hole Berkeley's philosophical position with convenient labels, the relevant terms are *idealism* and *phenomenalism*. It is appropriate to call Berkeley an idealist, since for him the ultimate ontic level was made up of ideas in the mind rather than matter. Mind had primacy. Phenomena are the direct objects of sensory perception — like the 'whiteness' that I perceive with my eyes and mind when I look at a snow-field; or the taste I have in my mouth when I suck a lollipop. If one holds that phenomena (or relations between phenomena) are the *only* objects of the mind's knowledge, then one is a phenomenalist. Berkeley was a leading representative of this tribe; but we shall meet others in the course of our journey, such as Ernst Mach.[41] Phenomenalism was an important component of nineteenth-century positivism, but it made an early entrance to the philosophical stage in the writings of Bishop Berkeley.

Phenomenalism, though probably implausible and unacceptable to most readers, at least has a sort of mad consistency to it, and one can see that it may have certain attractions. At the time when Berkeley began his work, the sensible world (ie the world of the senses) was regarded as perceivable and existing, but not corresponding with reality. The material world (ie the world of substance or matter) was regarded as real (if it existed), but it could not be perceived, and there were grounds for the empiricist doubting its existence. All this was rather disquieting. Berkeley was much tidier. The sensible world, for him, was real, perceivable and existent; the material world could not be perceived, it was unreal and didn't exist! But watertight consistency in philosophy doesn't always mesh with common sense. Admittedly, Berkeley could develop a self-consistent system on the assumption that objects were nothing more than bundles of sensations. But that does not make it any the less sensible to assert the existence of physical objects, upon the basis of 'clues' provided by the information received by the senses.

Berkeley's *Principles* also contained an attack upon Newton's ideas of absolute space and absolute time[42] for, like Leibniz, Berkeley could only envisage relative positions and motions. His attack was carried further in *De Motu*,[43] which work is also interesting for a substantial shift in his philosophical position. Berkeley now came to accept the proposal that concepts such as force, attraction or gravity, which formed such important features of the Newtonian doctrine, were admissible in so far as they worked successfully within a science and were useful in calculations:

> [J]ust as geometers for the sake of their art make use of many devices which they themselves cannot describe nor find in the nature of things, even so the mechanician [i.e. the natural philosopher concerned with mechanics] makes use of certain abstract and general terms, imagining in bodies force, action, attraction, solicitation, etc., which are of first utility for theories and formulations, as also for computations about motion, even if in the truth of things, and in bodies actually existing, they would be looked for in vain, just as the geometers' fictions made by mathematical abstraction.[44]

But absolute space and time were to be rejected because such notions didn't work; they were ineffective — useless in science. As we have seen, Newton, couldn't use them in the *Principia* and had to presume that the 'centre of the world' was fixed. That worked; God's sensorium didn't!

The position that Berkeley adopted in *De Motu* is today known as *instrumentalism*.[45] The term is not, perhaps, so very well chosen, but what it means is that scientific theories, or hypotheses may be used (if they work successfully) without any claims to the ontological reality of the entities they postulate or the truth of the theories or laws that they assert. Indeed, a hard-line instrumentalist would want to say that science cannot and does not provide a pathway to reality or truth. Berkeley, then, in his later writings, can be construed as an early representative of instrumentalism and positivism,[46] though the position he adopted in *De Motu* had much in common with earlier writings such as one finds in the mediaevals who were content to 'save the appearances' with their theories, not regarding them as more than calculating devices.[47]

There are also some interesting hints towards a different attitude towards science on Berkeley's part in *Alciphron*, even though this was chiefly a theological work. In the Seventh Dialogue, Berkeley developed further his theory of signs. He pointed out that a person performing arithmetical calculations could do so successfully by manipulating the Arabic numerals, without thinking directly of the numerical quantities that they represented. Likewise, concepts such as force could be used in science with good effect, even though no one could give a precise idea of the abstract notion of force.[48] As a further example, Berkeley drew attention to the utility of the mathematical symbol for the square root of minus one (i), which could serve to good effect in calculations, even though no one could form any idea at all of the nature of the square root of a negative number.[49] The same kind of argument could be extrapolated to encompass language as a whole.

There were, then, certain indications of new approaches in metascience in some of Berkeley's later writings; but they were not worked out in detail. The important point, for our present purpose, is Berkeley's criticism of Locke's empiricist theory of knowledge. Berkeley was able to make it consistent, but only at the expense of credibility. We shall now look at the work of Hume, and his development of empiricist doctrine, leading to the recognition of the great difficulty involved in giving a satisfactory account of cause-and-effect relationships and to what is customarily called the 'Problem of induction'.

Hume

David Hume (1711–1776), one of the great luminaries of the Scottish Enlightenment and the third of the celebrated trio of British empiricist philosophers, has a secure place as one of the leading figures in the history of philosophy; and in addition he was a distinguished historian. He wrote on ethics, political theory and philosophy of religion, but perhaps his most important work was in epistemology. At any rate, that is the area that is of particular relevance to the history of the philosophy of science. His work was so significant that it is still actively discussed amongst philosophers to this day (and not as a kind of philosophical museum piece),[50] and he had a profound influence on Kant, thereby producing a general reshaping of the whole philosophical scene.

Hume's first major philosphical work was his *Treatise of Human Nature*, first published in 1739–40.[51] It was not a publishing success, however, and Hume himself described it as having fallen still-born from the press. In an effort to retrieve the situation, he subsequently published a somewhat more popular version, the *Enquiry Concerning Human Understanding*, in 1748.[52] There was also a useful *Abstract* of the *Treatise*, which Hume himself issued as an explanatory pamphlet in 1740.[53] In what follows, material will be drawn from all three of these publications, but attention will be focussed chiefly on the *Treatise*.

Hume was profoundly influenced by Bacon, Newton and Locke, and also by the moralist Francis Hutcheson (1694–1746), who, however, need not enter our discussions here. Hume hoped that his philosophical investigations would allow a science of human nature to be established, analogous to the science of mechanics that had been constructed with such success by Newton. Indeed, as we shall see, Hume proposed an account of the association of ideas such that they came together rather like the coming together of massive bodies under the influence of gravity in the Newtonian theory. The Newtonian model was certainly not far below the surface of Hume's thought.

Hume was, as we have said, an empiricist and he rejected the doctrine of innate ideas.[54] But he rightly objected that Locke had been too loose in his employment of the word 'idea', using it to mean both a sense impression *and* a thought in the mind; for, as Hume said,[55] Locke conflated all perceptions. So right at the very beginning of his *Treatise*, Hume sought to remove this ambiguity, making a clear distinction between 'impressions' and 'ideas'.[56] If I look at the pencil in my hand, for example, I receive *impressions* of this object in my mind as a result of the 'information' transmitted through my sense organs. But I can also shut my eyes and conjur up a mental picture of my pencil; this, in Hume's terminology, would be an *idea*. Hume then went on to propose that there could be both simple and complex impressions in the mind, and also simple and complex ideas — but nothing else.[57] To be sure, some of the complex ideas might have no corresponding complex impressions. For instance, Hume stated that he could readily imagine the complex idea of a city paved with gold and with walls made of rubies.[58] Thus, his mind was capable of joining together two complex ideas that were not normally found in association. The mind was quite able to do *that*. But it could not form simple ideas for which there had never been any corresponding simple impression. To use Hume's own example: 'we cannot form to ourselves a just idea of the taste of a pine-apple, without having actually tasted it'.[59] This was so because, according to Hume's doctrine, there was always an exact resemblance between simple ideas and simple impressions. Thus, he held to the thesis that:

> [A]ll our simple ideas in their first appearance are deriv'd from simple impressions, which are correspondent to them, and which they exactly represent.[60]

It should be noted that this does not imply that all ideas are derived *directly* from impressions. We use words as symbols or signs for ideas, and some complex ideas — like the one expressed by the symbol 'envy', for example — do not have exactly correspondent impressions, even though there is plenty of envy in the world (unlike gold-paved, ruby-walled cities). But what empiricists such as Hume do claim is that symbols for complex ideas can be explicated by and ultimately reduced to symbols for simple ideas, the meaning of which can be understood through direct sensory experience. Symbols for all ideas, therefore, according to the empiricist program, are ultimately reducible to symbols for direct sensory impressions[61] — or to ostensive definitions.[62] This is not quite the same thing as saying that all ideas are derived from impressions.

In his examination of the workings of the human mind, Hume soon noticed that ideas are by no means randomly associated in the processes of thought. He suggested, therefore, that the factors normally determining the 'clustering' of ideas were: resemblance, contiguity in space and time, and cause and effect. He may, as suggested above, have had a kind of 'Newtonian' model for the working of the mind and its association of ideas, for he wrote:

> Here is a kind of ATTRACTION, which in the mental world will be found to have as extraordinary effects as in the natural, and to shew itself in as many and as various forms.[63]

Be that as it may, we find that Hume followed Locke in dividing the complex ideas, produced by association, into '*Relations, Modes,* and *Substances*'.[64] Hume was dismissive about modes and substances. They were mere collections of simple ideas, united by the imagination, the difference being that extra qualities could be added to the idea of a substance without upsetting the complex idea first entertained. For example, one might have a complex idea of gold, and then discover (what previously was not known) that this substance was soluble in *aqua regia*. This new 'quality' could simply be added to the known qualities of gold without any problem. But modes, according to Hume's usage, were complex ideas representing qualities that weren't united by contiguity or causation. (His examples were 'a dance' and 'beauty'.) Such complex ideas could not, apparently, receive newly discovered qualities. If new qualities were to be imposed on such complex ideas, then new names would have to be devised.

This was Hume's (somewhat arbitrary) distinction between modes and substances,[65] but the two were not so very important as in Locke's *Essay*, and we need not pursue them much further, beyond saying that, like both Locke and Berkeley, Hume rejected the suggestion that man could have any knowledge of substance *per se*. His discussion of the question was from the perspective of a nominalist, tending towards phenomenalism. But, unlike Berkeley, Hume didn't deny the possible reality of the category of substance.

Leaving aside, then, further consideration of Hume's treatment of modes and substances, let us say more about his treatment of relations. Seven types were identified: resemblance, identity, time and place, proportion (in number or quantity), degree (in any quality), contrariety, and causation[66] — this last being the sort to which we must give particular attention. But before doing so, we should refer to a famous dogmatic statement (occurring in the *Enquiry*, not the *Treatise*) sometimes called 'Hume's fork' –for it divided knowledge into two kinds, that which could be known *a priori* and that which could only be known *a posteriori*:

> All the objects of human reason or enquiry may naturally be divided into two kinds, to wit, *Relations of Ideas*, and *Matters of Fact*. Of the first kind are the sciences Geometry, Algebra, and Arithmetic; and in short, every affirmation which is either intuitively or demonstratively certain... Propositions of this kind are discoverable by the mere operation of thought, without dependence on what is anywhere existent in the universe. Though there never were a circle or triangle in nature, the truths demonstrated by Euclid would forever retain their certainty and evidence.
>
> Matters of fact...are not ascertained in the same manner; nor is our evidence of their truth, however *great*, of a like nature with the foregoing. The contrary of every matter of fact is still possible; because it can never imply a contradiction, and is conceived by the mind with the same facility and distinctness, as if ever so conformable to reality. *That the sun will not rise tomorrow* is no less intelligible a proposition, and applies no more contradition than, *that it will rise*. We should in vain, therefore, attempt to demonstrate its falsehood. Were it demonstratively false, it would imply a contradiction, and could never be distinctly conceived by the mind.[67]

This important passage may well be puzzling to the reader on first acquaintance. But what Hume was seeking to do was to draw a hard-and-fast distinction between the sciences of mathematics and logic and the empirical sciences such as mechanics and natural history. To make such a distinction — just like that — begged a large number of difficult questions. But what Hume was claiming was that the human mind could comprehend logical *relationships* without recourse to examination of the world of things. For example, one might accept the *forms* of the standard syllogisms as expressions of logical truths, without commitment to the particular truth of any given major or minor premiss. Likewise, we find it *inconceivable* that A is greater than B, and B is greater than C, but A is not greater than C. Also, according to Hume, mathematical reasoning may be conducted without knowledge of 'matters of fact'. It is made up entirely of 'relations of ideas'. By contrast, however, we have no difficulty in *imagining* a ball defying the laws of gravity, or a light ray passing through a block of glass in a way that contravenes Snell's law. We certainly don't expect such things to happen; but we are able to imagine them.

Such was Hume's way of carving up the cake of knowledge. His position certainly does not command general assent today, particularly amongst those contemporary theorists of the sociology of knowledge school who would argue that what is and what is not regarded as logically correct reasoning is a form of social custom or practice, dependent on numerous factors but particularly on education, rather than some, so to speak, transcendent truth.[68] And one might go further and question whether the 'rational' prong of 'Hume's fork' was compatible with his general rejection of innate ideas. But it is not necessary to press Hume on either of these points here. Rather, let us see what effect the empirical prong of Hume's fork had when applied without the allowance of any 'rational' element, to the notion of causality.

The notion of causality was obviously a difficult one for Hume, for it could not be reduced to ideas corresponding to simple impressions. Take a well-known example of a cause/effect relationship such as that of a billiard ball, *A*, moving up to a second billiard ball, *B*, striking it, and *causing* it to move.[69] Clearly, the relationship between the two balls is physical, not logical. Therefore, one does not say that the ball *B* has to move, of *logical* necessity, as a result of being struck by *A*. Obviously, there is a connection between the actions of *A* and *B*, but the connection is causal and describable in terms of the laws of physics, not the laws of logic — at least this was Hume's interpretation of the matter (or, rather, our gloss on Hume). But, Hume also said: 'all reasonings concerning matters of fact seem to be founded on the relation of *Cause* and *Effect*'.[70] Therefore, it behoved him to give a thorough account of the cause/effect relationship, consonant with his empiricist principles and his division of knowledge into 'matters of fact' and 'relations of ideas'. Otherwise his philosophical system would be a shambles of scepticism.

However, when Hume looked closely at examples of supposed cause/effect relationships, such as that of the billiard balls mentioned above, he found that all that his senses could tell him was that causes were always *prior* to effects, and *contiguous* to them.[71] On the basis of what he could observe, he could find no necessary connection, in the logical sense, between a cause and an effect. One could only observe contiguity, and temporal priority for the cause.[72] And since Hume rejected Locke's concept of '*causal power*',[73] he could really see no compelling reason — and certainly there was no logical reason arising from some inherent 'relation of ideas'[74] — to suppose that the same effects were always generated by the same causes. The 'empirical' side of the fork was also of no assistance in a search for some necessary connection between cause and effect. There were no sense impressions to which the word/symbol 'cause' might be attached, or to which it might be reduced.

Having reached this uncomfortable position, Hume tried to extricate himself by offering a psychological explanation of the commonly held belief that there is *some* kind of necessary connection between causes and effects. That is, Hume supposed that after repeated experiences of the association of two kinds of events — which we are accustomed to call the cause and the effect, respectively — we come to believe that there is a necessary connection, even if there is no basis in logic for such a belief, and even if all we have to rely on is the 'constant conjunction'[75] of what we are pleased to call causes and what we are pleased to call effects. Arguing in this manner, Hume eventually arrived at his well-known definition of a cause as:

> *An object precedent and contiguous to another, and so united with it in the imagination, that the idea of the one determines the mind to form the idea of the other, and the impression of the one to form a more lively idea of the other.*[76]

This account of the reason why we commonly believe that there are necessary connections between causes and effects (even if there are none such) was far from satisfactory, for it was entirely subjective in character, the connection residing in our beliefs rather than in the way things in the world actually behave.[77] It had the effect of making necessary-connection statements autobiographical, so to speak, which is odd, to say the least. For one would hardly suppose that one was talking about oneself when one said that billiard-ball A caused billiard-ball B to move. For reasons of this kind, subsequent philosophers have generally been dissatisfied with Hume's 'psychological' account of causality, and have looked for alternatives. Historically, however, Hume's examination of the problem of causality was of the greatest importance because it stimulated Kant to attempt his own account of the matter, effecting thereby a major reconstruction of philosophy.

But we have by no means done with Hume's problem of causality, for it flowed on to raise the question of whether or not it was reasonable to believe in the uniformity of nature. And it also led Hume to recognise and discuss the very important philosophical 'problem of induction'. Considering first the question of the uniformity of nature, it is clear that if nature *were* uniform, so that — quite regardless of the existence or otherwise of causal powers or anything of that kind — the same causes were *always* followed by the same effects, then there *would* be a kind of 'logical' connection between causes and effects. One could argue as follows. Cause A has occurred and has been noted to be followed by effect B. But nature is uniform in her operations; so the same causes are always followed by the same effects. *Therefore*, given a recurrence of cause A, effect B will necessarily follow; B is necessarily connected to A.

Hume, however, would have none of this. He denied the principle of the uniformity of nature, saying that:

> [T]he supposition, *that the future resembles the past*, is not founded on arguments of any kind, but is deriv'd entirely from habit, by which we are determin'd to expect for the future the same train of objects to which we have been accustom'd.[78]

Thus he gave a psychological account of our belief in the uniformity of nature, just as he did of our belief in the principle of causality. We are not further forward, therefore.

The third possible line of advance might be to proceed with the help of inductive reasoning or inference. But to proceed thus, Hume quickly found, led directly to the notorious 'problem of induction', though he did not himself refer to it in this way. Rather, he wrote:

> [T]here can be no *demonstrative* arguments to prove, *that those instances, of which we have had no experience, resemble those, of which we have had experience.*[79]

It will be appreciated that arguing from past experiences to ones that are to occur in the future is a form of inductive reasoning. It is like saying:

$A(1)$ has the property X
$A(2)$ has the property X
$A(3)$ has the property X

$A(p)$ has the property X
$A(q)$ has the property X
$A(r)$ has the property X

Therefore: All *A*s have the property *X*, where 'All *A*s' include future ones as well as those of the past and present.

However, if the matter is regarded from the perspective of a logician, such a generalised conclusion (or inductive generalisation) is not warranted on the basis of specific items of observational evidence ($A(1)$, $A(2), A(3), ..., A(p), A(q), A(r)$, etc), for the conclusion is 'wider' than the premises. Or, as Hume said, such arguments (or inductions) are not 'demonstrative'; and no matter how hard one may try, inductive arguments cannot somehow be twisted around, so to speak, to make them deductive. Inductive generalisations cannot be justified on the basis of experience without circularity of reasoning. (Or so our logicians tell us.)

The difficulty, however, is that science in our day, as in Hume's and at other times, does seem to place considerable reliance upon inductive inferences. To take a simple example, it does so whenever a physical constant is printed in a book of tables, and other physicists assume that it is perfectly 'safe' to use the result of some other person or persons in their own work. Induction is also used whenever someone plots points on a graph and draws the 'best-fitting' line through or between these points; for a general case is being inferred from individual experimental results.

I suggest that scientists are in fact happy to do this sort of things for a variety of reasons. One is social in nature. People place reliance on the work of others because it has received the seal of approval of other members of the scientific community. But this can hardly be the whole of the story. It says at the back of my *Cambridge Four-Figure Mathematical Tables* that the mechanical equivalent of heat is 4.185 joules per calorie; and recognising that this information derives from a highly reputable source (Cambridge University Press), I may feel confident in accepting its truth. But surely that isn't *all* that needs to be said on the subject? In fact, I have done some experiments earlier in my life that have led me to believe that the figure is certainly quite close to 4.185. And I know that physicists using this figure have found general 'coherence' in their practical and theoretical work. Moreover, the constant can be measured in a number of different ways, and the agreement of the various experimental results suggests that the figure is substantially correct, though some more refined experiments will allow additional significant figures to be added.

Yet, as I say, although the 'inductions' used in science are in some way (that is difficult to analyse) warranted by the social fabric of science, and its practical success, it is pretty obvious that this isn't the end of the matter. It would appear that there is *some* uniformity in nature, even though we cannot have certain knowledge as to precisely which features of nature are uniform and which are not. And the joining-up of points to draw the best-fitting line is a perfectly *rational* action, sensible to perform, even though it is contrary to what the canons of strict deductive logic will allow. I suggest, therefore, that science must settle for something less than absolute certainty in its enterprises — the certainty that is customarily associated with deductive logic. But it can, in fact, get along well enough without having to submit to the requirement of strict deducibility from phenomena when law-like generalisations are stated. And, as we shall see in Chapter 6 in our consideration of the work of Hans Reichenbach, it is possible to show that it is rational to employ induction in science, and that no other method can be found that is superior. Induction is, so to speak, the best bet.

It is important to emphasise that Hume's argument — that inductive reasoning does not, and cannot, have deductive certainty — appeared within a specific historical matrix: namely an attempt to base a satisfactory theory of knowledge on empiricist principles, and with an acceptance of the principle that there is an unbridgeable divide between 'matters of fact' and 'relations of ideas', or between *a posteriori* and *a priori* knowledge. It is, I suggest, quite possible that the 'problem of induction' would have *appeared* to be no problem at all if Western philosophy had had some other, quite different, internal history. (This is not to deny that there would always be a logical problem for those who might seek to make empirical observations that were *certain*, on the basis of finite observational data.) But the arguments *were* developed the way they were, and history cannot now be altered. So partly as a result of Hume's sceptical writings on induction, we find a distinctive philosophical tradition established on the Continent, with Kant trying to erase the difficulties generated by Hume's work. And in Britain the tradition of empiricism continued to be pursued through the work of philosophers such as Herschel, Mill, Russell and Ayer. When blended with the work of German mathematicians and logicians, this tradition generated the so-called logical-empiricist (or logical-positivist) movement[80] which has played such an important role in twentieth-century philosophy of science.

Some of the results of this have been rather curious. While, as I shall argue in my concluding chapter, much of the work of metascientists can best be understood as that of attempted legitimations of science, a considerable divide has opened up between scientists and metascientsts in the present century. Scientists do work that appears, *prima facie*, to involve inductive inference, and they extrapolate from the past into the future. Moreover, scientists claim to be 'logical' thinkers. Yet some modern philosophers of science (such as Karl Popper) have told scientists that it is 'illogical' for them to use induction so that in effect the edifice of science is constructed of crumbly material. As a result, scientists and metascientists have frequently looked at each other from separate camps, with mutual incomprehension.[81] This was particularly so when the metascientific citadel was fortified and commanded by philosophers of the logical-positivist school.

I think it is fair to say that David Hume had a leading hand in all this. He raised the 'problem of induction' within the context of an examination of the Lockian empiricist program, following it faithfully where'er it might take him — even if it were to the most unpalatable conclusions. Thereby he generated problems that have excited philosophical interest and inquiry from his own day right down to the present. One definite historical outcome of Hume's critique of induction was that he showed, to most people's satisfaction, that the structure of science cannot be represented simply as a great pyramid, such that one rises *deductively* from observations and experiments to laws and theories. The followers of Hume who have tried to sustain such a picture have either (like Mill, see p. 149) had to give a new interpretation of the word 'deduction', or (like Carnap in his *Aufbau*, see p. 238) have engaged in a philosophical lost cause.

As for Hume himself, he sought an escape route of his own, in Part III (Secton 15) of Book I of the *Treatise*, where the author listed eight 'Rules by which to judge of causes and effects'. For having presented his 'psychological' analysis of the problem of causality, Hume proceeded to give some account of the way in which scientific inquiries might be pro-secuted, regardless of the problem of the uniformity of nature and other awkward obstructions. And it was this 'methodological route' that subsequent British empiricists such as Herschel and Mill sought to exploit — with some advantage and success, as we shall see.

Hume's eight Rules were stated as follows:

1. The cause and effect must be contiguous in space and time.
2. The cause must be prior to the effect.
3. There must be a constant union betwixt the cause and the effect. 'Tis chiefly this quality, that consitutes the relation.
4. The same cause always produces the same effect, and the same effect never arises but from the same cause. This principle we derive from experience, and is the source of most of our philosophical reasonings. For when by any clear experiment we have discover'd the causes or effect of any phaenomenon, we immediately extend observation to every phaenomenon of the same kind, without waiting for that constant repetition, from which the first idea of this relation is deriv'd.
5. There is another principle which hangs upon this, viz. that where several objects produce the same effect it must be by means of some quality, which we discover to be common amongst them. For as like effects imply like causes, we must always ascribe the causation to the circumstance, wherein we discover the resemblance.
6. The following principle is founded upon the same reason. The difference in the effect of two resembling objects must proceed from that particular, in which they differ. For as like causes produce like effects, when in any instance we find our expectation to be disappointed we must conclude that this irregularity proceeds from some difference in the causes.
7. When any object encreases or diminishes with the encrease or diminution of its cause, 'tis to be regarded as a compounded effect, deriv'd from the union of the several different effects, which arise from the several different parts of the cause. The absence or presence of one part of the cause is here suppos'd to be always attended with the absence of a proportionable part of the effect. This constant conjunction sufficiently proves that the one part is the cause of the other. We must, however, beware not to draw such a conclusion from a few experiments. A certain degree of heat gives pleasure; if you diminish that heat, the pleasure diminishes; but it does not follow, that if you augment it beyond a certain degree the pleasure will likewise augment; for we find that it degenerates into pain.
8. The eighth and last rule...is, that an object, which exists for any time in its full perfection without any effect, is not the sole cause of that effect, but requires to be assisted by some other principle, which may forward its influence and operation. For as like effects necessarily follow from like causes, and in a contiguous time and place, that separation for a moment shews that these causes are not compleat ones.[82]

These rules of method may appear to be somewhat at variance with the sceptical tenor of Hume's other remarks concerning causation, elsewhere in the *Treatise*. Nevertheless, they do seem to provide some sensible suggestions as to how scientific inquiries might be prosecuted, regardless of worries that one might have about 'necessary connections' between causes and effects.[83] Rules 5 and 6 seem to be forerunners of John Stuart Mill's 'Method of agreement' and 'Method of difference',[84] but we have already encountered them in the writings of Duns Scotus and William of Ockham respectively.[85] Hume's Rule 7 seems to correspond roughly with Mill's 'Method of concomitant variations', employed by anyone looking for some causal relationship between two variables and hoping to display the relationship graphically. And very approximately Rule 8 was an ancestor of Mill's 'Method of residues'. The first three Rules might readily be expected, from what we already know of Hume's analysis of causality in terms of temporal and spatial contiguity, and the notion of 'constant conjunction'. But the fourth Rule may well cause eyebrows to be raised. For in effect Hume was saying here that for practical purposes it was quite all right to assume the uniformity of nature and make inductive generalisations, logical problems and inadequacies notwithstanding. In other words, he was quite willing to have science proceed as accustomed, regardless of philosophical objections that might be raised. Some subsequent metascientists have not been so generous!

Yet the problem may seem more acute, or Hume's position more incongruous, when we note another passage from Hume, this time from the *Enquiry*, which seems to recommend the efficacy and utility of Newton's 'Third Rule of Reasoning in Philosophy':[86]

> It is entirely agreeable to the rules of philosophy, and even of common reason; where any principle has been found to have a great force and energy in one instance, to ascribe to it a like energy in all similar instances. This indeed is Newton's chief rule of philosophy.[87]

To be sure, this passage occurs within the context of Hume's discussion of moral — not natural — philosophy. Nevertheless, it does seem strange that he should have chosen to give such high praise to Newton's 'Third Rule', in which induction was ostensibly legitimised — even from the macro- to the micro-world. Was Hume just philosophically inconsistent in all this? In a sense he was. Nevertheless, he had no wish that his philosophical scepticism should produce a paralysis of all action. We may suggest, therefore, that his eight rules were intended to help men to coordinate their thinking and to act in a rational manner, and conduct their affairs pragmatically and with success. But their employment does not mean that Hume supposed that his own philosophical arguments directed against induction, universal causation and the principle of uniformity, were themselves unsound. So his overall position does seem to have been somewhat incongruous or inconsistent. Nevertheless, his eight rules of method made a valuable contribution to the technique of 'arch-climbing', or to metascientific accounts of scientific methodology, a topic that was not given much attention in other parts of Hume's work.

There is one other topic in Hume's philosophy that we may mention just briefly, before we pass on to a consideration of Kant; that is, Hume's discussion of the distinction between primary and secondary qualities. Hume agreed with Locke that different observers may have quite different perceptions of the same objects; so the two philosophers were more or less in agreement about secondary qualities. But Hume thought that the doctrine of primary qualities was hopelessly confused. 'The idea of motion', he said, 'depends on that of extension, and the idea of extension on that of solidity'.[88] Yet a body could not just be solidity, and nothing more! So like Berkeley, Hume concluded that:

> [A]fter the exclusion of colours, sounds, heat and cold from the rank of external existences, there remains nothing, which can afford us a just and consistent idea of body.[89]

Besides, Hume added,[90] there can be no simple impression corresponding with solidity.

So Hume's scepticism was extended to a consideration of the adequacy of Locke's distinction between the primary and secondary qualities, and all that depended thereon so far as the mechanical philosophy was concerned. And the distinction was found wanting. We may note further that Hume implicitly rejected the notion of absolute motion when he wrote; "tis evident... [motion] is a quality altogether inconceivable alone, and without a reference to some other object'.[91] It appears, therefore, that although he did not specifically attack the science of his day, as did Berkeley, Hume's philosophy was really incompatible with some of its leading suppositions. In other words, the philosophical warrant for science that Locke had sought to provide was further shaken.

So let us now turn our attention to Kant, who specifically described himself as having been awoken from his 'dogmatic slumber' by the influence of Hume.[92] We must try to see how Kant sought to reconstitute philosophy in a way that would take account of criticisms such as those of Berkeley and Hume, and yet at the same time provide a firm metaphysical basis on which science might rest, securely and comfortably. Kant's program was by no means carried through with complete success. Nevertheless, its overall importance in the history of philosophy and the history of Western thought can hardly be stressed sufficiently. We must, therefore, attempt to come to grips with at least some of the leading ideas of this philosopher, his complexities and obscurities notwithstanding.

Kant

Immanuel Kant (1724–1804) taught at the University of Königsberg on the Baltic coast in East Prussia for the greater part of his life, and he is well known for the absolute regularity with which he conducted his affairs, so that, for example, it is said that the citizens of Königsberg were able to set their watches by him as he passed by on his daily walk. This passion for order, system and general tidiness can readily be discerned in Kant's philosophical writings, even though the coherence of the whole may in fact be more apparent than real.

Kant was an extremely prolific writer, but only two of his works will attract our attention here: the celebrated *Critique of Pure Reason* (1781, 2nd edn 1787)[93] and *Prolegomena to any Future Metaphysics* (1783).[94] The *Prolegomena* (meaning 'Introductions') were intended as a simplification and clarification of the ideas presented in the *Critique of Pure Reason*, which Kant reckoned (no doubt correctly) had been ill understood on its first presentation to the public. There were considerable differences between the two editions of the *Critique of Pure Reason*. But fortunately there is a 'composite' version in the standard modern English edition of Kemp Smith[95] which we shall be referring to here.[96] So, let us attempt the formidable task of trying to make some sense of the leading features of the *Critique of Pure Reason* within the compass of just a few pages.

Kant was impressed and disturbed by the fact that the rationalist ('dogmatic') philosophers with which he was familiar (notably Descartes, Leibniz and Wolff) were noticeably *unsuccessful* in increasing the sum of human knowledge, whereas, by contrast, science was having considerable success in this enterprise. This was so, despite the fact that Hume had apparently shown that scientists could make no claims to certain knowledge of the causes of phenomena — which might suggest that scientific knowledge was not securely grounded. (Kant was not much impressed with Hume's psychological account of induction and causation.) The reason for the lack of success of rationalist philosophy was, Kant believed, due to the fact that it sought to generate 'pure' knowledge, without any taint of empirical input. To put the matter in a homely way, if philosophers were content to sit in their philosophers' chairs and devise accounts of the nature of things, without going out into the world to examine its features with the help of observation and experiment, then they could not be expected to say anything of interest; they would make no contribution to *knowledge*; they would be mere Baconian spiders, or dogmatists.

We can see, then, why Kant called his book a *Critique of PURE Reason*. By 'pure' he meant *a priori*.[97] And in his view any philosophical system that was wholly 'pure' or rationalistic, making no allowance for the input of information through the senses, was doomed to disaster. Metaphysics as a speculative science, isolated from empirical input, was useless. As Kant himself said in one of the most famous sentences in the *Critique*: 'Thoughts without content are empty; intuitions[98] without concepts are blind'.[99] To illustrate the philosophical problems that could arise through reliance on 'pure reason' alone, Kant devised four famous 'antinomies' (that is, contradictions). These appear quite late in the *Critique*, but we may choose to deal with them first in illustration of the evils of indulgence in 'pure reason', and by way of introduction to some of the other major themes of Kant's book.

What Kant sought to show was that the 'purely' rationalist philosopher could perfectly well sit down and 'prove' statements that were in total contradiction to one another. Thus he could 'prove' that the world *does*, and *does not*, have a beginning; and he could 'prove' that it *is*, and *is not*, infinitely extended. Likewise he could 'prove' that the world both *is*, and *is not*, made up of simple parts; that the human will *is*, and *is not*, free; and that there *is*, and *is not*, an absolutely necessary being (namely, God)!

We shall not examine all of Kant's antinomies, but by way of illustration let us take the question of whether the universe did or did not have a beginning. Consider the notion of an infinite sequence of regular time intervals, which for convenience we may represent as in Figure 15.

Figure 15

Such a sequence, according to Kant, could never be *completed*. It would be a contradiction in terms to have an end to an infinite sequence. However, the world *must* have had a beginning in time, otherwise an infinite number of years must already have elapsed ('passed away'). So the world *did* have a beginning in time.

Now we may start again. Suppose, for the moment, that the world *did* have a temporal beginning (as represented in Figure 16).

Figure 16

Universe created at this point in time

One would presume that the time intervals before the creation of the world were all identical. (Clearly, Kant's idea of time was quasi-Newtonian here. It would make little sense to a Leibnizian to talk about time before the creation of the cosmos.) However, the time interval immediately before the Creation would be different from the ones that preceded it, for it came just before a unique event. But this should not be so if all points in time before the Creation were identical. One is forced to conclude, therefore, that the very idea of a Creation at a point in time is unintelligible. Therefore the universe is infinitely old; it did not have a beginning.[100]

These arguments are not, I think, particularly impressive. For example, there seems to be no very obvious reason why an infinite length of time should not elapse, or come to an end. It could still very well stretch to infinity 'in the other direction', so to speak. Our task here is not, however, to find objections to Kant's arguments, but to understand his method of approach. To repeat, he believed that the antinomies demonstrated the philosophical problems that arise if one relies solely on 'pure reason'. His argument was that statements made about matters that transcend all possible experience — for example, which purported to say something about the universe as a whole — would inevitably be fallacious. 'Pure reason', therefore, was to be utterly eschewed. All good arguments were based upon information received via the senses.

Next, we may look at the question of what are called 'synthetic *a priori*' propositions, with which Kant was much concerned in both the *Critique* and the *Prolegomena*. Two general classes of knowledge may first be considered. There is knowledge that one may (arguably) have without recourse to experience — which may be known *a priori*. For example, one could (arguably) recognise the truth of the syllogisms of Aristotle's *Prior Analytics* without climbing out of one's philosophical chair to go and have a look at the world outside. One could achieve this kind of 'knowledge' simply by mental introspection — by examining the *form* of the propositions. Likewise statements such as 'The whole is greater than the part' could (arguably) be known *a priori*. More widely, the truth of any 'analytic' proposition could be known without recourse to experience. Here we introduce a new element of the Kantian vocabulary. According to Kant, an analytic proposition is one in which the (grammatical) predicate is 'contained within'[101] the (grammatical) subject. Such propositions are knowable *a priori*.

Of course, one may well wonder what Kant meant by 'contained within'. If we say 'A green apple is an apple' there really isn't any difficulty, and one might readily agree that this proposition can be known *a priori*, for the predicate is genuinely 'contained within' the subject, and can be found by 'analysing' it — by chopping it into its parts. Stretching things a little further, we can perhaps agree that 'All triangles have three angles' is analytic (and therefore knowable *a priori*). When we go just a little further, however, we may be a bit less sure. For example, is the proposition 'A rainy day is a wet day' analytic? It would be if one agreed that the notion of wetness is 'contained within' or implicit within the notion of raininess.[102] But one could, I think, argue about this. Kant's own example was 'All bodies are extended'. For him, the notion of extension was part of the very meaning of body. Therefore, one could say, *a priori*, that the proposition was true: it was an analytic proposition, knowable *a priori*. On the other hand, Kant maintained that the statement 'All bodies are heavy' was not analytic. One would not find the notion of heaviness 'within' the notion of body. One could discuss all this at considerable length, but I do not propose to do so here. The point I wish to make is that Kant believed that analytic propositions could all be known *a priori*. And he thought he could recognise an analytic proposition when he met one — though we might not be so sure about this.

There were also synthetic propositions that were not knowable *a priori*, but only *a posteriori* (by getting out of that philosophical chair and having a look at the world); for in synthetic propositions the predicates were not 'contained within' the subjects. The predicates added something to the subjects. For example, the proposition 'The apple is rotten' may be classified as 'synthetic *a posteriori*'. The predicate adds something to (or says something about) the subject. Therefore, the truth or otherwise of the proposition cannot be known by mental intro-spection — *a priori*. This synthetic proposition, and others like it, is knowable only *a posteriori*.

One might suppose that no synthetic proposition could be known *a priori*. But Kant thought otherwise. For example, he claimed that the proposition 'The shortest distance between two points is a straight line' was a synthetic proposition (in that the predicate is not 'contained within' the subject) but one which nevertheless was knowable *a priori*. It was for us necessarily true. Its converse was unthinkable. Kant wrote:

> That a straight line between two points is the shortest, is a synthetic proposition. For my concept of *straight* contains no notion of quantity, but only a quality. The concept of the shortest is therefore wholly an addition, and it cannot be drawn [derived] by any analysis from the concept of the straight line.[103]

Nevertheless, this addition or synthesis could (Kant believed) be performed *a priori*.[104] So we have an example of a synthetic *a priori* proposition. Other examples proposed by Kant were: '7 + 5 = 12'; 'All men are free'; and 'Every event has a cause'. It is with this last that we shall be particularly concerned.

Accepting for the moment that there are such creatures as synthetic *a priori* propositions,[105] we may wish to inquire, with Kant, how this might be possible. It would not be so if one followed standard empiricist doctrine according to which the mind conforms to objects and the impressions received from them as it makes its judgments. If this epis-temology were correct, no synthetic propositions could be known without examination of the world in some way. How, then, was it that Kant felt able to say that men could know, *a priori*, that every event has a cause, particularly when he was also saying that 'pure reason' led one into grievous philosophical errors, exemplified by the antinomies?

Speaking very approximately for the moment, we may say that Kant's answer was that objects in some sense conform to the operations of the mind, rather than the other way round as empiricism would suggest. This striking suggestion was at the heart of what he called his Copernican Revolution in philosophy.[106] Kant did not mean that the mind actually created objects or was endowed with innate ideas. Rather, he was proposing that the mind brings something to the objects which it experiences, in the process of their cognition. Unlike Locke (say), Kant envisaged the mind as an active agent, in some way imposing itself on the objects of which it forms conceptions, rather than being a mere 'sheet of paper' for the reception of impressions. It was as if everything we perceived were 'coloured' through the faculty of our thought. Inevitably, we think according to the particular structures of our mental apparatuses. If we wish to offer a very child-like illustration, we might say that the person who wears rose-tinted spectacles inevitably sees the world from a 'rosy' perspective. Of course, spectacles can be taken off, but a man cannot divest himself of his whole cognitive apparatus and still have means of gaining knowledge of the world.

However, putting the matter this way is very simplistic. If we wish to pursue the matter further, we must now make a more strenuous effort to enter the grim and forbidding castle of the *Critique of Pure Reason*[107] — to try to form some idea of the line of argument according to which Kant sought to persuade his readers of the justice of his Copernican Revolution, the acceptability of (some) synthetic *a priori* propositions, and in particular the truth of the statement that 'every event has a cause'.

It may be helpful first to give some further details concerning Kant's philosophical vocabulary. We have already seen his usage of *a priori*, *a posteriori*, analytic and synthetic, though it should be noted that the term *a priori* was sometimes used in the sense of 'logically necessary', rather than in a 'psychological' sense of 'prior to experience'. We have also seen that Kant sometimes used the word 'pure' to mean '*a priori*' in the 'psychological' sense), as in the very title of his book: *Critique of Pure Reason*. And as if this were not enough, he also made frequent use of the term 'transcendental' to mean *a priori* in a psychological sense. A good way of interpreting the whole Kantian enterprise is to see it as an examination of what Strawson has called the 'limiting framework of ideas and principles the use and application of which are essential to empirical knowledge'[108] — that is, the conditions that must necessarily be fulfilled in order that the human mind may have experience and knowledge of the world. So Kant's inquiry, in so far as it was transcendental, was concerned with locating and describing the *a priori* elements in cognition.[109] The word 'transcendental' should not be confused with 'transcendent'. A transcendent entity, such as God or a Platonic Idea, is one that lies wholly beyond human empirical experience, or is outside the world altogether, being an an altogether different plane of existence, so to speak. (Incidentally, 'transcendent', not 'transcendental', is the converse of 'immanent'.)

A Humean 'impression' (or sense-datum, in modern jargon) was called by Kant an 'intuition'.[110] He called a mental event or a mental experience (a thought, if you like) a 'representation'. He also used the term 'concept' in more or less the manner in which we are accustomed today, as when we speak of the 'concept' of a triangle. The faculty governing intuitions (or sensations) Kant called the 'sensibility'; and the faculty of concepts he called the 'understanding'. The 'pure' (that is, *a priori*) concepts, which handled the 'manifold of intuitions' provided by the sensibility to the mind, were called 'categories' (the meaning of which within the Kantian vocabulary we shall seek to explicate below). So in essence, then, Kant's theory of cognition was that *both* sensibility and understanding were needed for knowledge. The sensibility produced a 'manifold' out of the intuitions; and the understanding brought the manifold of intuitions under concepts. Kant's well-known dictum is worth restating here: 'Thoughts without content [that is, 'intuitions'] are empty; intuitions without concepts are blind.'[111] His general account of the cognitive process is of course debatable. One might, for example, question whether the faculties of sensibility and understanding can in fact be treated separately, as Kant supposed. But at least his system did allow for a kind of middle way between unalloyed (no longer dare I say 'pure') empiricism and untramelled rationalism.

One other Kantian word may be mentioned here — 'judgment'. It was Kant's term for 'proposition', as used above in our discussion of the analytic/synthetic distinction. It need cause us no particular difficulties. There is also Kant's term, 'schema'; but we cannot explain this until we have got a little further with our exposition of his system. Explanation of this will therefore be deferred for the moment.

Let us now try to give an indication of the general structure and content of the *Critique of Pure Reason*. The Introduction directed attention to the problem of synthetic *a priori* propositions and indicated that there were three kinds of 'science' where they characteristically found employment: mathematics, physics and metaphysics. But as we have seen already, Kant wanted to reject metaphysics — or any speculations that ran beyond the 'boundaries of sense'. The major division of the book was into the 'Transcendental Doctrine of Elements' and the 'Transcendental Doctrine of Method'; but really the battle was nearly all over by the time the second half was reached, it being concerned with an exposition of the general method of philosophical inquiry, already deployed in the first part of the book. So we shall not pursue our examination beyond the first part.

The 'Transcendental Doctrine of Elements' is subdivided into a so-called 'Transcendental Aesthetic' and a 'Transcendental Logic'; and the latter is divided into a section entitled 'Transcendental Analytic' and another called 'Transcendental Dialectic'. It may be easier, however, to think of a three-fold division, each section being of roughly equivalent status, namely:

1. Transcendental Aesthetic
2. Transcendental Analytic } being two forms of
3. Transcendental Dialectic } Transcendental Logic.

This has the advantage of meshing with Kant's intention of dealing successively with the three branches of 'science' in which synthetic *a priori* propositions ostensibly find employment: mathematics, physics, and metaphysics. However, Kant's actual discussion was by no means as tidy as the foregoing might suggest.

The 'Transcendental Aesthetic' did not (as one might be forgiven for supposing) have to do with a theory of aesthetics or taste. It was, in fact, concerned with Kant's theory of the way in which notions of space and time were involved in forming a manifold out of the sensory intuitions (or Humean 'impressions'). It dealt with Kant's 'doctrine of sensibility'.

The 'Transcendental Analytic' was basically concerned with the doctrine of the categories — that is, the *a priori* concepts that were supposedly applied to the intuitions (perceived as a manifold in space and time) in the process of understanding or cognition. And the 'Transcendental Analytic' was itself divided into an 'Analytic of Concepts' and an 'Analytic of Principles'. The first of these set out Kant's twelve celebrated 'categories of the understanding' and sought moreover to prove [*sic*] that the list was both correct and complete. This task was attempted in the latter part of the first chapter of the 'Analytic of

Concepts', in a section popularly known as the 'Metaphysical Deduction of the Categories'.[112] There was also a 'Transcendental Deduction of the Categories' in Chapter 2. The 'Analytic of Principles' contained Kant's theory of 'schematisation' (of which more in a moment), leading to the statement of the various 'Synthetic Principles of Pure Understanding', really the outcome of all his immense philosophical labour. At this stage, the reader, probably exhausted before we have even begun, may find it helpful to be provided with an 'anatomy' of the *Critique* (see Figure 17). It is incomplete, but sufficient for our purposes.

Figure 17

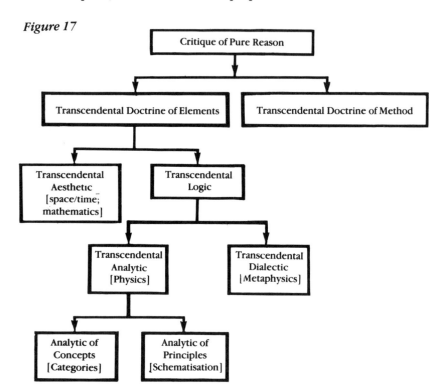

These preliminaries can now be put aside, and we may proceed further to have a look at some of the specific ideas propounded in the various parts of the *Critique*. The 'Transcendental Aesthetic' was chiefly concerned with the notions of space and time, which Kant regarded as the *a priori* aspects or components of the faculty of intuition (or sensation) in the overall process of understanding. He called them 'pure intuition[s]'[113] or 'pure form[s] of sensible intuition'[114] and emphasised that they were not themselves concepts. A good deal can be, and has been, said about this doctrine, but in essence it boiled down to the claim that all objects of experience must — in order to be perceived — be perceived in space and time, which are, so to speak, 'logically prior' to the objects actually recognised in experience. But objects themselves must also be spatial and temporal in order that they be experienced.

Thus, Kant maintained, there has to be a kind of congruence between objects and our perceptions of them, in order that perception may occur; but the mental aspect is 'prior'. However, it is a *necessary condition of experience* that objects be extended in space and time. In this sense, one may say that space and time are the *a priori* aspects of sensuous intuition.

Space and time were not, however, treated as being exactly on a par. Kant distinguished between 'inner sense' and 'outer sense', space supposedly being the 'form of all appearances' of the latter[115] while time fulfilled the same role for both inner and outer senses.[116] In other words, time was held to be the *a priori* form of sensuous intuition for all experiences (including memories, for example) while space had this role only for 'outer' experiences: of the external world outside ourselves.

This distinction between space and time seemingly underlay Kant's rather curious attempt to maintain that geometry was the mathematics of space[117] and arithmetic was the mathematics of time[118] — which would explain (if the general theory sketched above were found acceptable) why the propositions of geometry and arithmetic could be said to be 'synthetic *a priori*'.

Kant's doctrine of space and time can usefully be distinguished from that of Newton and that of Leibniz. Newton, it will be recalled, adopted an 'absolute' theory of space and time. Leibniz's theory was 'relational', space and time being regarded as functions of the relations between objects. Kant's theory, however, which proposed that space and time were 'forms of sensuous intuition', made them *functions of perceivers*. They were, so to speak, buried in man's psyche. The fact, for example, that we (or most of us!) can only imagine or entertain three-dimensional Euclidean space would, for Kant, be a necessary feature of *experience*, not of the world.

This point brings us to another important feature of Kant's doctrine that is also to be found in the 'Transcendental Aesthetic' — his so-called 'Transcendental idealism'. If, for example, we can only know the objects of the world through the 'necessary forms of sensuous intuition', namely space and time, then we can never know things as they are 'in themselves'. What we know of them must always and necessarily be shaped by the 'forms of intuition' (and also by the categories — of which more anon). So the whole realm of objects 'in themselves' is wholly inaccessible to us.[119] Kant called it the realm of *noumena*.[120] All we can know are *phenomena*; but phenomena are necesarily apprehended in determinate ways, according to the forms of space and time. This doctrine of transcendental idealism clearly has a strong Berkeley-like flavour.[121] And in a sense it was somewhat similar to Locke's view that there *were* objects in the micro-world, possessed of primary qualities; but we could never know them. The whole doctrine of transcendental idealism in Kant, though used by him with good effect in his criticism of 'pure reason' (speculative metaphysics), had a baneful influence in nineteenth-century philosophy through the work of Fichte and Hegel, and led to some untoward political and social consequences.[122] But we need not attend to such matters in this book.

We must now pass on to the even more forbidding regions of the

Critique and say something about the awesome 'Transcendental Analytic', notorious for its philosophical difficulty and obscurity. Following the order of treatment in the *Critique*, we encounter first the 'Analytic of Concepts', about which Kant wrote with pleasing clearness and distinctness:

> By 'analytic of concepts' I do not understand their analysis, or the procedure usual in philosophical investigations, that of dissecting the content of such concepts as may present themselves, and so of rendering them more distinct; but the hitherto rarely attempted *dissection of the faculty of the understanding* itself, in order to investigate the possibility of concepts *a priori* by looking for them in the understanding alone, as their birthplace, and by analysing the pure [ie *a priori*] use of this faculty.[123]

And when he analysed the 'faculty of the understanding itself' Kant claimed that he found therein just twelve categories, neither more nor less; and he attempted moreover to 'prove' that his list was both correct and complete.

To grasp the notion of a Kantian category, we may consider a simple example. Suppopse we look at what we call a rose. We receive certain 'intuitions' (that is, sensations or impressions) such as redness, fragrance, softness, and 'rose-shapedness'. These 'intuitions' may be correlated and coordinated by envisaging them as collectively located in an *object*, which we think of as a member of a class of things that can exist independently of ourselves. The concept of a 'thing' is not itself a thing or an object. It may be regarded as a 'mental frame' or 'pattern' into which a large number of bundles of intuitions (perceived in space and time) may be fitted, so that they are comprehended as single objects or things. Such a mental frame or pattern Kant called a '*category*'. It would not itself be objective, but external objects could only exist for an observer by virtue of being perceived or understood through the employment of this category.[124]

In order to obtain his list of categories, Kant first provided a 'Table of Judgments' (a judgment being for Kant a proposition, as we have already noted), which was supposed to give a list of all possible logical forms of propositions. Then he sought to abstract from each proposition the most general idea contained therein, and this supposedly would be the corresponding category. For example, one could supposedly abstract the category of 'allness' (that is, 'totality') from the proposition 'All *A*s are *B*' (which was called the 'universal' judgment). Thus Kant felt able to give his list of twelve categories:[125]

Judgments	Categories
1. *Quantity*	1. *Quantity*
Singular: 'This is *B*'	Unity
Particular: 'Some *A*'s are *B*'	Plurality
Universal: 'All *A*'s are *B*'	Totality
2. *Quality*	2. *Quality*
Affirmative: '*A* is *B*'	Reality
Negative: '*A* is not *B*'	Negation
Infinite: 'All *A* is non-*B*'	Limitation

3. *Relation*
   Categorical: 'A is B'

   Hypothetical: 'If A, then B'

   Disjunctive: 'A is either B or C'

4. *Modality*
   Problematic: 'A may be B'
   Assertoric: 'A is B'
   Apodeictic: 'A must be B'

3. *Relation*
   Inherence and Subsistence
   (Substance and Accident)
   Causality and Dependence
   (Cause and Effect)
   Community
   (Reciprocity between agent
   and patient)

4. *Modality*
   Possibility/Impossibility
   Existence/Non-existence
   Necessity/Contingency

One may well wonder how Kant obtained his list of judgments. Somewhat ingenuously, he stated that it was the one 'ordinarily recognised by logicians', with some departures;[126] but to my knowledge no one has been able to find such a list in the logical texts of Kant's predecessors. And in fact it has been suggested by de Vleeschauwer that Kant's *actual* procedure was quite the reverse of that which is implied in the *Critique*.[127] That is to say, he may have made up his list of categories first and then constructed the list of judgments to fit it! Or he may have worked on the two together. In any case, one may be highly sceptical about any suggestion that the categories *did* emerge from the judgments as Kant would have us believe. For one thing, the similarities between the items in the parallel lists are by no means obvious. If, for example, you were to present the 'Table of Judgments' to a hundred philosophers it would, I think, be extremely unlikely that they would all come up with the same list of categories as that suggested by Kant. Indeed, very likely none of them would come up with the same list. And many of them would, I suspect, refuse to accept the suggestion that *any* category could be extracted from a judgment. The judgments are examples of logical concepts, while the categories are, we understand, psychological entities — the 'pure' concepts of experience. To suppose that the one can be deduced (or somehow educed) from the other is what Kant himself would very likely have called a metaphysical illusion! We may, therefore, be sceptical about Kant's supposed 'Metaphysical deduction of the categories' from his list of forms of statements (that is, from 'judgments').

So we now turn our attention to the next section of the 'Analytic of Concepts': the notorious 'Transcendental Deduction of the Categories'.[128] Here Kant took up the problem of *self*-consciousness and its seemingly necessary unity; and although he was ostensibly *deducing* the categories (all twelve of them, presumably), actually all he succeeded in doing (at best) was to show that there have to be categorical concepts, or forms of thought, in order for each of us to have unified and coherent thoughts. That is, he felt able to conclude, on the basis of the requirement of a unified self-consciousness, that there had to be categories as necessary conditions of experience. But the 'Transcendental Deduction' did not show that there were just twelve categories, of the kind previously derived in the 'Metaphysical Deduction'. All it showed was that they might reasonably provide the rule-governed connections of experience that were required by, or were a necessary condition for, the unity of self-consciousness. So although Kant had some important things to say about the nature of self-hood, and similar matters, and one may (or may not) be persuaded of his argument to the necessity of categories, he really made no further progress towards showing that there were precisely the twelve categories ostensibly generated by the 'Metaphysical Deduction'. So in large measure the 'Transcendental Deduction' was rendered useless by the insufficiencies of the 'Metaphysical Deduction'.[129]

Next we may move briskly on to the 'Analytic of Principles' where we encounter Kant's doctrine of 'schematism' and the establishment of the 'Principles of the Pure Understanding'. Kant was aware of a difficulty in his system, as presented so far, in that there seemed to be a disparity between the objects perceived and the conceptual categories according to which they were supposed to be comprehended:

> [P]ure concepts of understanding being quite heterogeneous from empirical intui-
> tions...can never be met with in any intuition. For no one will say that a category, such as
> that of causality, can be intuited through sense and is itself contained in appearance.
> How, then, is the *subsumption* of intuitions under pure concepts, the *application* of a
> category to appearances, possible?[130]

In answer to this self-inflicted puzzle, Kant proposed that time 'mediated' between objects and categories in the process of schematisation. Thus, the categories were to be given a temporal interpretation, yielding a list of 'schemata' corresponding to the list of categories: 'A schema is, properly,...the phenomenon, or sensible concept, of an object in agreement with the category'.[131] Kant did not, however, carry the procedure of temporalising the categories through to completion, so we can only offer the following partial list:[132]

Categories	Schemata
1. *Quantity*	*Number*
Unity	–
Plurality	–
Totality	–
2. *Quality*	[*Degree of Intensity*]
Reality	Being in time
Negation	Not being in time
Limitation	Time both filled and empty? [discreteness?]
3. *Relation*	–
Inherence and Subsistence	Permanence of the real in time
Causality and Dependence	Causality
Community	Co-existence
4. *Modality*	–
Possibility/Impossibility	Possibility in time
Existence/Non-existence	Existence in some determinate time
Necessity/Contingency	Existence of an object at all times

But schematisation was still not the end of the story. Even the schemata, by which the mind might comprehend phenomena, were not yet in the form of statements of principles that might find application in concrete cases. For this purpose, the schemata had to be developed into the form of specific propositions, such as the principle of causality. If, then, we proceed directly to Kant's results in this endeavour, we have the following 'Principles of the Pure Understanding', which allegedly could be stated as a result of the schemata being developed into propositions:[133]

Schemata (General concepts)	Principles of the pure understanding (Statements of general principles)
Number	*Axioms of Intuition*: 'All appearances are, in their intuition, extensive magnitudes.' [Or, we might say, 'All phenomena are measurable, being extended in space and time.']
–	–
–	–
–	–
[*Degree of intensity*]	*Anticipations of Perception*: 'In all appearances sensation, and the *real* which corresponds to it..., has an *intensive magnitude*, that is, a degree.' [Or, we might say, 'All phenomena have a certain degree of intensity.']
Being in time	–
Not-being in time	–
Time both filled and empty? [discreteness?]	–
–	*Analogies of Experience*: 'All appearances are, as regard their existence, subject *a priori* to rules determining their relation to one another in one time.' [Or, we might say, 'All phenomena are connected in law-like ways.']
Permanence of the real in time	1. 'In all change of appearances substance is permanent; its quantum in nature is neither increased nor diminished.' [Or, we might say, 'In all observed phenomenal changes, matter is conserved, that is, neither created nor destroyed'.]
Causality	2. 'All alterations take place in conformity with the law of the connection of cause and effect.'
Co-existence	3. 'All substances, so far as they coexist, stand in thoroughgoing community, that is, in mutual interation.' [Or, we might say, 'All things mutually interact with one another.']

	Postulates of Empirical Thought: ' — '.
– Possibility in time	1. 'That which agrees with the formal conditions of experience, that is, with the conditions of intuition and of concepts, is *possible*.' [Or, we might say, 'Only those phenomena are possible that conform with the *a priori* conditions of space/time and the categories.']
Existence in some determinate time	2. 'That which is bound up with the material conditions of experience, that is, with sensation, is *actual*.' [Or, we might say, 'All actual phenomena are verifiable by sensation.']
Existence of an object at all times	3. 'That which in its connection with the actual is determined in accordance with the universal conditions of experience, is (that is, exists as) *necessary*.' [Or, we might say, 'Phenomena are *necessary* in so far as they are determined by the *a priori* conditions of space/time and the categories.']

It should be noted that in the 'Analogies of Experience' Kant was, in effect, giving expression to the Newtonian notions of matter, force, and action and reaction, approximately corresponding, therefore, to the three laws of motion of the *Principia*. It is for this reason, therefore, that we felt empowered to say that the 'Transcendental Analytic' was concerned with physics, just as the 'Transcendental Aesthetic' was concerned with mathematics.[134]

The 'Principles of the Pure Understanding', and perhaps above all the second 'Analogy of Experience' ('All alterations take place in conformity with the law of cause and effect'), represent the little grain of gold that may be extracted from Kant's extraordinary philosophical enterprise. And we can finally see how Kant presented his answer to Hume's critique of causality. The causal principle was not to be discovered by induction from observations of phenomena. It was a *necessary condition of experience*, which Kant located in the mental apparatus of the human observer — who is utterly unable to conceive phenomena as operating in any other way than in accordance with the principle of causality.[135]

Yet we have not yet done with Kant's system. There is still the 'Transcendental Dialectic' to contend with. The reader may recall that we actually started here, with our discussion of the antinomies. Metaphysics deployed only 'pure reason', without empirical input (that is, appropriate 'intuitions'). And when this happened, Kant claimed, one could easily run into all sorts of philosophical absurdities. Yet it seems that he still hankered after metaphysics! For example, although he showed that the philosophical arguments in favour of the existence of God were wanting, he still wished to retain the idea of God as a 'regulative principle'. In other words, some ideas of pure reason, such as God, soul, final causes, etc., though not constitutive to the human mind (there being no corresponding categories or schemata), might nevertheless have considerable practical utility. For example, the ideas of God and the soul might be of considerable significance in society, and might be crucial for the purposes of adherence to moral law. So Kant thought it appropriate to admit them, with a 'regulative' rather than a 'constitutive' function. Thus, on the distinction between 'regulative' and 'constitutive' principles, he wrote:

[T]ranscendental ideas without any basis in sensuous intuition never allow of any constitutive employment. When regarded in that mistaken manner, and therefore as supplying concepts of certain objects, they are but pseudo-rational, merely dialectical concepts [of pure reason]. On the other hand, they have an excellent, and indeed indispensably necessary, regulative employment, namely that of directing the understanding towards a certain goal upon which the routes marked out by all its rules converge, as upon their point of interaction.[136]

Thinking in this way, therefore, it might be appropriate to regard the cosmos *as if*[137] if were a divinely created entity, working out some divine purpose, even if there were no constitutive constraints, within the human mental apparatus, that made such suppositions inescapable. Or in science it might be a worthwhile regulative principle to hold to the atomic theory or the principle of uniformity of nature, even though there was no constitutive requirement so to do. And in the *Critique* Kant made some suggestions as to regulative principles that could usefully find employment in taxonomy.[138] (By contrast, as we have seen, Kant thought that the principles of Euclidean geometry, arithmetic, and Newtonian mechanics were inescapable — that is, constitutive — for us.)

This aspect of Kant's philosophy is all most unsatisfactory. If he was going to allow such play to regulative principles, one wonders really why there was any need to work so hard to establish the constitutive ones. All could have been represented as regulative, and we could have got on with the job of doing science! In all this, the sociologists of knowledge can, no doubt, make a great deal of scope. They could say, for example, that Kant's society generally adhered to the doctrines of God and the immortal soul, and Kant brought them into his philosophy through the regulative back door, since they had been unable to gain entrance through the front by courtesy of either the Metaphysical or the Transcendental Deductions. But more significantly, it has been ably argued by Wilkerson that the constitutive/regulative distinction cannot satisfactorily be sustained.[139] If I behave 'as if' God exists, I must know what the concept of God is. Yet according to Kant, the idea of God must be a metaphysical illusion, since there are no sensory intuitions by means of which I might hope to know of His existence!

Personally, therefore, I am little inclined to accept the regulative/constitutive distinction, as propounded by Kant. This does not mean, however, that regulative principles have no useful role to play in science. (Consider, for example, Newton's 'Rules of Reasoning'.) And one may well agree with Kant that 'white-paper empiricism' is wholly unsatisfactory as an epistemology for science, and that our processes of understanding do involve the bringing of information received by the senses under suitable concepts. In point of fact, *this* was the message that some of the nineteenth-century philosophers of science took from Kant. And *this* will form one of our major foci of attention in the following chapter.

But before we continue our inquiry into the nineteenth century, the reader may be curious as to whether Kant's extraordinary endeavour may be seen as some extravagant version of the art of arch building. I think it can. For (disregarding for the moment the role of regulative principles within Kant's system) we can see that the major principles of Newtonian science were supposedly generated (in the 'Analogies of Experience') as a result of Kant's lengthy philosophical journey. And likewise the synthetic *a priori* principles of geometry and arithmetic were supposedly arrived at (in the 'Transcendental Aesthetic'), though not quite so explicitly. Such principles could, one might suggest, properly stand at the apex of any self-respecting 'arch of knowledge'. And by means of his 'Copernican Revolution' Kant might claim to have demonstrated their necessity — locating them within the very cognitive apparatus of the human psyche. Since, however, his 'proof' of the categories in the 'Metaphysical Deduction' depended upon the examination of the characteristic structures of language (in the 'Table of Judgments'), one might say that no sort of *a priori* necessity had been demonstrated for the particular 'Principles of the Pure Understanding' that Kant arrived at. (It might be, for example, that the examination of non-European languages would not yield anything like Kant's claimed list of twelve Judgments, or types of proposition.) Nevertheless, Kant may well have put his finger on a profound truth. Twentieth-century linguistic research has strongly suggested (though, to be sure, not to everyone's satisfaction) that people of different cultures may have utterly different cosmic world views, according to the structures of their languages.[140] This finding is certainly consonant with the overall results of Kant's inquiry, intellectual categories seemingly being grounded in the structure of the language that one happens to employ.

But, returning to the question of our arch, are there any indications within Kant's *Critique of Pure Reason* that he sought to make any use of the traditional methodological pathways of analysis and synthesis (which we must distinguish carefully from Kant's usage of the terms 'analytic' and 'synthetic')? Well, he certainly made some use of the arch metaphor, proper, as when he wrote:

> The transcendental ideas thus serve only for *ascending*, in the series of conditions to the unconditioned, that is, to principles. As regards the *descending* to the conditioned, reason does, indeed, make a very extensive logical employment of the laws of under-standing, but no kind of transcendental employment.[141]

However, something curious has happened to Kant's terminology in the *Critique*, *vis-à-vis* earlier 'arch language'. This becomes clear when we examine the following passage, where Kant was discussing criteria for the use of hypotheses (which elsewhere[142] he stated could certainly play a useful regulative role in science):

> [T]he criterion of an hypothesis consists in the intelligibility of the assumed ground of explanation, that is in its *unity* (without any auxiliary hypothesis); in the *truth* of the con-sequences that can be deduced from it (their accordance with themselves and with

experience); and finally in the *completeness* of the ground of explanation of these consequences, which carry us back to neither more nor less that [what?] was assumed in the
hypothesis, and so in an *a posteriori* analytic manner give us back and accord with what
had previously been thought in a synthetic *a priori* manner.[143]

This passage is not easy to construe. But it seems to refer to hypotheses
— which are synthetic propositions, generated *a priori*. (That is, they are
'guesses' or 'conjectures'.) Certain consequences may flow from the
hypotheses, which consequences are thereby explained by the
hypotheses — receiving a ground of explanation. Thus the truth of the
consequences (determined by experience) allows an *a posteriori*
analytic argument for the hypotheses, previously generated *a priori*. But
it is really very difficult to know if this is what Kant *did* mean, for at one
moment one seems to be climbing the arch by synthesis and the next by
analysis.[144] And I am not exactly clear what he meant by 'an *a posteriori*
analytic manner'. But the difficulty that one may encounter in
interpreting this passage is by no means surprising. I have already
commented several times on the linguistic slipperiness of these terms,
and it has been noted that at one time Hooke and Newton chose to use
them in diametrically opposite ways.[145] But in the section headings of his
Critique, Kant used 'analytic' to mean 'logic'; and in general his 'analysis'
of propositions seems to have been a process that was unambiguous, if
the predicate was 'contained within' the subject. No one would deny that
a green apple is an apple! But this usage is really at odds with the
traditional association of synthesis with deduction and analysis with
induction. Things have got in a muddle, in fact. Indeed, if we are to follow
the persuasive argument proferred by C.M.Turbayne, two metaphors had
inadvertently become switched. As he points out:

> Kant unhappily decided to use the names 'analytic' and 'synthetic' both in the old way and
> in a way diametrically opposed to the old. As a result it is now fashionable to call
> deduction 'analytic' and induction 'synthetic'.[146]

Thus the muddle — which, as we have seen, was a product of the
conflation of mathematical and methodological traditions, Aristotle's use
of the word 'Analytics' for both his 'deductive' and his 'inductive' treatises,
an insufficiently clear distinction drawn between induction and
deduction, and the unrecognised use of metaphorical language — finally
manifested itself as a terminological confusion of considerable magnitude,
to which Kant probably made a significant contribution. What success
there was in sorting out some of these matters in the nineteenth century
we shall hope to see in some measure in the succeeding chapters. But we
shall have many other issues to deal with besides.

NOTES
1   On this, see A R Hall *Philosophers at War: The Quarrel Between Newton and Leibniz* Cambridge
    University Press Cambridge 1980
2   S Clarke *A Collection of Papers which Passed Between the Late Learned Mr.Leibniz and Dr.Clarke
    in the Years 1715 and 1716 Relating to the Principles of Natural Philosophy and Religion* James
    Knapton London 1717; H G Alexander ed *The Leibniz–Clarke Correspondence Together with
    Extracts from Newton's Principia and Opticks* Manchester University Press Manchester 1956
3   R Latta ed *Leibniz: The Monadology and other Philosophical Writings* Oxford University Press
    Oxford 1898 p 235
4   G W Leibniz 'That corporeal phenomena cannot be explained without an incorporeal principle, that
    is God [1669]' in L E Loemker ed *Gottfried Wilhelm Leibniz: Philosophical Papers and Letters* 2nd
    ed Reidel Dordrecht 1969 pp 109–12
5   Leibniz *op cit* (note 3) p 239
6   *Ibid* pp 247–8
7   In Leibnizian language, 'compossible worlds'.
8   G W Leibniz 'Selections from the Paris Notes [1676]' in Loemker ed *op cit* (note 4) pp 157–60 (at
    p 157)
9   G W Leibniz 'First truths [c. 1680–84]' *ibid* pp 267–70 (at p 268)
10 The relationship of the principles one to another, may be summarised as in Figure 18.

Figure 18

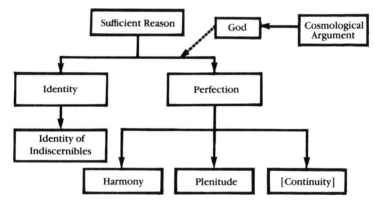

It may be noted that the whole system is beholden to the cosmological argument, which is partly *a
posteriori* in character, since it depends upon a knowledge of the existence of things. Hence it
cannot be said that Leibniz's system succeeded in being wholly rationalistic. It will be noticed also
that I have drawn on several of Leibniz's writings; this was necessary in order to present a synoptic
account of his metaphysical first principles. Not even in the systematically presented *Monadology*
did Leibniz put all his arguments together in a neat and coherent form.

11 *The Leibniz–Clarke Correspondence* (note 2) p 37. As we have seen above, p 83, Newton himself
    had to have recourse to relational space, when he assumed the centre of the world, the Sun for his
    purposes, to be a fixed point of reference.
12 *Ibid* p 38
13 For a more detailed exposition of Leibniz's theory of space and time, see: K E Ballard 'Leibniz's
    Theory of Space and Time' *Journal of the History of Ideas* 1960 Vol 21 pp 49–65
14 *The Leibniz–Clarke Correspondence* p 64 (note 2)
15 *Ibid* p 69. This is simply a gloss on Leibniz's own way of representing the argument.
16 Such a view is by no means without its problems. If space is a system of relations between objects,
    then it can be described by specifying all the distance relationships between all the objects in the
    universe. One might think that, in principle, this would be possible but it breaks down when one
    considers objects that are mirror images of one another. Consider, for example, two mirror-image
    triangles, as shown in Figure 19.
    The distance relationships, AB, AC, BC and A′B′, A′C′, B′C′ are all the same, yet the 'spaces' 'mapped'
    by the two figures are not the same. Extrapolating to the whole world, Leibniz's relational theory of
    space would be unable to distinguish between our world and its mirror image. Yet two different
    spaces would be mapped. Also, Leibniz might want to specify (by convention) one point in his
    cosmos as a 'marker' with which other places or positions could be compared. However, this would
    not require him, Newton-like, to accept the doctrine of absolute space; likewise for time.

Figure 19

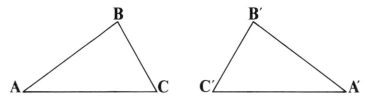

17 F Cajori ed *Sir Isaac Newton's Mathematical Principles of Natural Philosophy and his System of the World Translated into English by Andrew Mott in 1729...* University of California Press Berkeley 1934 pp 10–11

18 *The Leibniz–Clark Correspondence* p 48 (note 2)

19 I refer here to Newton's celebrated 'bucket experiment', intended to show that there could, in principle, be empirical evidence for absolute rotational motion. Newton claimed that the curvature of the surface of water in a bucket was an indication of its absolute rotational motion, the surface being flat when the water was stationary. Likewise, one can imagine two spheres connected together by a spring. If such an apparatus were moving in a linear manner, there would be no tension in the spring; but if it were rotating there would be a stretching of the spring and the two spheres would move apart, indicating the 'absolute' rotation of the apparatus. All this, Newton would have us believe, would be detectable in common human sensoria, not those of the Divine kind only. On Newton's arguments about absolute space, see R Laymon 'Newton's Bucket Experiment' *Journal of the History of Philosophy* 1978 Vol 16 pp 399–414. (It should be noted here that Laymon maintains that Newton's 'thought experiment' was not intended to *prove* the existence of absolute space. Rather, its existence was assumed in the argument presented in the bucket thought experiment.)

20 G W Leibniz 'First truths [c 1680–84]' in Loemker ed *op cit* (note 4) pp 267–70 (at p 269)

21 *The Leibniz–Clarke Correspondence* p 44 (note 2)

22 Leibniz *op cit* (note 9) pp 270–1

23 *The Leibniz–Clarke Correspondence* p 36; and 'On nature itself, or on the inherent force and action of created things [1698]' in Loemker ed *op cit* (note 4) pp 498–507 (at p 506)

24 G W Leibniz 'Critical thoughts on the general part of the principles of Descartes [1692]' in Loemker ed *op cit* (note 4) pp 383–410 (at pp 406–7). Newton himself sought to account for cohesion and chemical affinity in terms of his 'ultimate' – gravity. However, this would not explain how the different parts of a single atom would hold together, which presumably was necessary since the Newtonian atoms were thought to occupy volume. (Leibniz's monads, by contrast, were mere points.)

25 Leibniz *op cit* (note 3) p 217

26 For a simple exposition of Leibniz's theory of monads, see, for example, R Saw *Leibniz* Penguin Harmondsworth 1954

27 Leibniz *op cit* (note 3) p 219

28 G W Leibniz 'On universal synthesis and analysis, or the art of discovery and judgment [1679?]' in Loemker ed *op cit* (note 4) pp 228–33. It is worth noting the way in which Leibniz conflated here the vocabulary of scientific methodology — analysis and synthesis — and that of the rhetorical tradition — discovery and judgment.

29 Loemker ed *op cit* (note 4) p 21

30 G W Leibniz 'On the elements of natural science [c 1682–84]' in Loemker ed *op cit* (note 4) pp 277–89

31 G Berkeley *An Essay Towards a New Theory of Vision* Pepyat Dublin 1709. We have not space to consider this work here, but for a discussion of Berkeley's theory of vision in relation to his account of the nature of space, see R Gray 'Berkeley's Theory of Space' *Journal of the History of Philosophy* 1978 Vol 16 pp 451–4

32 All Berkeley's writings may be found in A A Luce & T E Jessop eds *The Works of George Berkeley Bishop of Cloyne* 9 vols Nelson Edinburgh 1948–57

33 *Ibid* (Vol 2 1949) p 41 (#1)

34 See p 86

35 Berkeley *op cit* (note 33) p 42 (#3)

36 See J Boswell *Boswell's Life of Johnson* ed G B Hill 6 vols Clarendon Press Oxford 1934–50 Vol 1 p 471: 'I [Boswell] observed, that though we are satisfied his [Berkeley's] doctrine is not true, it is impossible to refute it. I never shall forget the alacrity with which Johnson answered, striking his foot with mighty force against a large stone, till he rebounded from it, "I refute it *thus*."'

37 Berkeley *op cit* (note 33) pp 61 & 74 (##48 & 76). Berkeley did not give this argument much emphasis in the *Principles*. In the *Dialogues*, however, it formed what the author believed to be an important new argument for the existence of God.

38 *Ibid* p 33

39 *Ibid* p 31
40 *Ibid* p 45
41 See p 176
42 Berkeley *op cit* (note 33) pp 81–94 (##110–17)
43 G Berkeley *De Motu; sive, De Motus Principio & Natura, et de Causa Communicationis Motuum* (no place or publisher stated) 1721 in Berkeley *op cit* (note 32 Vol 4) pp 11–30; English translation pp 31–52
44 *Ibid* p 41
45 For more on this, see below, p 188, 191 & 195.
46 Several writers have noticed this anticipation. See V I Lenin *Materialism and Empirio-Criticism; Critical Comments on a Reactionary Philosophy* Foreign Language Publishing House Moscow 1947; J Myhill 'Berkeley's *De Motu*: An Anticipation of Mach' in *George Berkeley: Lectures Delivered before the Philosophical Union of the University of California* University of California Press Berkeley 1957 pp 141–57; K R Popper 'A Note on Berkeley as precursor of Mach' *British Journal for Philosophy of Science* 1953–54 Vol 4 pp 26–36. However, such interpretations have been questioned by R J Brook in his study *Berkeley's Philosophy of Science* Nijhoff The Hague 1973. Brook points out that in his critique of Newton's bucket experiment Berkeley did accept that there was a legitimate distinction between real motion and apparent motion, in which case the argument for his being a 'true' relativist and precursor of Mach and Einstein is less satisfactory.
47 See p 195
48 Berkeley *op cit* (note 32) Vol 3 p 304 (#11)
49 *Ibid* p 307 (#14)
50 See, for example, G P Morice ed *David Hume: Bicentenary Papers* Edinburgh University Press Edinburgh 1977
51 D Hume A *Treatise of Human Nature: Being An Attempt to Introduce the Experimental Method of Reasoning into Moral Subjects* 2 vols John Noon London 1739–40. The standard modern edition of this was edited by L A Selby-Bigge and has recently been reissued, edited by P H Nidditch (Clarendon Press Oxford 1978).
52 The title of the first edition was different from that which is now used. It was *Philosophical Essays Concerning Human Understanding. By the Author of the Essays Moral and Political* (no place or publisher stated) 1748. The standard modern edition is L A Selby-Bigge ed *An Enquiry Concerning the Human Understanding, and an Enquiry Concerning the Principles of Morals…* 2nd ed Clarendon Press Oxford 1902, rev 3rd ed edited by P H Nidditch 1975.
53 D Hume *An Abstract of a Book lately Published; Entitled, A Treatise of Human Nature, &c. Wherein the Chief Argument of that Book is further Illustrated and Explained* C Borbet London 1740. The text of this *Abstract* may be found accompanying the Nidditch edition of the *Treatise* mentioned in note 51.
54 Hume *op cit* (note 51 1978) p 7
55 *Ibid* p 2
56 *Ibid* p 1
57 *Ibid* p 3
58 *Ibid*
59 *Ibid* p 5
60 *Ibid* p 4
61 Hume made one curious exception. He imagined (*Treatise* p 6) someone who was familiar with most shades of blue, but whose experience was such that there was one shade of blue that he had never seen. Nevertheless, Hume supposed, in apparent contradiction to his normal empiricist thesis, that the person *would* be able to imagine the shade of blue that was missing from his experience. This admission on Hume's part is very curious. We shall not explore it further here, but readers are invited to consult B E Rollin 'Hume's blue patch and the mind's creativity' *Journal of the History of Ideas* 1971 Vol 32 pp 119–28
62 An ostensive definition is one achieved by directly showing that which is to be defined. For example, one might ostensively define an elephant for a child by taking him or her for a visit to the zoo, or one could give an ostensive definition of the colour yellow by showing a lemon.
63 Hume *op cit* (note 51 1978) pp 12–13
64 *Ibid* p 13
65 *Ibid* pp 15–17
66 *Ibid* p 69
67 Hume *op cit* (note 52 1975) pp 25–6
68 Hume himself gave this example in his *Abstract* (*op cit* note 51 1978) p 649
69 Hume *op cit* (note 52 1975) p 26
71 The relationship of adjacency or contiguity has long since been abandoned. Action-at-a-distance theories abound in modern physics, and everyone is familiar with the wire-less. But we still are reluctant to accept causes following effects. It is strange, perhaps, that Hume seemed to discount gravitational 'actions at a distance'.
72 Hume *op cit* (note 51 1978) pp 75–8
73 *Ibid* pp 160–1

74 So far as logic was concerned in this matter, Hume was undoubtedly correct. But of course, the requirements of logic were not necessarily appropriate, or-were much too stringent.

75 Hume *op cit* (note 51 1978) p 88

76 *Ibid* p 172

77 Hume himself wrote: 'upon the whole, necessity is something, that exists in the mind, not in objects' (ibid p 165); and 'the efficacy of causes lies in the determination of the mind' (ibid p 167).

78 *Ibid* p 134

79 *Ibid* p 89

80 See below, p 231

81 If I may be permitted an autobiographical remark here, I well remember my first encounter with philosophy of science (at University College, London, in the late 1950s), when our class was informed by the lecturer that there was 'no logical justification' for drawing a straight-line graph through a set of plotted experimental data that lay nicely on a straight line! The other students in my group were equally enraged and there was a good deal of disturbance in the class that evening. I suggest, with hindsight, that the difficulty might well have been avoided if an *historical* approach to the philosophy of science had been employed.

82 Hume *op cit* (note 51 1978) pp 173–5

83 In his daily life, it appears that Hume had little regard for concerns generated by his sceptical analysis of causality.

84 See below, p 150

85 See above, p 35

86 See above, p 82

87 Hume *op cit* (note 52 1975) p 204

88 Hume *op cit* (note 51 1978) p 229

89 *Ibid*

90 *Ibid* p 231

91 *Ibid* p 228

92 I Kant *Prolegomena to any Future Metaphysics that Will be Able to Present Itself as a Science* trans P G Lucas Manchester University Press Manchester 1953 p 9

93 I Kant *Kritik der reinen Vernunft* J F Hartknoch Riga 1781 1787

94 I Kant *Prolegomena zu einer jeden kunftigen Metaphysic die als Wissenschaft wird auftreten können* Riga 1783

95 *Immanuel Kant's Critique of Pure Reason* trans Norman Kemp Smith 2nd ed Macmillan London 1933 (lst ed 1929, 2nd ed reprinted many times)

96 We may mention here that there were two other 'critical' works of Kant: the *Critique of Practical Reason* (1788) which contained statements of Kant's moral philosophy; and the *Critique of Judgment* (1790), which was concerned particularly with teleology. There are, besides, a great number of writings on politics, history, law, geography, anthropology, and so on. It should be noted that Kant's 'critical' writings were published relatively late in his career, after he had reacted against the rationalist philosophies that provided the staple diet in German universities in the eighteenth century. Among Kant's 'pre-critical' writings his *Universal Natural History of the Heavens* (1755), in which he set out a theory of the origin of the solar system, was of particular interest and importance, and showed Kant's strong adherence to the principles of Newtonian mechanics. As we shall see, the influence of Newtonian philosophy is manifest to a considerable degree in the *Critique of Pure Reason*.

97 The word is still used in this sense today when we speak of 'pure' mathematics.

98 As we shall see shortly, the word 'intuition' had a special meaning within the Kantian philosophical vocabulary. It meant what Hume called an 'impression', or what twentieth-century philosophers have called a 'sense datum'.

99 Kant *op cit* (note 95 1933) p 93

100 The two opposing arguments are set out on p 397 of Kant *op cit* (note 95 1933). As presented there they are not particularly easy to understand and I have therefore tried here to clarify and make more intelligible the two arguments. (Kant's antinomies are discussed at length in S J Al-Azm *The Origins of Kant's Arguments in the Antinomies* Clarendon Press Oxford 1972; this author shows that the first antinomy is best interpreted with the historical context of the Leibniz-Clarke debate carefully in mind.)

101 *Ibid* pp 48–9; Kant *op cit* (note 92) p 20

102 Kant also defined analytic propositions as ones that 'say nothing in the predicate that was not already thought in the concept of the subject' (ibid 1953 p 66). And according to this sense, the proposition 'A rainy day is a wet day' could reasonably be said to be analytic.

103 *Ibid* p 20

104 *Ibid* pp 39 & 43–4

105 This assumption is, to be sure, highly questionable. It is an issue over which there has been intense philosophical debate, over two hundred years. Even the distinction between analytic and synthetic propositions is regarded as suspect by many. See W V O Quine 'Two dogmas of empiricism' in *From a Logical Point of View: 9 Logico-Philosophical Essays* Harvard University Press Cambridge Mass 1953 pp 20–46

106 Kant *op cit* (note 95 1933) pp 22 & 25
107 My own attempts at entry have been greatly facilitated by T E Wilkerson *Kant's Critique of Pure Reason: A Commentary for Students* Clarendon Press Oxford 1976. This admirable and amusingly written work does great service to modern readers by showing clearly the extraordinary inadequacies of certain parts of Kant's text. On reading Kant, one may often feel that the difficulties that one has in understanding must be due to one's own intellectual inadequacies. Wilkerson gives great encouragement to the reader by showing that Kant himself may often be held responsible. There are, of course, a great many commentaries on Kant's *Critiques*, some of them almost as opaque as the originals. Strawson's *Bounds of Sense* (see note 108) has been particularly influential. See also W H Walsh *Kant's Criticism of Metaphysics* Edinburgh University Press Edinburgh 1975
108 P F Strawson *Bounds of Sense: An Essay on Kant's Critique of Pure Reason* Methuen London 1966 p 18
109 His formal definition was: 'I entitle transcendental all knowledge which is occupied not so much with objects as with the mode of our knowledge of objects in so far as this mode of knowledge is to be possible *a priori*' (*op cit* [note 95 1933] p 59).
110 Kant's intuitions were not of the kind commonly associated with the female sex; nor were they anything to do with Sherlock Holmesian hunches.
111 See note 99
112 This section was not headed in this way by Kant. But later (*ibid* p 170) he did refer to it in this way.
113 *Ibid* p 69
114 *Ibid* p 71 (The word 'form' here can be taken in a traditional Aristotelian sense without difficulty.)
115 *Ibid* p 71
116 *Ibid* p 77
117 *Ibid* pp 68–9 (1st ed), 70–1 & 80
118 *Ibid* pp 75 & 80 (Kant did not make the connection with arithmetic *explicitly* in this passage, though one feels that he might like to have done so. The connection being, to say the least, somewhat implausible, only an implicit or imputed link is to be found in the text. The connection was made more explicitly in the *Prolegomena*.)
119 *Ibid* p 82
120 The terminological distinction between *noumena* and *phenomena* only appeared later in Chapter 3 of the 'Analytic of Principles'; but the philosophical distinction was made earlier in the 'Transcendental Aesthetic', under discussion here.
121 Kant, however, did not wish to be construed as a Berkeleian and he wrote somewhat patronisingly of his Irish predecessor. See Kant *op cit* (note 95 1933) p 89
122 See K E Popper *The Open Society and its Enemies* 2 vols Routledge London 1945
123 Kant *op cit* (note 95 1933) p 103
124 Kant's own list of categories did not include 'thing' or 'quality', but he did include 'substance' and 'accident'. For the purposes of exposition here I have used the term 'thing', which may perhaps be more readily understood than the rather archaic philosophical term, 'substance'.
125 Kant *op cit* (note 95 1933) pp 107 & 113 (The examples, such as 'This *A* is *B*', etc., were not given by Kant himself. It should be noted that Kant's published 'Table of Judgments' gave the order of the first three elements of the table as Universal, Particular, Singular. This does not mesh with the published order of the first three categories. I have therefore taken the liberty of reversing the order of the first three Judgments to give: Singular, Particular, Universal.)
126 *Ibid* p 107
127 H J de Vleeschauwer *The Development of Kantian Thought: The History of a Doctrine* trans A R C Duncan Nelson Edinburgh 1962 p 75
128 Kant *op cit* (note 95 1933) pp 129–75
129 For a very helpful reconstruction of the argument of the 'Transcendental Deduction', see Wilkerson *op cit* (note 107) pp 47–57.
130 Kant *op cit* (note 95 1933) p 180
131 *Ibid* p 186
132 Kant's text for the schemata of quantity and quality (pp 183–5) is very sketchy, and one can hardly be certain of his intentions at this point. One has to work backwards to the schema corresponding to the category of 'quality' from the corresponding 'Principle of the Pure Understanding' (see p 131). That is why it is placed in square brackets here.
133 In drawing up this synopsis, I have drawn indiscriminately on passages from both editions of the *Critique*, selecting the ones that appear to be most intelligible for the present expository purpose.
134 It may be noted that in another work, the *Metaphysische Anfangsgrunde der Naturwissenschaft* (1786), Kant sought to elaborate the science of mechanics from the categories and the 'Analogies of Experience' ostensibly established in the *Critique of Pure Reason*. Corresponding to the main categorial divisions of 'Quantity', 'Quality', 'Relation' and 'Modality', were four scientific divisions which Kant named 'Phoronomy' (i.e. kinematics) 'Dynamics', 'Mechanics' and 'Phenomenology' (i.e. the science of the different *kinds* or modes of motion, such as 'absolute' or 'relative'). And from statements of universal metaphysics, akin to the three 'Analogies of Experience' in the *Critique*, Kant sought to infer three laws of mechanics. But they were not the same as Newton's. The first and third laws of Newton were 'distilled' from the second and third Analogies. The first Analogy yielded the

principle of conservation of mass. Newton's second law remained unaccounted for in this way. See E B Bax trans *Kant's Prolegomena and Metaphysical Foundations of Natural Science* Bell London 1883 pp 145–7 & 220–3.

135 Should any reader be finding all this rather hard to take, we may invite him or her to consider a thought-provoking example due to W H Walsh ('Categories' in R P Wolff ed *Kant: A Collection of Critical Essays* Anchor Books New York 1967 pp 54–70). Imagine yourself driving along a road in a car with a friend, and the car suddenly stops. Suppose further that your friend asserts that there is *no reason* why this has happened. Needless to say, you will reject this as an absurd assertion. It will be inconceivable to you that there is no cause whatsoever for the breakdown. Thus *our minds will not allow us* to envisage events occurring without causes. It is in this kind of a way, with suitable homely examples, that one may try to come to terms with the Kantian doctrine of categories. (Walsh also has a useful article on the Kantian art of schematisation in the same volume.)

136 Kant *op cit* (note 95 1933) p 533

137 For a statement of Kant's philosophy of '*as if*', see *ibid* p 551.

138 *Ibid* pp 539–43

139 Wilkerson *op cit* (note 107) p 156

140 See, for example, Benjamin Lee Whorf (ed J B Carroll) *Language, Thought and Reality: Selected Writings of B L Whorf* Wiley New York and Chapman & Hall London 1956. However, if one follows the ideas of Durkheim and Mauss, and certain later social anthropologists, social structures should be given primacy, essential language categories (or ways of classifying the world) supposedly being traceable to social formations. See E Durkheim and M Mauss *Primitive Classification* (trans and ed R Needham) Cohen & West London 1963.

141 Kant *op cit* (note 95 1933) p 325

142 *Ibid* p 535

143 *Ibid* p 119

144 It is noteworthy that Kant described his method in the *Prolegomena* as analytic, and in the *Critique of Pure Reason* as synthetic (*op cit* [note 92 1953] pp 13 & 29). But what this actually meant is hard indeed to determine.

145 See above, p 81

146 C M Turbayne *The Myth of Metaphor* Yale University Press New Haven 1962 p 30

CHAPTER 4

FACTS AND THEORIES: HERSCHEL, MILL & WHEWELL, & THE WHEWELL–MILL CONTROVERSY

The problem in metascience of making a clear distinction between induction and deduction, has been referred to on a number of occasions. In looking at some leading British philosophers of science of the first half of the nineteenth century, we shall now face this recurring difficulty even more directly, with the domain of induction being considerably extended in philosophical usage, encroaching into the territory normally regarded as the province of deduction. We shall also see a continuation of the debate that had been to the fore in the second half of the eighteenth century, of which there was some account in the previous chapter. As we have seen, the eighteenth-century metascientists of the empiricist tradition regarded knowledge as a product of sensations, worked upon by the mind, which (so to speak) had to 'conform' to the dictates of experience. By contrast, Kant and his followers maintained that mental activity was, as it were, 'prior' to experience and that understanding of the world could only be achieved in accordance with certain 'mental frames' which preceded experience. In Britain in the nineteenth century, the empiricist tradition was actively pursued by the scientist John Herschel and the philosopher John Stuart Mill. The Kantian position (or a modified version thereof) was presented by the philosopher/scientist/historian, William Whewell. The debate between Mill and Whewell is particularly instructive in illustration of the chief differences between the empiricist position and that of Kantian idealism. Also, in the work of all three of these writers we find some interesting suggestions about the actual methods of inquiry which might usefully be followed in science. In fact, we find here a considerable addition to the literature of arch building.

Herschel

John F.W.Herschel (1792–1871), son of the celebrated astronomer William Herschel, was one of Britain's most distinguished scientists in the first half of the nineteenth century. He was a brilliant mathematician, achieving the coveted position of Senior Wrangler for his year in the mathematics examinations at Cambridge. He made major contributions in astronomy and also had interests in optics, crystallography, chemistry, metrology, mineralogy, geology, meteorology — indeed virtually all the physical sciences. He also contributed to the new technique of photography. But his interests were wider still: he also published a verse translation of Homer's *Iliad*. Herschel's major claim to scientific fame was based on his use of his father's observations of the motions of double stars to show that they moved around each other in elliptical orbits according to the inverse square law, just as the Newtonian theory required. This was no small achievement, for it supported the view that the laws of Newtonian physics were of general application throughout the universe, not being restricted in their range of applicability to the solar system alone. As a result of this triumph, Herschel became *the* 'man of science' in early nineteenth-century England, paralleling the chemist, Humphry Davy, in public esteem.

It was as a general authority on science (or natural philosophy) that Herschel was invited to write an introductory work on the philosophy and methodology of science, as a contribution to a series of semi-popular books, collectively called the *Cabinet Cyclopaedia* and edited by one Dionysius Lardner. Herschel's volume was entitled *Preliminary Discourse on Natural Philosophy* and was published in 1830.[1] It proved a considerable success, running through several editions. Indeed it is still in print today.[2] On examining the *Preliminary Discourse*, we immediately notice that it bears an engraving of Bacon on the title page, and in this we can, perhaps, see some of the background to the success of Herschel's work. Accompanying the rising interest in science and technology associated with the Industrial Revolution, there was a significant revival of interest in Bacon's work in the early nineteenth century, and (as D.E.Allen has shown) natural history collecting became increasingly popular as the century progressed.[3] Of course, the Baconian methods of natural history were hardly apt for the complex mathematical calculations with which Herschel engaged in the course of his work as a professional astronomer. Yet at the level of metascience, the Baconian description was still esteemed. Indeed, as I say, it was coming into renewed favour in the early nineteenth century, so that much of the amateur science and metascience of the period was consciously Baconian. Moreover, as we shall see, even one such as William Whewell, whose work contained strong Kantian elements, made explicit obeisance to Bacon.

The *Preliminary Discourse* was important as being the first book in English specifically concerned with philosophy of science to have been written by an eminent scientist. However, Herschel's own reading in the existing literature on the philosophy of science was probably not very extensive, and may not have gone much beyond Bacon and Hume. We find, then, that Herschel engaged in a little 'amateurish' epistemology near the beginning of his account of scientific procedures,[4] but it was not very successful and we shall not consider it here[5] beyond noting Herschel's adoption of an essentially empiricist stance, with the influence of Hume clearly in evidence.

As we shall see, Herschel's account of the general 'shape' of scientific inquiry lay very much within the tradition of the 'arch', which we have been tracing in this book. Thus, a clear distinction was made throughout his text between the two 'sides' of science: discovery, and verification or justification. Moreover, Herschel had some worthwhile things to say about the process of discovery or 'arch-ascent', with which most previous writers had been noticeably unsuccessful in dealing.

According to Herschel, then, the first step in the inductive ascent consisted on the 'analysis of phenomena'.[6] It is not easy to know exactly what Herschel had in mind for this. There is one passage[7] which may be construed as requiring the scientist to make a preliminary analysis of the phenomena, picking out those aspects that were capable of being specified in law-like form. And this seems eminently sensible, though how exactly one might go about such an 'analysis' could be a puzzle to Herschel's readers, for no general rules could be specified for the process. In fact, his explanation of the process was skimpy and not very satisfactory, and the example he gave hardly meshes with the simple focussing of attention of the relevant aspects of a complex whole, which is what his instructions might lead one to expect. Thus, Herschel said, the phenomena of sound might be analysed in a preliminary way into:

> 1st, The excitement of a motion in a sounding body. 2ndly, The communication of this
> motion to the air or other intermedium which is interposed between the sounding body
> and our ears. 3dly, The propagation of such motion from particle to particle of such
> intermedium in due succession. 4thly, Its communication, from the particles of the
> intermedium adjacent to the ear, to the ear itself. 5thly, Its conveyance in the ear, by a
> certain mechanism, in the auditory nerves. 6thly, The excitement of sensation.[8]

Of course, one might reasonably object that in presenting this 'preliminary' analysis of the phenomenon of sound Herschel had in fact taken for granted his solution to the particular scientific problem with which he had chosen to concern himself. This was hardly a preliminary analysis of phenomena. Such an objection seems to me to be well founded — so all we can do, I fear, is ignore it and proceed with our exposition. We may agree with Herschel, however, that some kind of preliminary analysis — or perhaps scientific training — is called for. One cannot usefully go straight into the world or into the laboratory and simply begin by observing.

It should be remarked that Herschel chose to call the preliminary stage of the inductive ascent 'analysis'; and it involved (as in the example of the sounding body) a dividing up of one type of phenomenon into a number of distinct contributory phenomena. In fact, in Herschel's example the outcome was not unlike Bacon's explanation of the phenomenon of heat in terms of the phenomenon of motion. But so far we have little indication from Herschel as to *how* one might carry out the process of analysis. So let us proceed.

Herschel noted that there appeared to be two *causes* emerging from the preliminary analysis of sound, which could not be analysed further, namely *motion* and *sensation*.[9] He believed that these could not be analysed (or explained) in terms of causal powers, but that they could be treated in terms of scientific *laws of nature*; that is 'statement[s] in words of what will happen in such and such proposed general contingencies'.[10]

So the inquiry was to be directed towards an examination of the laws of nature. Sound, for example, was to be analysed into motion; so the investigation of sound could proceed through an investigation of the laws of motion. This also was quite Baconian in approach.

We should note, however, Herschel's more extended description of a law of nature:

> A law of nature, being the statement of what will happen in certain general contingencies, may be regarded as the announcement, in the same words, of a whole group or class of phenomena. Whenever, therefore, we perceive that two or more phenomena agree in so many or so remarkable points, as to lead us to regard them as forming a class or group, if we lay out of consideration, or *abstract*, all the circumstances in which they disagree, and retain in our minds those only in which they agree, and then under this kind of mental convention, frame a definition or statement of one of them, in such words that it shall apply equally to them all, such statement will appear in the form of a general proposition, having so far at least the character of a law of nature.[11]

The example offered was: 'Doubly refracting substances exhibit periodic colours by exposure to polarised light.'[12] We see, therefore, that Herschel's account of the methodology of science was still deeply influenced by the Aristotelian logic of classes. Wearing his metascientific hat, Herschel thought of the process of scientific investigation as one in which facts were grouped into general facts, which were then classified into statements about laws; and eventually 'axioms of the highest degree of generality of which science is capable'[13] were to be *raised*. Also, with a Humean touch, Herschel spoke of laws as expressions of '*constant association*'.[14]

The various processes involved in the establishment of scientific laws were together to be called induction. But this upward inductive ascent was to proceed in two stages. The first led to the establishment of 'laws' and the second to 'theories'. And in his Chapter 6 (entitled 'Of the first stage of induction. — The discovery of proximate causes, and laws of the lowest degree of generality, and their verification') Herschel gave his readers some very much more specific indications as to how scientific inquiries were to be conducted, drawing on the suggestions made by Hume for acertaining the relationships between causes and effects[15] and ideas that in fact may be traced back to Bacon.[16] Later, as we shall see,[17] they were to emerge as the celebrated 'Canons of induction' of John Stuart Mill's *System of Logic* (1843).

In summary, Herschel's 'general rules for guiding and facilitating our search, among a great mass of assembled facts, for their common cause'[18] (that is, a search for cause/effect relationships as a basis for discovering the laws of nature) were to look for:

1. A constant conjunction of cause (antecedent) and effect (consequent);[19]
2. The absence of an effect (consequent) in the absence of a cause (antecedent);[20]
3. A proportionality between cause (antecedent) and effect (consequent);[21]
4. The same as '3', expressed in different words;

5. A reversal of the effect (consequent) upon a reversal of the cause (antecedent).[22]

Herschel also gave a statement of what Mill was later to call the 'Method of residues'.[23] The overall use of these methods was intended to allow the successive discovery of laws of ever-increasing generality,[24] as Bacon would have approved.

Next, Herschel offered a number of examples of the application of such rules in successful scientific practice, and a more extended description of the work of one William Wells (1757–1817) on the cause of the formation of dew,[25] in illustration of the supposed workings of 'scientific method'. However, Herschel went on to say that the discovery of laws could, on occasions, be achieved by the formulation of hypotheses and their testing, rather than by direct inductive ascent:

> In the study of nature, we must not... be scrupulous as to *how* we reach to a knowledge of... general facts [i.e. laws]: provided only [that] we verify them carefully when once detected, we must be content to seize them wherever they are to be found.[26]

Clearly, Herschel did not suppose that scientific research could be successfully undertaken simply by employing his rules for inductive research in a mechanical manner. When employing hypotheses, verification was to be achieved by the experimental or observational testing of the consequences that might logically be derived from the hypotheses.[27]

In his following Chapter 7, Herschel dealt with 'The higher degrees of inductive generalization, and of the formation and verification of theories'. Theories he described as 'creatures of reason rather than of sense'.[28] They were to be arrived at by the 'raising' of 'higher inductions'.[29] Herschel had in mind such things as the atomic theory of matter, the wave theory of light, the kinetic theory of heat, and so on. Here, science would pass beyond the region that could be penetrated directly by the senses, even with the assistance of instruments, and into more speculative domains, for which, however, testing and verification were still feasible.

Theories should be arrived at by the formulation of hypotheses, and perhaps by a 'fair inductive consideration of general laws',[30] which they should at the same time be capable of explaining. Moreover, theories should be capable of experimental test and verification.[31] However, Herschel's account of the inductive ascent from laws to theories was vague, though he was inclined to think that the use of analogies was important. It would, however, have been remarkable if he *had* been able to suggest any general rules for the successful production of theories. If he had done so satisfactorily, the discussion of metascientists would very likely have been brought to an early end! In fact, one would not expect to find any universally applicable set of rules for the prosecution of scientific inquiry. If there were such, the difficult act of turning induction into deduction would somehow have been achieved; or the mysteries of the psychological processes of the creative mind would by some means have been penetrated.

Following the language of Newton's 'First Rule'[32] Herschel urged that scientists should hunt for 'true causes' (*'verae causae'*)[33] rather than speculative fictions. It is sometimes possible to find fortuitous correlations between phenomena, where there is in fact no causal connection at all. Also, on occasions, some true causal connection may be suspected, but cannot be identified with certainty — as in the supposed connection between smoking and lung cancer. And sometimes in the history of science certain theoretical entities (such as the aether) may have seemed *bona fide* causal agents in one generation, only to be rejected in another. We can, of course, understand why Herschel should have thought it desirable to look for 'true causes' in science, and to deploy them in theories; but unfortunately his suggestions for distinguishing the true from the false were necessarily inadequate, and no exact criteria for making the distinction have been found to this day. As a general guide, though, a *vera causa* should be feasible or equal to the task that it is required to perform, and should be comprehensible with the help of some intelligible analogy.

We may now attempt to summarise Herschel's description of scientific method by means of a diagram, such as Figure 20.[34]

Figure 20

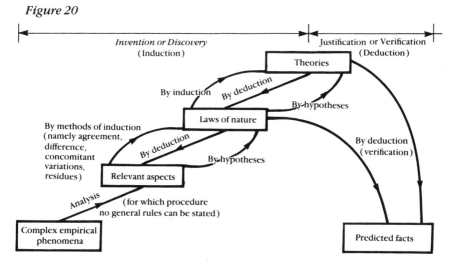

However, this picture is something of an oversimplification, for like Bacon and Hooke before him,[35] Herschel was inclined to see a whole hierarchy of laws of increasing degrees of generality, and the ascent to high-level laws and theories would consist of alternate inductive and deductive steps:

> The path by which we rise to knowledge must be made smooth and beaten in its lower steps, and often ascended and descended, before we can scale our way to any eminence, much less climb to the summit.[36]

So we should envisage a whole series of laws at different 'levels' interconnected by numerous inductive and deductive pathways, and all capable of verification by deductive connection with empirical facts.

Clearly, what is important in all this is the developing picture of the 'structure' of science, based upon an empiricist epistemology, and usefully 'informed' (in an Aristotelian sense of the word) by Herschel's personal familiarity with the actual practices of scientists and his own experience of scientific research.[37] The feature that is of more than special interest is to be found in the 'general rules for research', and one may wonder whether Herschel had here found the touchstone for the successful conduct of scientific inquiry. We may, however, ask this with greater advantage when we see the rules in their fully-fledged form in Mill's 'Canons of induction'. We shall therefore defer this question for the moment and return to it shortly.[38]

Mill

So, then, let us proceed forthwith to a consideration of the contributions to metascience of John Stuart Mill (1806–1873). Mill was unquestionably one of England's great Victorians. In addition to being a distinguished philosopher, he was also eminent as politician, economist, editor, writer, and spokesman for women's rights. And although we shall only be concerned with his contributions to philosophy of science, it should be recognised that this was only a part of his total philosophical output, which was chiefly concerned with moral and political questions. His father was James Mill (1773–1836), the social reformer and friend of the utilitarian philosopher Jeremy Bentham (1748–1832). James Mill wished to use his son as subject for a remarkable educational 'experiment', and accordingly he submitted him to a rigorous — or cruel — regimen of learning. The young boy was put to Greek at the age of three and Latin at eight, and subsequently he became proficient in several modern languages, as well as ancient history, Hebrew, mathematics and logic. However, his knowledge of experimental science was largely derived at second hand from books.

Doubtless as a result of the intellectual pressures to which he was subjected in his youth, Mill suffered a nervous breakdown, but fortunately he passed through this crisis safely. At the age of twenty three he fell in love with a remarkable 'blue-stocking', Mrs Harriet Taylor, and when her husband died she and Mill were eventually able to be married. Their relationship constituted one of the great Victorian romances and excited considerable public interest. The couple collaborated in their later writings, and many of Mill's thoughts on the emancipation of women were derived from his wife.

Max Lerner has divided eminent writers into those whose originality revives each time we read him, and those — not necessarily more original or profound — whose 'influence spreads imperceptibly over the generations, so that his originality has become part of the intellectual air we breathe'.[39] Mill was of this second type, but none the less significant for that: we still breathe air that is strongly permeated by Millian doctrines, both in political theory (metapolitics) and in metascience. And his single major work on philosophy of science, the *System of Logic* (1843),[41] must therefore be of considerable interest to us. Certainly, Mill laboured at it mightily for thirteen long years; and it was his first book.

It is usual to think of Mill as a pivotal figure in the long history of British empiricism: Bacon, Locke, Berkeley, Hume, Herschel, Mill, Jevons, Russell, Ayer,... and many others besides. Mill, however, did not like to accept the label of empiricist (which he associated with unwarranted generalisations and surmises), preferring to be identified with the 'School of Experience' or 'sensationalism'; but the terminological distinction that he sought does not appear particularly significant today. In fact, as we shall see, Mill's empiricism was of such an 'extreme' kind that he has had few followers in the totality of his argument, the pervasive nature of his influence notwithstanding. It seems reasonable enough, nevertheless, to call him an empiricist.

It should be noted at the outset that Mill's book was entitled *A System of LOGIC*; yet we are considering it here within the context of the history of the philosophy of science, not the history of logic or of mathematics. This apparent anomaly arises (in part) because of the somewhat unusual meaning which Mill attached to the word 'logic'. He did not mean by it deductive reasoning, the use of the syllogism, formal logic, or anything of that kind, but rather the processes of reasoning as a whole — though not treated psychologically or physiologically. He was concerned with how we can reason successfully from one statement of fact to another. This was possible — he supposed — in the kind of reasoning employed in science. It was not found in, for example, the logic of the syllogism. Indeed, Mill did not regard the syllogism as a mode of inference at all.[41] It yielded nothing in the conclusions that was not already given in the premises. It was a logic of consistency, not of truth; it was, so to speak, interpretation rather than inference. Indeed, for Mill, there was no such thing as deductive *inference*!

Mill was the antithesis of a Kantian idealist, and he specifically sought to combat such views.[42] It will be recalled that for Kant the axioms of geometry (say) were necessary, being an inescapable product of the human mind's way of seeing the world through the 'spectacles' of space and time. But Mill found this metaphysically repugnant, being, he thought, incompatible with the basic consideration of the freedom of the human will,[43] which stood at the core of his political liberalism. Mill also objected to Hume's treatment of 'necessary' connection in relation to causation.[44] As we shall see, the independent line of argument which he created to give a basis to his own system, in opposition to that of the idealist camp, was to represent *all* knowledge — even that of mathematics — as *empirical* in character. And induction was the procedure whereby general propositions were established; the syllogism merely allowed the results of induction to be 'interpreted' — rather as a judge interprets the law.[45]

If, then, logic and deduction[46] were to be accorded this highly novel, but apparently subordinate, status, we may wonder what Mill had to say about induction, and the processes of discovery; for evidently a great deal had to be made to rest on them. This brings us immediately to Mill's celebrated '*Canons of induction*', which have already been mentioned in our discussions of Herschel, and from whose work they were in large measure derived.[47] The Canons were as follows, where *A, B, C,* etc, stand for particular antecedents to certain consequents, *a, b, c,* etc.

First Canon ('The Method of Agreement')

> If two or more instances of the phenomenon under investigation have only one cir-
> cumstance in common, the circumstance in which alone all the instances agree, is the
> cause (or effect) of the given phenomenon.[48]

Thus, if A, B, C, ..., are followed by a, b, c, ...,
and A, D, E, ..., are followed by a, d, e, ...,
then one may infer that A is the cause of a (or a is the effect of A).

For example, suggested Mill, the factor which (he believed) crystalline
substances have *in common* (equivalent to A) is that they are all deposited
from a liquid. That is, the solidification of a substance from a liquid is an
invariable antecedent, and consequently a cause (or a contributory factor
to) the process of crystallisation.

Second Canon ('The Method of Difference')

> If an instance in which the phenomenon under investigation occurs, and an instance in
> which it does not occur, have every circumstance in common save one, that one
> occurring only in the former; [then] the circumstance in which alone the two instances
> differ, is the effect, or the cause, or an indispensable part of the cause, of the
> phenomenon.[49]

Thus, if A, B, C are followed by a, b, c
and B, C are followed by b, c
then one may infer that A is the cause of a (or a is the effect of A).

For example, suggested Mill, if a man in the 'fullness of life' suddenly
died, and if the death was immediately preceded by a gunshot through the
heart, one might conclude that the gunshot caused the death, for it was the
circumstance of the gunshot wound that was *different* before and after the
death.

Third Canon ('The Joint Method — of agreement and difference')

> If two or more circumstances in which the phenomenon occurs have only one cir-
> cumstance in common, while two or more instances in which it does not occur have
> nothing in common save the absence of that circumstance; the circumstance in which
> alone the two sets of instances differ, is the effect, or the cause or an indispensable part
> of the cause, of the phenomenon.[50]

Thus, if A, B, C are followed by a, b, c
           A, D, E are followed by a, d, e
           A, F, G are followed by a, f, g
and B, C are followed by b, c
           D, E are followed by d, e
           F, G are followed by f, g etc,
then one may infer that A is the cause of a (or a is the effect of A).

Fourth Canon ('The Method of Residues')

> Subduct [i.e. subtract] from any phenomenon such part as is known by previous
> inductions to be the effect of certain antecedents, and the residue of the phenomenon is
> the effect of the remaining antecedents.[51]

Thus, if *A, B, C* are followed by *a, b, c* and if, by previous investigations, it has been shown that *A* causes *a*, and *B* causes *b*, then:

> [S]ubtracting the sum of these effects [*a* and *b*] from the total phenomenon, there remains *c*, which now, without any fresh experiments we may know to be the effect of *C*.[52]

Or, *C* is the cause of *c*.

Mill's language is not specially transparent here, but his meaning is easily made clear with the help of an example. For instance, in the discovery of the planet Neptune, certain perturbations in the orbit of Uranus were known, and could, for the most part, be accounted for by the gravitational influences of the large neighbouring planets, Jupiter and Saturn. (That is, schematically *A* and *B* accounted for *a* and *b*.) However, there was still some small perturbation of Uranus's orbit (*c*) that was not accounted for by the influence of Jupiter and Saturn. Consequently, the existence of another planet (*C*), which might be the cause of the residual perturbations, was proposed; and as is well known it was eventually discovered.

Fifth Canon ('The Method of Concomitant variations')

> What ever phenomenon varies in some particular manner, is either a cause or an effect of that phenomenon, or is connected with it through some fact of causation.[53]

Thus, if as *A* varies *a* varies concomitantly[54] (*B, C,* etc, and *b, c,* etc, all remaining constant) then one can infer that *A* is the cause of *a*. For example, variations in the position of the Moon are regularly and proportionably accompanied by variations in the times and places of high tide. Thus, the Moon is (wholly or partially) the cause of the tides.[55]

We may well wonder what efficacy these Canons might have in relation to the problem of arch climbing. Would they, for example — if deployed in their new, polished-up Millian condition — be of great service in Herschel's methodology for rising to laws and theories? We have already seen in an earlier discussion[56] that the 'Method of Agreement', if used in isolation, could lead to absurd conclusions. Also, when we look at Mill's own examples, we may have some scepticism as to the efficacy of the Canons. For example, crystalline material may readily be produced by sublimation, which involves no precipitation from a liquid. Again, in the case of the man shot through the heart, Mill was simply wrong in saying that the conditions before and after agreed in all respects except that of life and death. The circumstances *b, c,* etc, could not and would not be the same after the event as before. Then, in the case of the 'Method of Concomitant Variations', there have been many reported correlations between phenomena which are quite fortuitous, or more often tenuous and uncertain, and providing a hint of causal connection, but certainly not secure scientific knowledge. As for the 'Method of Residues', it might certainly lead to discoveries in some instances, and did so in the case of the discovery of Neptune, but on other occasions it might lead to all manner of wild goose chases.

It is generally agreed, therefore, that Mill's Canons do *not* succeed in making inductive arguments deductive (in the traditional sense of this word, not Mill's). They provide no way of achieving certain knowledge, and no certain method for achieving knowledge. But it is undoubtedly the case that they are very widely used in the course of scientific investigations. For example, the testing of drugs by pharmaceutical companies customarily employs the Methods of Agreement and Difference. Anyone in the business of looking for functional relationships between variables is likely to use the Method of Concomitant Variations, though scientists usually tend to eschew the philosophers' language of cause and effect.[57]

However, even if the Canons may have certain practical use and are widely deployed in scientific researches, it does not follow that they can be employed straight away by any novice entering upon a scientific investigation for the first time. This was clearly recognised by Mill's critic, William Whewell (or whom more anon), when he wrote:

> Upon these methods [of Mill], the obvious thing to remark is, that they take for granted, the very thing which is most difficult to discover, the reduction of the phenomena to formulae such as are here presented to us.[58]

Whewell objected further that an examination of the history of science did not reveal Mill's Canons in common use. Whewell's point was this. By the time a particular scientific inquiry has been reduced to A, B, C, and a, b, c, etc, the problem has virtually been solved and only 'mechanical' manipulation is needed. The difficult thing is to get the inquiry into a form such that it can be treated by Mill's Canons. For, said Whewell, when one begins an investigation it is by no means obvious which of the many possible phenomena that might be taken into consideration are relevant. So Mill was bringing his Canons into the fray far too early in his account of the processes of scientific investigation. Much more attention had to be given to the processes involved in the preliminary analysis of the phenomena in order to determine the aspects that were relevant to the inquiry — the A's, B's, C's, a's, b's, c's, etc. Herschel, we have seen, had recognised the need for such a preliminary analysis — whatever form it might take — but his account of the matter was not at all satisfactory.

It may be further remarked that Mill's assumption that the phenomena could be reduced to discrete A's, B's, C's, a's, b's, c's, etc, meshed well with — though I would not wish to say was directly caused by — his empiricist epistemology. Locke, we remember, had supposed that sensations entered the mind in single file, so to speak, and imprinted themselves on the mind. Mill's methodology seems to have reflected this epistemological tradition. Later, as we shall see, Russell was to develop his so-called 'logical atomism', which can, I suggest, be regarded as a descendant of this particular epistemological tradition.

It will be recalled that Hume had drawn attention to the problems of induction, universal causation and the uniformity of nature, so that even if the quasi-psychological/methodological problem raised for Mill by Whewell could be solved satisfactorily there would still remain the

logical difficulty of using the Canons of induction as a basis for the establishment of *general* scientific laws, theories, principles, axioms or
whatever. Mill was, of course, not unaware of the problem of induction,
and the fact that inductive generalisations might easily be mistaken or
erroneous. As he said, most pertinently:

> To Europeans, not many years ago, the proposition, All swans are white, appeared an...
> unequivocal instance of uniformity in the course of nature. Further experience has
> proved ... that they were mistaken; but they had to wait fifty centuries for this
> experience. During that long time, mankind believed in an uniformity of the course of
> nature where not such uniformity really existed.[59]

However, Mill was not dismayed by this, believing that the inductive
ascent involved when his Canons were in use was warranted by virtue of
the truth of the wide, over-arching principle of the uniformity of nature.
So while he acknowledged that some 'low-level' inductive generalisations
might turn out to be false, he claimed that very general principles could
be accepted, for they covered a variety of instances and had never been
found wanting. He wrote:

> [T]he precariousness of the method of [induction by] simple enumeration is in an
> inverse ratio to the largeness of the generalization. The process is delusive and
> insufficient, exactly in proportion as the subject-matter of the observations is special
> and limited in extent. As the sphere widens, this unscientific method becomes less and
> less liable to mislead; and the most universal class of truths, the law of causation for
> instance, and the principles of number and geometry, are duly and satisfactorily proved
> by that method alone, nor are they susceptible of any other proof.[60]

So, the law of universal causation, for example (that the same causal
antecedent circumstances always give rise to the same consequent
effects), is of very wide application and has never been found to be
disobeyed in nature. So we can assume its truth and use it as a major
premiss, serving as a warrant for the ascent of the inductive ladder by
means of the Canons of induction.

Unfortunately, however, it is generally agreed by subsequent commentators that Mill's argument here was unsatisfactory. The overarching
principles of the uniformity of nature and of universal causation rest on,
or have to find their warrant in, many lower-level inductions. To try to
use the grand principles as logical justification for the low-level
generalisations is to argue in a circle. However, it must be acknowledged
that the generality of 'subsequent commentators' have held a different
view of logic from that of Mill, who was interested in the general
processes of reasoning that *are* employed in science, as much as the
niceties of formal logic (which, as we have seen, he thought was a vastly
overrated enterprise). So the objections of critics who believed in the
standard account of deduction would not have impressed Mill greatly.
Also, there is nothing to stop scientists 'holding to' such general
principles as that of universal causation as general guides to research
should they wish to do so, using them as regulative principles — to use
Kant's terminology — or methodological directives. Whether this makes
for successful or unsuccessful science, one might have to judge by some

kind of pragmatic criterion.[61] Or, one might eventually be forced to forego some of the high-level principles, as occurred when the principle of causality was abandoned in the twentieth century by the workers on the quantum theory of the atom.

Besides the inductive ascent utilising the Canons of induction, Mill also considered what he called the 'deductive method', which was to be employed when it proved impossible to use the direct methods of observation and experiment. It was what we today would normally call the 'hypothetico-deductive method'. Its three stages were, according to Mill:

1. ascertaining a set of laws relevant to the phenomena under investigation, and which might together be deployed in an explanation of the phenomena — to be found with the help of the Canons, or by formulating hypotheses;
2. deducing certain statements or conclusions from these laws;
3. ,testing or verifying the deductions by experiment and further observation.[62]

Mill preferred not to employ hypotheses, although he acknowledged that they were necessary on occasions and were justified if their consequences could be well confirmed or verified. But a 'verified' hypothesis should be the *only* one capable of accounting for the phenomena under investigation. All other possible hypotheses had to be excluded — an impossibly stringent requirement we would think, given that there can (theoretically speaking) always be an infinite number of possible hypotheses to account for any given phenomenon.

Mill did give a useful example in illustration of his 'deductive method'. He pointed out that the inverse square law of the Newtonian theory accounted satisfactorily for the observed motions of the heavenly bodies, and was the *only* law which could do this with sufficient precision. In Mill's day, this must surely have seemed to have been correct; but as is well known, at the beginning of the twentieth century the new relativity theory of Einstein, which envisaged variations of mass with velocity, offered an alternative to the Newtonian mechanics. This well-known historical fact, therefore, indicates that Mill was mistaken in believing that the Newtonian theory had been satisfactorily *proved* to be true by means of the 'deductive' — or hypothetico-deductive — method.

One of the most remarkable and interesting features of Mill's philosophy of science was his belief that the principles of mathematics were inductive generalisations; or, in the language of Hume, they were 'matters of fact' rather than 'relations of ideas'. The principles of mathematics, including those of arithmetic, were, for Mill, empirical laws which men had come to know by observations of the world. They were not arbitrarily selected axioms, or starting points for deduction, nor were they in some way products of the human mind as the philosophers who followed Kant supposed. So, in reference to the axioms of geometry, Mill wrote that they were:

> experimental truths; generalizations from observation. The proposition, Two straight lines cannot enclose a space — or in other words, Two straight lines which have once

met, do not meet again, but continue to diverge — is an induction from the evidence of our senses.[63]

This is, I suggest, one of the most 'extreme' forms of inductivism/empiricism that one will encounter in the philosophical literature, and even Mill admitted that 'there is probably no proposition enunciated in this work [the *System of Logic*] for which a more unfavourable reception is to be expected'.[64] He was indeed right in this prognostication. Few have accepted his view of mathematics as an empirical science, on a par with (say) physics and chemistry. The usual line of argument taken against Mill is something like this.[65] We allow no empirical falsification to *count against* the laws of arithmetic or geometry. If we find an apparent empirical falsification of some arithmetical principle, we immediately assume that our countings or calculations have been in error in some way or other. We do not suppose that it is the laws of arithmetic that are at fault. Again, it is claimed that there *cannot be* any possible empirical method for showing the falsity of geometrical principles, for any measurements that might be used to falsify the principles do themselves rely upon the acceptance of those principles.

However, such criticisms of Mill do not seem *quite* as solid and secure today as they did a few years ago. One must, after all, think of the processes whereby each and every one of us comes to know such mathematics as we do know. And it is clear that all of us have come to know our mathematics, at least in its early stages, in an empirical way — by counting counters, by drawings, measurements, cut-outs, and so on. So although the (pure) mathematicians would claim that one can have mathematical axiom systems, with postulates, rules of inference, chains of deductions and theorems, all quite independent of the physical world, even they would not be likely to be able to engage in the art of pure mathematics if they did not make use of an empirical grounding for their thoughts in the early stages of their mathematical education. Further, it is possible to envisage a situation such that mathematical principles originally, in the history of human thought, had an empirical or experiential origin, and subsequently they became 'hardened' or conventionalised, so that no empirical refutation would be countenanced.[66] Finally, noting the very important role of education in the whole process of the acquisition of mathematical knowledge, both for elementary students and for those engaged in mathematical research, one may wonder whether the fact that we regard certain mathematical propositions as axiomatic and necessarily true is to be accounted for in *social* terms as much as anything else.

Such views have recently been advanced by the sociologist of knowledge, David Bloor, drawing heavily on the work of Wittgenstein, and further mention of such ideas will be made in the closing pages of this book.[67] What is interesting to us at this juncture is that in order to press the sociological argument Bloor has had to have recourse once again to an empiricist view of the nature of mathematics, such as that of Mill. To be sure, Bloor's views have not been received with anything like universal acclaim. The idea that there might be several 'kinds' or 'qualities' of mathematics, according to the social system in which they happen to arise, may well seem very strange. Nevertheless, such views are attracting increasing attention at present, and in order for them to get off the ground, so to speak, they have had to abandon the 'necessary' character of mathematical knowledge and claim that it is in some way empirical in character. Mill might have been glad to have seen these recent developments in metascience.

What relationship, if any, did Mill's account of science bear to the 'arch of knowledge'? It will be clear that he envisaged a markedly asymmetrical structure. It had one greatly overgrown leg (the inductive side), and one that was severely emaciated (the deductive side). Mill believed that there was no such thing as deduction, as commonly understood. There could only be reasoning (on the basis of general laws, which for Mill functioned as 'inference warrants') from 'particulars to particulars', even in the syllogism, and *a fortiori* in the process of inductive inquiry, with the aid of the Canons of induction. However, as we have seen, there was also the so-called 'deductive method', called into play, apparently, when all else failed. This was a hypothetico-deductivist methodology and as such it was possessed of the arch-like structure with which we have become familiar. It must be acknowledged, however, that the form that Mill preferred was that of a solidly based inductive pillar.

It was mentioned above that Mill's ideas have tended to become 'part of the intellectual air we breathe'. This is so, I think, despite the fact that so many of his basic ideas, and particularly his representation of mathematics as an empirical science, have not found general acceptance. But the role of induction as the process of paramount importance in the prosecution of scientific inquiry was widely esteemed in the nineteenth century, Hume's identification of the 'problem of induction' notwithstanding. Indeed, 'induction' came to be almost synonymous with 'doing science', and when someone produced scientific ideas that for some reason were displeasing it was common to berate him for having 'abandoned the true path of induction'. (This happened to Darwin.) Central to Mill's inductive methodology were his celebrated Canons, which do indeed give valuable advice for the conduct of inquiry — but which are not sufficient in themselves for the successful conduct of scientific research.

The strengths and weakness of Mill's account of science, and the way it depended upon empiricist principles in disregard of Kant's 'Copernican Revolution', is shown with considerable clarity in the controversy in which Mill engaged with his redoubtable adversary, William Whewell. And therefore we shall want to look at some aspects of this interesting controversy. But first we must examine Whewell's own views on, and contributions to, metascience.

Whewell
Whewell (1794–1866) was one of the most interesting and perceptive of nineteenth-century philosophers of science. A man of extraordinary breadth of learning and intellectual power, he was a distinguished classicist and moralist, as well as historian of science, mathematician, and

scientist in his own right. He did important work in mathematics, physics, astronomy, geology, architecture and economics. He was Master of Trinity College, Cambridge, and University Vice-Chancellor. For short time he held the chair of mineralogy. Whewell is also remembered for his contributions to scientific nomenclature. He suggested to Faraday the terms 'anode', 'cathode', 'ion', 'electrode', etc. He coined for geology the words 'catastrophism' and 'uniformitarianism'; and the very term 'scientist' (as opposed to 'natural philosopher') was proposed by Whewell.

Whewell was also an ardent theist and ordained Anglican minister. He opposed the Darwinian theory, though much of his philosophy of science had strong 'developmental' or 'evolutionary' aspects. From our point of view, Whewell's philosophy of science is particularly interesting because of the way in which it linked together certain 'Kantian' concepts and elements of the tradition of British empiricism. Thus, one of Whewell's avowed aims was a suitable refurbishment of Baconian ideas for nineteenth-century readers in the light of Kant's 'transcendental idealism'.

Whewell's views on metascience (unlike Mill's) were thoroughly grounded in actual scientific research experience and detailed researches in the history of science. His major writings with which we shall be concerned were: *History of the Inductive Sciences* (1837),[68] *The Philosophy of the Inductive Sciences* (1840),[69] and *Of Induction* (1849).[70] The third edition of the *Philosophy* was issued as three separate books: *The History of Scientific Ideas* (1858),[71] *Novum Organon Renovatum* (1858),[72] and *The Philosophy of Discovery* (1860).[73]

The very title, *Novum Organon Renovatum*, is a clear indication of Whewell's wish to establish a refurbished Baconian philosophy, consonant with the results of nineteenth-century historical and philosophical researches, and it is interesting that Whewell also chose to present a synopsis of his argument in the *Philosophy of the Inductive Sciences* in a series of aphorisms, adopting the literary device that Bacon himself had favoured. The particular 'refurbishment' that Whewell had in mind was, of course, the inclusion of some of the results of Kant's 'Copernican Revolution' in a new metascience. Also, there was a very significant historical dimension to Whewell's account, which had been noticeably absent in earlier philosophies of science. This gave a typically nineteenth-century 'historicist' character to his scheme.

It is not possible to give a precise estimate of Whewell's knowledge of Kant's writings, but Kant's ideas had been available in English since the end of the eighteenth century[74] and had made a considerable impact through the interest of the poet Coleridge, who was also concerned with trying to develop a personal world-view through the spectacles of German idealism.[75] Certainly Whewell referred to the *Critique of Pure Reason* in his *Philosophy* and showed evident familiarity with it. He stated specifically that he had adopted Kant's views on the nature of space and time, though (he claimed) his views differed widely from those of 'that acute metaphysician'.[76] This was true in that, for example,

Whewell did not adopt Kant's distinction between the ideas of the 'Transcendental Aesthetic' and the 'Transcendental Analytic'. That is to say, he did not separate off the notions of space and time from the categories of the understanding.[77] Also, Whewell had no truck with Kant's attempted 'deduction' of the categories. But the Kantian vision of the mind as an active agent, in some manner 'imposing itself' on the world during the process of its apprehension, was very much to the fore in Whewell's system.

As we know, Kant believed that the human mind achieved its understanding of the external world, and performed its cognitive processes according to a 'framework' of twelve categories — neither more nor less. Corresponding to this, Whewell believed that there were certain *'Fundamental Ideas'* necessary for the prosecution of science, and he believed that a list of such 'Ideas' might be determined by examining the history of science, not by some metaphysical deduction. Whewell was not, perhaps, as clear as one might wish as to exactly what the Fundamental Ideas were supposed to be; but he had this to say:

> [T]he necessity and universality of the truths which form a part of our knowledge, are derived from the *Fundamental Ideas* which these truths involve. These ideas entirely shape and circumscribe our knowledge; they regulate the active operations of our mind, without which our passive sensations do not become knowledge. They govern these operations, according to rules which are not only fixed and permanent, but which may be expressed in plain and definite terms; and these rules...may be made the basis of demonstrations by which the necessary relations imparted to our knowledge by our ideas may be traced to their consequences in the most remote ramifications of scientific truth.[78]

The Ideas, then, were 'certain general forms of apprehension, or relations of our conceptions'.[79]

The actual list of Fundamental Ideas, which Whewell thought he could discern as active within the history of scientific inquiry, was: *Space*; *Time* (including *Number*[80]); *Cause* (including *Force* and *Matter*); the *Outness* or *Externality* of objects; the *Media* for the perception of secondary qualities; *Polarity* (or contrariety); chemical *Composition*; *Affinity*; *Substance*; *Likeness* (or resemblance) and natural *Affinity*; *Means* and *Ends*; *Symmetry*; *Vital Powers* (*Assimilation* and *Irritability*); *Final Cause*; and *Historical Causation*.[81] Such Fundamental Ideas were supposed to 'inform', guide, shape, or regulate the sensations in a quasi-Kantian way. They gave coherence and sense to a person's understanding of the world.

Now the various sciences, and mathematics, all had their several axioms such as those of Euclidean geometry, which — according to Whewell — could be 'unfolded'[82] from the various Fundamental Ideas. The Ideas were somehow responsible for the necessary character of the various scientific axioms and principles. For example, the Idea of space lay at the back of our 'intuitive' feeling that it was *necessarily* the case that the shortest distance between two points is a straight line. Also, the various 'conceptions' of the several sciences could be construed as corresponding to their appropriate Fundamental Ideas. For example:

Fundamental Ideas
   Space
   Number
   Cause
   Composition
   Resemblance
Corresponding Conceptions
   Line, circle, square, etc.
   Square number
   Accelerating force
   Natural combination of elements
   Genus, species, and other taxonomic groupings

It might seem natural to object against all this that Aristotle had no notion of the Newtonian conceptions of accelerating force and inertia. Indeed, Aristotle's physics was based upon wholly different principles. For example, he held that bodies move only when they are impelled in some way — either by the 'Unmoved Mover'or some derivative source of motion. How, then, could 'accelerating force' be a Fundamental Idea, if some people possessed it while others did not? Well, in point of fact this difference between the physics of Aristotle and that of Newton actually supported Whewell's account — grounded in the *history* of science — rather than refuting it. For Whewell maintained that the intuition of scientific truths was progressive, or was slowly developing and evolving. The whole process of scientific inquiry was one in which the Fundamental Ideas were gradually elucidated and made clear to scientists. So, once a particular Fundamental Idea had been properly clarified, its associated axioms could be formulated appropriately and would then seem to have the quality of philosophical necessity; for their contraries would be incompatible with the Fundamental Idea. (By contrast, Mill said that the scientific axioms that we use are necessary for us because their contraries clash with experience. And unlike Descartes' first principles, Whewell's axioms were the end-product of scientific inquiry, not the starting point.)

The process of clarification of the Fundamental Ideas was to be achieved in the course of the development of science, with the help of dialectical interchange between scientists and philosophers, evaluating and discussing the information acquired during the processes of experimental investigation.[83] In placing such emphasis upon the processes of dialectical discussion within the scientific community, Whewell was, I suggest, one of the first to recognise the importance of social factors within the construction of knowledge (though, one could also, I suppose, see the long arm of Plato influencing the nineteenth-century philosopher's depiction of the scientific process).

However, unlike some modern exponents of the 'sociology of knowledge' doctrine, Whewell believed that the overall trend of scientific research involved a progressive approach towards the successful establishment of Fundamental Ideas and the discovery of 'the truth'. In his view, the reason that one might have some confidence in the progressive character of science lay in the fact that from time to time successful 'consiliences[84] of inductions' might be achieved. To illustrate this process, Whewell constructed two interesting tables, one for optics and one for astronomy, to show the way in which, during the history of science, what were initially disparate formulae, isolated facts or theories, gradually came to be subsumed under laws and theories of increasing levels of generality. For example, such apparently distinct and separate phenomena as the observed motions of the planets, the tides and the fall of apples from apple trees, might all eventually be explained and accounted for in terms of the Newtonian theory of universal gravitation. So, wrote Whewell:

> The Consilience of Inductions takes place when an Induction obtained from one class of facts, coincides with an Induction, obtained from another different class. This Consilience is a test of the truth of the Theory in which it occurs.[85]

And by induction he meant either the process by means of which laws or theories were established or the propositions that stated such laws or theories.[86] His terminology here was, therefore, not unlike that of Bacon. But the underscoring of the importance of consilience was something new, although no doubt it had been noticed by others before Whewell, without any special emphasis being given to it. In fact, it was, in a sense, related to Bacon's idea of science as a kind of hierarchy or pyramid,[87] made up of laws or axioms of ever-higher degree of generality.

As previously stated, Whewell's epistemology was essentially Kantian in character. He did not regard general conceptions as in any way 'real' (as had the 'realists', following Plato); nor were they merely notions corresponding to boundaries marked out by names (as the 'nominalists' had supposed).[88] Conceptions were gradually established from Fundamental Ideas in the course of the history of science, as described above. Then, when one examined the external world, what one knew of it was a product of both the sensations received by courtesy of the sense organs and the ideas or conceptions which the mind brought to bear in its attempt to form its understanding. There was, said Whewell (drawing on elements of German idealist philosophy), always a subjective and an objective element in the process of knowing or understanding;[89] or the sensations supplied the 'matter' and the mental conceptions or ideas supplied the 'form'[90] for the cognitive process, giving rise to what we are pleased to call facts.

But for Whewell fact and theory were not essentially different, as they had been for Mill. Whether a particular notion might be regarded as a fact or a theory depended upon its relative familiarity and the particular stage of development that the history of science might have reached. Thus, the theories of one generation could serve as the facts of the next, so that there was always a step-wise progression in scientific knowledge.[91] As a social phenomenon, describing the way in which the human scientific community behaves in its discourse concerning facts and theories, this contention seems to me to be essentially true, although Whewell, of course, was not specially concerned with the sociological perspective.

To understand the 'Kantian' element in Whewell's epistemology more clearly, we should look at his remarks about 'induction', which, as we have seen, meant for Whewell the processes by means of which new facts, laws and theories are established — with a nice 'consilience' if possible. During induction:

> [T]he ideal conception which the mind itself supplied is *superinduced* upon the facts as they are originally presented to observation. Before the inductive truth is detected, the facts are there, but they are many and unconnected. The conception which the discoverer applies to them gives them connexion and unity.[92]

So the scientist does not and cannot survey the world as if his mind were a blank piece of paper. It is customarily well stocked with certain Fundamental Ideas and derivative conceptions and in the process of discovery the observational data are linked together — *colligated*[93] in Whewell's language — to establish new facts. For example,[94] Kepler was presumed to have combined the conception of an ellipse (the subjective element in the process of knowing) with the observational data collected by his mentor Tycho Brahe (the objective component) to yield the 'fact' (or hypothesis or theory, if you prefer) that the planet Mars moves round the Sun in an ellipse. In this way, the combination of the subjective and the objective elements allowed an induction to be made.

This was a nice example, well illustrating the 'Kantian' elements of Whewell's philosophy, for it revealed the role of the mind as a kind of active agent in the process of cognition. Moreover, it served as a convenient focus of debate in the controversy that began to develop between Mill and Whewell on the question of the way in which the scientist proceeds in the course of his investigations.[95] According to Mill, the empiricist Kepler merely summed up Tycho's observations when he asserted that the orbit of Mars was elliptical, thereby giving a 'description' of the phenomena.[96] This was a rather curious usage of the word on Mill's part, and one might well have expected him to have employed the term 'induction' in accordance with his usage only a few pages previously.[97] Whewell, as we may imagine, was not satisfied, and in 1849 brought out a small booklet specially devoted to an attempted refutation of Mill's account of science and his treatment of 'induction'.[98]

Whewell pointed out[99] that the distinction that Mill sought to make between 'induction' and 'description' was spurious. For Mill seemed to think that when particular facts were joined together, or colligated, by consideration of spatial relationships (as in the case of Kepler and the elliptical orbit of Mars), a 'description' was involved. Only when other colligating concepts such as time or cause were employed did one have an 'induction'. It seems to me that Whewell's rejection of the distinction that Mill sought to make was perfectly justified. We do colligate phenomena with the help of such concepts as we may possess, including spatial ones, and in *that* simple sense the Kantian philosophy was and is to be preferred to that of the empiricists. Today, it is a commonplace of metascience that 'concepts influence percepts'. How we see and make sense of the world is necessarily shaped by the concepts that we possess. This we can say without committing ourselves to the full-blown Kantian doctrine of categories, the 'Metaphysical Deduction', and so forth. But the doctrine of the theory-dependence of observations doesn't make much sense in terms of Mill's empiricism, and Whewell was, I believe, in this matter more in tune with twentieth-century developments in metascience than was Mill. Yet it was Mill's influence that chiefly prevailed through the nineteenth century, and it is perhaps only in the past twenty years or so that the merit of Whewell's metascientific work has been sufficiently recognised.

There were other important differences between Mill and Whewell, such as the latter's rejection of Mill's claims that all reasoning, even 'induction', involved arguing from 'particulars to particulars'. But such matters, though important and interesting, were subsidiary to the major issue of the extent to which the mind played an active role in cognition and understanding, 'superinducing' concepts on data in the process of attempted understanding. One can, of course, see why one such as Mill might have found Whewell's metascience unacceptable. It might appear to open the door to, or admit the existence of, all kinds of subjectivities and speculations in science. And Whewell's idealism was, of course, part of his wider moral philosophy and his Anglican theology, to neither of which was Mill specially sympathetic. Nevertheless, as I say, Whewell's 'Kantianism' has found considerable subsequent favour. But it took a long time for this favourable assessment to be achieved. Finally, in discussing Whewell's writings on metascience, we may say something further about his discussions of method, analysis and synthesis, and his picture of the 'arch of knowledge'. We shall refer here chiefly to his most fully developed account of these matters in the *Novum Organon Renovatum*.

As preliminary steps, Whewell suggested that the conceptions which were to be deployed had to be 'explicated' and the facts 'decomposed' (or analysed). Like Herschel, Whewell recognised that scientists had to decide which factors were relevant to their inquiries, rather than collecting facts holus bolus. The process of 'explication of conceptions' was one that we have already discussed — involving the gradual, 'historical' clarification of the Fundamental Ideas and the elucidation of their correspondent conceptions or concepts. But Whewell's account of the 'Decomposition of Facts' was, I fear, of little more help than that of Herschel's discussion of the preliminary analysis of phenomena. In the *Novum Organon Renovatum* we read:

> We resolve the complex appearances which nature offers to us, and the mixed and manifold modes of looking at these appearances which rise in our thoughts into limited, definite, and clearly-understood portions. This process we may term the *Decomposition of Facts*.[100]

In this preliminary phase, phenomena were to be observed and classified, with such terminology as might be required on each occasion being introduced as necessary.[101] But as I say, Whewell's account of the process was pretty thin.

Next in the process of induction, there was the very important 'colligation of facts'. This involved the 'Selection of the [appropriate] Idea', the 'Construction of the Conception', and the 'Determination of the Magnitudes'.[102] There was to be no bar against the use of hypotheses in these procedures. Indeed, Whewell's methodology might, in a nutshell, be described as one of hypothetico-deductivism. He acknowledged, however, that no general rules could be given for the creative processes involved in the formulation of hypotheses. Even the process of selecting the appropriate Fundamental Idea had to be carried out by something like trial and error![103]

The process of testing of the Ideas, the appropriate application of the conceptions and the correct determination of the magnitudes was to be achieved by the prediction of facts which were further to be tested by observation and/or experiment, the consilience of inductions (as previously described), and the simplification of a conception as it was extended to new cases. So as hypotheses were verified by their testing and the successful consilience of inductions, they might be clarified, gradually achieving the status of necessary truths.

As we have seen above,[104] Whewell could see difficulties in the direct application of Mill's Canons of induction in the process of scientific inquiry. He claimed that Mill was inclined to bring them into the story much too early. Whewell himself formally adopted only one of Mill's Canon — the Method of Residues. But Whewell did have some suggestions as to specific scientific procedures that one might find useful: the 'Method of curves', the 'Method of Means', and the 'Method of Least Squares'.[105] Essentially, these were commonplace statistical methods, and methods for fitting curves to data plotted graphically. They had no special philosophical significance though they betoken the recognition on Whewell's part of the need to realise that all experiments are subject to experimental error. Whewell also made some suggestions as to particular techniques of observation, such as double weighing,[106] which might be useful towards improvements in accuracy and precision. Again, they need not detain us here.

Whewell did not give much attention to the language of analysis and synthesis, although he did write:

> Geometrical deduction (and deduction in general) is called *synthesis*, because we introduce, at successive steps, the result of new principles. But in reasoning on the relations of space, we sometimes go on *separating* truths into their component truths, and these into other component truth; and so on; and this is geometrical *analysis*.[107]

This appears to have been largely a nineteenth-century gloss on the confused terminological tradition to which nineteenth-century writers were heir. Certainly, Whewell made no particular use of the distinction between analysis and synthesis in his description of the methods and structure of science. So far as he was concerned, the overall shape of science — when well established with satisfactory consilience of inductions — was pyramidal, with laws of ever-increasing generality towards the apex. The steps for moving from one level to the next were such as I have sought to describe in the foregoing account. Yet the inductive pyramid, if so we may call it, was only the final product of a whole range of scientific activity, which might in practice take thousands of years to complete. And in the establishment of the finished product a vast number of supporting 'arches' might have to be constructed — arch-like because of their employment of induction, the formulation of hypotheses, and their deductive testing.

In all this, Whewell's chief interest was in the many processes that contributed to the 'upward' inductive ascent. His system, while primitive in certain respects, particularly in the account of the 'unfolding' of conceptions from 'Fundamental Ideas', was nurtured in ground rich in Baconian nutriment. Yet included within the whole was the essential Kantian insistence on the central role of 'pre-existing' concepts, actively employed by the mind in the colligation of data. In this respect, Whewell's philosophy of science was more perceptive than that of his contemporary Mill; and he provided important anticipations of views that have been accorded detailed discussion by twentieth-century philosophers. Further, Whewell's emphasis on the historical dimension to scientific knowledge has much to recommend it.

The trio of writers that we have considered in this chapter were all actively concerned, one way or another, with methodological issues — with suggestions as to the proper way to conduct scientific inquiries. And the answers they gave to methodological questions were in many ways remarkably similar, despite differences stemming from different epistemological perspectives. All were chiefly interested in the inductive, ascending limb of the arch of knowledge, and in the case of Mill at least, the usual distinction between induction and deduction became very blurred. Indeed, induction began to 'take over' from deduction, with the 'relations of ideas' side of Hume's Fork coming to be regarded as within the domain of empirical, *a posteriori* knowledge. Herschel and Whewell did not go as far as this but, for all three, induction became loosely equivalent to 'doing science'. Thus the firm distinction between induction and deduction that we hoped we might find in the nineteenth-century was by no means clearly displayed in the three methodologists we have considered here. It was, however, the modern terminological distinction that was chiefly lacking; it was not that there was some special species of mental confusion on the part of these writers. They used the language of their day in the manner that was usual or which seemed appropriate to them. It is not that there is some transcendent distinction between induction and deduction that we moderns are privileged to see, but which was veiled to the obscurantist minds of nineteenth-century metascientists.

NOTES

1   J F W Herschel *Preliminary Discourse on the Study of Natural Philosophy* Longman Rees Orme
    Brown & Green London 1930. (I have used the 'New edition' of 1833.)
2   Johnson Reprint Corporation New York 1967
3   D E Allen *The Naturalist in Britain: A Social History* Lane London 1976
4   Herschel *op cit* (note 1 1833) pp 83–4
5   For a critique of it, see C J Ducasse 'John F W Herschel's Methods of Experimental Enquiry' in R M
    Blake *et al Theories of Scientific Method: The Renaissance through the Nineteenth Century*
    University of Washington Press Seattle 1960 pp 153–82
6   Herschel *op cit* (note 1 1833) p 85
7   *Ibid* p 97
8   *Ibid* p 88
9   *Ibid* p 89
10 *Ibid* p 90 (There are, of course, many such laws in science, such as for example, Hooke's law, Boyle's
    law, Snell's law, the law of inverse squares, and so on. They are statements — usually given alge-
    braically today, in terms of functional variables, rather than words — of generalised descriptions of
    phenomena, which are thought to be of universal application if certain limitations are complied
    with. For example, Boyle's law, concerning the inverse relationship that obtains between the
    pressure and volume of a fixed mass of gas, only holds at fixed temperatures, and even then only
    approximately.)
11 *Ibid* pp 98–9
12 *Ibid* p 99
13 *Ibid* p 102
14 *Ibid* p 101
15 See above, p 117
16 For example, Bacon's 'Table of Degrees or Comparison in Heat' was in a sense a forerunner of what
    Mill was to call the 'Method of Concomitant Variations' or Herschel's third rule. See also Bacon's
    fourth Aphorism to Book II of the *New Organon*, quoted above, p 63
17 See below, p 150
18 Herschel *op cit* (note 1 1833) pp 151–2
19 Equivalent to Mill's 'Method of agreement' and (approximately) to Hume's third, fourth and fifth
    rules.
20 Equivalent to Mill's 'Method of difference' and to some extent to Hume's sixth rule.
21 Equivalent to Mill's 'Method of concomitant variations'.
22 Equivalent (very roughly) to Mill's methods of 'Agreement', 'Difference' and 'Concomitant
    variations' rolled into one.
23 See below, p 150
24 Herschel *op cit* (note 1 1833) p 159
25 W Wells *An Essay on Dew, and Several Appearances Connected With It* Taylor & Hessey London
    1814. In this work, Wells collected a number of instances in which dew was observed to form, and
    set up experiments (with pieces of wool to absorb the moisture) to discover the meteorological
    conditions when dew was formed. He was able thereby to establish the 'cause' of the formation of
    dew — a nice little scientific investigation carried through most successfully. It is interesting the
    extent to which this particular piece of work has been used by writers on scientific method as a
    paradigm case. For all I know, it may still be used by some teachers today. Certainly it was used in
    1960 (when I was attending lectures on philosophy of science at University College, London) as an
    illustration of the canons of scientific method. Thus did the shades of Herschel and Mill linger long
    in London!
26 Herschel *op cit* (note 1 1833) p 164
27 *Ibid* pp 164–81
28 *Ibid* p 190
29 *Ibid*
30 *Ibid* p 196
31 *Ibid* p 197
32 See above, p 82
33 Herschel *op cit* (note 1 1833) p 198
34 This follows suggestions made by J Losee (*A Historical Introduction to the Philosophy of Science*
    Oxford University Press Oxford 2nd ed 1980 p 116), but some modifications of his synopsis are
    made here.
35 See above, pp 62 & 67.
36 Herschel *op cit* (note 1 1833) p 175
37 One must acknowledge, however, that Herschel's reading of Bacon was as much in evidence as his
    own practical experiences.
38 See below, p 152
39 M Lerner ed *Essential Works of John Stuart Mill* Bantam Books New York 1961 p vii
40 J S Mill *A System of Logic, Ratiocinative* [i.e. deductive] *and Inductive Being a Connected View of
    the Principles of Evidence and the Methods of Investigation* 2 vols Parker London 1843. I have used

the version edited by J M Robson Mill's *Collected Works* (Vols 7 and 8 University of Toronto Press Toronto & Buffalo and Routledge & Kegan Paul London 1973–74). For a history of Mill's composition of his Logic, see the Introduction to Vol 7 of his *Collected Works*; or O A Kubitz 'Development of John Stuart Mill's *System of Logic' Illinois Studies in the Social Sciences* 1932 Vol 18 No 1

41   *Ibid* (1973–74) p 196
42   *Ibid* pp 224–5
43   This is not to say that Kant himself was disposed to deny the freedom of the human will; quite the contrary.
44   Mill *op cit* (note 40 1973-74) p 841
45   *Ibid* p 194
46   I use the terms here in their traditional senses. As we have seen, Mill identified logic with the general reasoning process, and emphasised its inductive character.
47   We have, however, also noted that Herschel's 'Rules' were themselves partly indebted to Hume and Bacon; and the methods of agreement and difference appeared in Duns Scotus and William of Ockham, respectively.
48   Mill *op cit* (note 40 1973–74) p 390
49   *Ibid* p 391
50   *Ibid* p 396
51   *Ibid* p 398
52   *Ibid* p 397
53   *Ibid* p 401
54   Concomitant going together, accompanying, concurrent, attending' (*Shorter Oxford English Dictionary*).
55   It may be noted that the 'Method of Difference' cannot be applied here: one cannot simply take away the Moon in order to gauge what effect this might have.
56   See above, pp 35 & 36
57   It may be noted here that in his statement of his Canons, Mill wrote 'or the effect' every time after he wrote 'the cause'. He was, in the Canons, establishing functional relationships between *A* and *a*, etc. But until the element of time was introduced (in terms of antecedent and consequent), one could not determine which of *A* and *a* was the 'cause' and which the 'effect'.
58   W Whewell *Of Induction, With Especial Reference to Mr J.Stuart Mill's System of Logic* Parker London 1849 p 44
59   Mill *op cit* (note 40 1973–74) pp 311–12. Mill's notion of the age of intelligent European man — 5000 years? — is worthy of remark here. Also, it was, I believe, this passage of Mill that set philosophers on the pathway, on which they have been muttering 'All swans are white' ever since.
60   *Ibid* p 569
61   On this, see below, p 183
62   Mill *op cit* (note 40 1973–74) pp 454–63
63   *Ibid* p 231
64   *Ibid*
65   See, for example, K Britton *John Stuart Mill* Harmondsworth Pelican 1953 p 134
66   For a more extended discussion of conventionalism, see below, p 191
67   See below, p 351
68   W Whewell *History of the Inductive Sciences, from the Earliest to the Present Times* 3 vols Parker London and Deighton Cambridge 1837
69   W Whewell *The Philosophy of the Inductive Sciences* 2 vols Parker London and Deighton Cambridge 1840
70   Whewell *Of Induction* (note 58)
71   W Whewell *The History of Scientific Ideas. Being the First Part of the Philosophy of the Inductive Sciences* Parker London 1858
72   W Whewell *Novum Organon Renovatum. Being the Second Part of the Philosophy of the Inductive Sciences* Parker London 1858
73   W Whewell *On the Philosophy of Discovery, Chapters Historical and Critical,...Including the Completion of the Third Edition of the Philosophy of the Inductive Sciences* Parker London 1860
74   *Elements of the Critical Philosophy: Containing a Concise Account of its Origin and Tendency; a View of all the Works Published by its Founder Immanuel Kant; and a Glossary...To which are Added: Three Philological Essays, Chiefly from the German of I.C.Adelung... By A.F.M.Willich* London 1798
75   See G N G Orsini *Coleridge and German Idealism: A Study in the History of Philosophy with Unpublished Materials from Coleridge's Manuscripts* Southern Illinois University Press Carbondale 1969; and T H Levere *Poetry Realized in Nature: Samuel Taylor Coleridge and Early Nineteenth-Century Science* Cambridge University Press New York 1981
76   Whewell *The Philosophy of the Inductive Sciences* (note 69) Vol I p x
77   Whewell did, however (*ibid* p 77), suggest that the sciences of space and time (that is, geometry and arithmetic) were 'pure', not 'inductive'.

78   *Ibid* pp 63–4
79   *Ibid* p xiii
80   It is interesting that Whewell followed Kant some way here by connecting the truths of arithmetic with time. (See above, p 127.)
81   Whewell *op cit* (note 69) Vol I p 77
82   *Ibid* p 71; Vol 2 p 172
83   *Ibid* Vol 1 p xxxvi; Vol 2 pp 172–3
84   'Consilient' means literally 'jumping together'; hence 'concurrent' or 'accordant' (Latin *salire* 'to jump').
85   Whewell *op cit* (note 69) Vol 1 p xxxix
86   *Ibid*
87   See above, p 62
88   Whewell *op cit* (note 69) Vol 1 p xviii
89   *Ibid* p 32
90   *Ibid* p 34 (Such distinctions between matter and form, objective and subjective, real and ideal, etc., Whewell thought formed the 'Fundamental antithesis in philosophy'.)
91   *Ibid* pp 46–7
92   *Ibid* p 42
93   Latin *colligere* 'to bind together'
94   Whewell *op cit* (note 69) Vol 1 pp 41–7
95   For a more extended treatment than that which we have room for here, see, for example: E W Strong 'William Whewell and John Stuart Mill: Their Controversy about Scientific Knowledge', *Journal of the History of Ideas* 1955 Vol 16 pp 209–21; H T Walsh 'Whewell and Mill on Induction' *Philosophy of Science* 1962 Vol 29 pp 279–84; G Buchdahl 'Inductivist versus Deductivist Approaches in the Philosophy of Science as Illustrated by Some Controversies Between Whewell and Mill' *The Monist* 1971 Vol 55 pp 343–67
96   Mill *op cit* (note 40 1973–74) pp 292–3 (Mill likened Kepler's discovery to that of a navigator sailing along a coast line and taking bearings, etc. Eventually, he might recognise that he had been sailing round an island. But for Mill this was a 'description of a complex fact', not an 'induction'.)
97   *Ibid* p 286. Here Mill wrote: '[S]uch facts as the magnitudes of the bodies of the solar system, their distances from one another, the figure of the earth, and its rotation, are proved indirectly, by the aid of inductions founded on other facts which we can easily reach.'
98   Whewell *op cit* (note 58)
99   *Ibid* p 21
100 Whewell *op cit* (note 72) p 56
101 *Ibid* pp 57–8
102 *Ibid* p 186
103 *Ibid* p 191
104 See p 152
105 Whewell *op cit* (note 72) p 202
106 *Ibid* p 158
107 Whewell *op cit* (note 69) Vol 1 p xxiii

CHAPTER 5

NINETEENTH-CENTURY POSITIVISM*

A species of philosophy of science to which many nineteenth- and twentieth-century metascientists have belonged is that known as positivism. Or perhaps we should refer to it as a genus, order or class rather than a species, for it was and is a very large, amorphous and ill-defined philosophical taxon. But for this very reason, we cannot evade such an important component of the philosophical landscape, viewed historically. The positivist movement began fairly tidily with the French philosopher Auguste Comte introducing the term, and attempting to set up a 'positivist school' (or indeed, as we shall see, even a 'positivist religion'!), in which the methods of the physical sciences would be extended to the study of society. Then in the twentieth century we have the logical-positivist movement, whose members sought to create a new philosophy of science, integrating results obtained in mathematics and logic with the empiricist tradition of nineteenth-century positivism. The logical-positivists formed a relatively coherent group, but between them and Comte we have a number of moderately distinct schools or 'isms', such as pragmatism, conventionalism and instrumentalism, which may nonetheless be classified more or less satisfactorily as different manifestations of positivism. So we shall deal with them in this chapter, recognising, however, that it is only with some stretching of the historical and philosophical imagination that they can all be represented as particular versions of the nineteenth-century positivist movement. It may be mentioned here that it would not be inappropriate to refer to Mill as a positivist; but we have chosen to discuss him in the previous chapter because of the historical relation of his work to that of Herschel and Whewell.

One of the most useful books on the history of the positivist movement is Kolakowski's *Alienation of Reason: A History of Positivist Thought*, in which the author makes a valiant effort to set down some general identifying marks to enable the student to recognise the leading features of positivist philosophy. Kolakowski's four criteria are: (i) The rule of phenomenalism; (ii) The rule of nominalism; (iii) The rule that denies

---

*Inevitably in this chapter, we will have to carry over into the twentieth century, but all the figures in this chapter were old enough to have formulated their basic views in the nineteenth century.

cognitive value to value judgments and normative statements; (iv) The rule (or belief) that there is an essential unity of scientific method.[1] The positivist, then, emphasises the importance of empirical evidence as the source of knowledge, and would wish to eliminate from science or philosophical discourse any kind of entities that lie beyond the reach of empirical investigation — such as gods, souls, entelechies, essences, or whatever. And in its most thoroughgoing and pure form, positivism subscribes to a phenomenalist epistemology.[2] So according to the positivist view, hypothetical entities such as atoms or genes, that might lie beyond the reach of observational evidence, should be eliminated from the the theoretical armoury of science. As we shall see, this was the position adopted by that 'hard-line' positivist, Ernst Mach.

But not all positivists are necessarily phenomenalists. There is, after all, a certain relativity in the matter of what is and what is not accessible to sensory experience. What might be thought of as an 'occult cause' for one generation might well be clearly visible under an electron microscope in the next. However, as Kolakowski points out, positivists generally unite in their desire to eliminate 'essences' from science, and they take the nominalist view of language. Moreover, they believe it possible to distinguish between facts and values, and maintain that values should be kept out of science.[3] They usually repudiate the claims of traditional theology, and hold that there is no transcendent world of value. Or if there is, it is a matter quite beyond the concern of science. They think it futile to seek to determine the ultimate causes and origins of things. By restricting inquiries to empirical 'matters of fact' — to the determination of the 'laws of nature' or the regularities in the relationships between observable phenomena — they believe that one can have a certain and secure basis for knowledge, and hence for action.[4]

Comte

Given these preliminaries, we may now look at the work of the 'official' founder of the positivist school, Isidore Auguste Comte (1788–1857). It should be noted, however, that earlier philosophers — including one no less than Newton himself — have sometimes been referred to as positivists.[5] And Comte certainly did not suppose that positivism sprang forth, fully fledged, when he published his celebrated *Course of Positive Philosophy*.

Comte was born in Montpellier of a strict Catholic family, son of an obscure clerk. He studied science and mathematics in Paris at the celebrated Ecole Polytechnique, acquiring there republican ideals and a belief in the great power and worth of science. He was also influenced by the utopian doctrines of Saint-Simon, precursor of the socialist movement. Later, Comte became on intimate terms with Saint-Simon and served as his secretary. Comte married quite young — to a woman who had to earn a living as a prostitute, and on whose 'immoral earnings' he sometimes found it necessary to rely. It appears that the relationship between the two was never very satisfactory, though Comte's wife stood by him during a period of his life when he lost his sanity (partly due to overwork, it seems), and she admired his intellect and philosophical achievements. Later, Comte became enamoured of a certain Clothilde de Vaux. But sadly this lady died only a year after they became acquainted; and as a result the love-sick Comte, hitherto the austere positivist concerned to purge the sciences of all values, came to worship her and all womankind in the latter part of his life. This middle-age infatuation was linked to a change on Comte's part from his early efforts to promulgate a positivist philosophy to the later desire to establish a positivist religion. We shall say a little more about this anon.

Much of Comte's life was spent in dire poverty, for he had no official teaching position, having quarrelled with the authorities when a young man. So he eked out a precarious living, giving private lessons, and a private course of lectures, in Paris. It was this lecture-course that became the basis of Comte's celebrated *Course of Positive Philosophy*, published in six volumes between 1830 and 1842.[6] Comte's lectures were based on an extensive study of the history of science, so like Whewell he was giving philosophy of science a characteristically nineteenth-century historicist slant. Comte supposed that a proper understanding and due appreciation of science could best be attained by a study of its history.[7]

If we look at Comte's published synopsis of his lectures, we find that he treated the various sciences in the following order:

Mathematics
Astronomy
Physics
Chemistry
Physiology
Sociology

Thus, sociology (or 'social physics' as Comte also called it) became as it were the 'queen of the sciences'. It was Comte who consciously 'invented' the new science of society and gave it the name to which we are accustomed. He thought that it would be possible to establish it on a 'positive' basis, just like the other sciences, which served as necessary preliminaries to it. For social phenomena were to be viewed in the light of physiological (or biological) laws and theories[8] and investigated empirically, just like physical phenomena. Likewise, biological phenomena were to be viewed in the light of chemical laws and theories; and so on down the line. For, as Comte said near the beginning of his lecture course:

> The end to be kept in view is the arrangement of the sciences in their natural sequence, *i.e.* according to their dependence on one another, so that they can be successively expounded without turning in a vicious circle.[9]

Thus, Comte regarded the several sciences as logically related, and dependent one on another, both historically and conceptually. This does not mean, however, that he envisaged that the whole of biology (say) would one day be 'reduced' to chemistry, so that (for example) all biological phenomena might eventually be explained wholly in chemical terms. There was undoubtedly a reductionist tendency in Comte's system, but he did not suppose that there would only be need for a single kind of knowledge — that of mathematics — to have a complete scientific understanding of phenomena. Indeed, he specifically denied any suggestion that all might ultimately be explained by one universal law.[10] Comte's system, it should be emphasised, was chiefly designed for pedagogic purposes. He held that in teaching science it was essential to begin with the most fundamental branch first. If one began (say) with physiology, without knowing any physics or chemistry, the result would be hopeless confusion.[11] There is, to be sure, some truth in this; but it should not be carried to excess. Obviously, one does not and cannot master the whole of mathematics before embarking on the study of elementary astronomy or physics.

Comte stated[12] that there were two essentially different ways of teaching science: the historical and the 'dogmatic'. That is, it could be taught either historically or theoretically. In the early stages of the development of a science, the historical mode would be appropriate, but gradually it would be superseded by the 'dogmatic' approach as the science achieved a mature condition. Thus, by considering the history of the sciences and their 'logical' relations one to another, one might be able to construct 'the encyclopaedic ladder of the fundamental sciences'.[13] For, although there would inevitably be some overlap and loose ends,[14] Comte supposed that the general order of the sciences, considered in their theoretical dependence one to another, and in their order of historical development, could be established satisfactorily. And this was to be the basis of his pedagogy. Comte's study of science and its history, then, led him to the hierarchy of the sciences that has been given above — which was what he taught in his lectures.

Since Comte's day, there has been a great deal of metascientific discussion as to the possibility and desirability of a general 'reductionist' program in science, such as might seem to be implied in the hierarchy of the sciences given in the *Course of Positive Philosophy*. That is to say, metascientists have wondered whether the laws and theories of one branch of science might be fully explained in terms of the laws and theories of some other, more fundamental branch. If this were possible, then the one branch might be 'reduced' to the other, with consequent economy of thought and increase of understanding.

In fact, there has been a strong reductionist tendency throughout the history of science, and it was given considerable impetus by the positivist movement. However, it does not follow from this that Comte's own intention was to reduce all the sciences to mathematics — the subject at the top rung of his 'ladder'. Neither does it mean that the 'logical' order of the sciences, in so far as there is one, is actually the same as the historical one. New areas of inquiry constantly emerge in the 'fundamental' physical sciences just as much as in the 'derivative' life sciences; and few, if any, areas of science seem to become fully and finally established as a firm base, towards which the other sciences may be 'reduced' with confidence and security. Moreover, even if some degree of 'reduction' is achieved in some domain of science, it obviously does not follow that all research in this area necessarily becomes superfluous thereafter. It can perfectly well continue, generating new information which may or may not be successfully explained in terms of ideas at some more fundamental level or levels of knowledge. Also, separate branches of science may well proceed independently for long periods of time without achieving consilience with knowledge at other levels of the scientific hierarchy. However, if and when such linking of two previously separate domains is achieved, it is usually supposed that progress has been made.[15]

But to return to Comte. One of the most famous features of his work was the well-known law of the three stages, according to which every branch of learning passes successively through three phases:

1. the theological stage;
2. the metaphysical stage; and
3. the positive stage.

His descriptions of these were as follows:

> In the theological state, the human mind, directing its search to the very nature of being, to the first and final causes of all the effects that it beholds, in a word, to absolute knowledge, sees phenomena as products of the direct and continuous action of more or less numerous supernatural agents, whose arbitrary intervention explains all the apparent anomalies of the universe.
>
> In the metaphysical state, which at bottom is a mere modification of the theological, the supernatural agents are replaced by abstract forces, veritable entities (personified abstractions) inherent in the various types of being, and conceived as capable in themselves of engendering all observed phenomena, the explanation of which consists in assigning to each its corresponding entity.
>
> Finally, in the positive state, the human mind, recognising the impossibility of attaining to absolute concepts, gives up the search for the origin and destiny of the universe, and the inner causes of phenomena, and confines itself to the discovery, through reason and observation combined, of the actual laws that govern the succession and similarity of phenomena. The explanation of the facts, now reduced to its real terms, consists in the establishment of a link between various particular phenomena and a few general facts, which diminish in number with the progress of science.[16]

To illustrate this thesis of the historical development of each of the sciences in three stages, it may be helpful to consider a simple example. Suppose, for instance, the phenomenon of lightning is the topic under investigation. According to Comte's model, in the 'theological' stage of the history of meteorology, explanations might be given in terms of the anger of the gods or in terms of the activities of deities of one kind or another. Subsequently, in the 'metaphysical' stage, explanations might be given in terms of some hypothetical non-tangible entity such as 'phlogiston'[17] or by postulating the existence of hypothetical 'nitro-aerial particles'.[18] Finally, in the 'positive' stage of the inquiry, scientists would (and should) be content to provide an accurate statement of the laws and conditions according to which the phenomena of lightning occur. One might, for example, seek to determine the time of year when lightning is most frequent, the usual prevailing barometric conditions, or the particular effects that it is able to produce. But no further inquiry below or beyond the level of phenomena would be necessary or appropriate. The objective of science in its 'positive' stage — the level to which all sciences should aspire — would simply be the exact determination of the laws of nature. 'Theological' and 'metaphysical' modes of explanation should definitely be eschewed, being marks of scientific immaturity.

It will be appreciated that Comte held a very narrow view of the nature of science, and he failed to recognise the important role (as it appears to us) of explanatory entities (albeit fictitious, on occasions) such as phlogiston, genes, atoms, nervous fluids and so on in the successful prosecution of scientific research and the development of scientific theory. And one might well wonder how explanation of facts at one level of the ladder of the sciences might be given in terms of scientific ideas at another level if all that the various levels contained was nothing more than laws.[19] However, although we may think that a 'mature' science is not at all as Comte envisaged it, or would have liked it to have been, it is certainly arguable that the positivistic stance of much of nineteenth-century French philosophy of science was a beneficial antidote to the more excessive forms of speculation that were rife in some circles, such as those associated with German Nature Philosophy.[20] And as we shall see, positivist critics such as Ernst Mach, who were unwilling to allow the speculations of science to proceed further than that which was required for the exact determination of the relations between phenomena (that is, the precise determination of the laws of nature), succeeded in drawing attention to some fundamental problems within the corpus of 'orthodox' science (for example, Newtonian mechanics).

So the austerities of positivistic metascience can — with hindsight — be seen to have been of benefit within science itself to a considerable degree. Besides, while there were many who payed lip-service to the speculative restraints that positivism sought to impose, not so very many actually practised what they preached in this matter, and as a consequence what we might consider to be the advantages of free-ranging speculation and hypothesis formation were not by any means wholly submerged, even at the highest tide of positivist thought. Incidentally, some interesting forms of science were generated under the aegis of the positivist movement. The chemist Wilhelm Ostwald (1853–1932), for example, wrote chemical texts[21] without any mention of the atomic theory, even though it was recognised in his day as being a grand unifying doctrine, extremely useful in the colligation of a great mass of phenomena. One could not 'see or touch' atoms; so they were not allowed any place within Ostwald's system. However, he was able to proceed a remarkable distance without them, purely at the level of phenomena. Those important theoretical entities, atoms, are by no means essential to all forms of chemical doctrine. One may mention here also that the eminently respectable nineteenth-century science of thermodynamics was (and is) positivistic in character. Its laws 'work', no matter whether heat be thought a fluid or a form of motion.

Comte's *Course* was chiefly an exercise in pedagogy rather than epistemology, but it is clear that the positivist science that he described and advocated was based on phenomenalist principles, and hence was grounded in an empiricist theory of knowledge. Comte maintained that science was concerned with the discovery of the *laws* connecting facts, not merely the facts themselves. He did not see this as raising fundamental epistemological problems. So Comte seems to have been largely untouched by Kant's 'Copernican Revolution'.

There are in Comte's text various references to the 'positive method', but apart from the desire to eliminate theology and metaphysics from science, concerning oneself only with the search for laws of nature, it is by no means easy to understand exactly what he meant by it. Indeed, Comte does not seem to have formulated a general theory of method. Nevertheless, there was supposed to be a unity of method for all the sciences. John Stuart Mill, who wrote quite a sympathetic critique of Comte's system[22] (and also gave money to Comte when he was in a particularly severe state of financial distress), drew attention to this deficiency. Mill pointed out that although Comte's work contained a great deal of incidental information about scientific practices, culled chiefly from the study of the history of science, it really had no general account of method to offer, other than the recommendation of observation, experimentation and comparison. According to Comte, in fact, method should be learned by actually practising science or by studying the history of science. But for Mill, this wasn't good enough. For example, he wanted more information than Comte provided on 'proof' — or the testing of scientific ideas. Comte was willing to see the employment of hypotheses in science, but unwilling to regard them as demonstrated if they successfully accounted for observed facts. So, lacking either the method of orthodox hypothetico-deductivism or the codification of inductive method that Mill himself gave in his Canons, it seemed to Mill that Comte's account of scientific method was seriously deficient. Mill[23] wanted Comte to invoke some high-level principle such as that of universal causation as a warrant for a general scientific method. But then, we might say that Mill's complaint was no more than that Comte did not propose the same overall account of science as did Mill himself. Perhaps Comte was right: the only way to learn anything about the methods of science is actually to practise them for oneself.[24]

Science, for Comte, could never know ultimate and absolute causes of the innermost essential natures of things. He seems to have supposed that science could only attain an approximate understanding of reality, but he did not work out in any detail an account of the nature of the relationship that obtains between scientific 'knowledge' and the real world that science purports to describe. However, in recognising that knowledge is dependent on the stage of cultural development that has been reached, Comte was, in effect, recognising a social component to knowledge.

Comte's sociology incorporated areas such as psychology (to which he did $_{not}$ allot a special position in the hierarchy of the sciences),[25] political economy, ethics and philosophy of history. He believed that like the other sciences it had 'static' and 'dynamic' aspects. The study of social statics involved inquiry into the forms of social structures, political and social conventions and constitutions, ethical systems, networks of authority, and so on. Comte was certainly no anarchist; he believed that well-structured social systems were conducive to happiness, and he greatly admired the remarkable social structure which he associated with the influence of the Catholic Church in the Middle Ages.

Yet Comte was also interested in social dynamics and he hoped to discern the laws of change to which society was subject. His law of the three stages was supposedly one such. Also, he believed that militarism corresponded historically to the theological stage, a juristic system to the metaphysical stage, and the industrial system to the positivistic. The progress of society towards the industrial positivist phase was supposedly linked with increased social cohesion and a rise in altruistic behaviour. As has been shown by J.C.Greene,[26] Comte's ideas on both the statics and dynamics of society were strongly influenced by his reading in the biological literature of his day, particularly the studies in comparative anatomy of Georges Cuvier (1769–1832).

In the latter part of his life, following the period of insanity and also the middle-aged infatuation with Clothilde de Vaux,[27] Comte tried to develop his system into a whole new secular 'religion of humanity'. His new 'Church' was to have many of the outward forms of religious expression such as temples, services, holy writings, a calendar of saints and so on; but it was humanity itself that was to be worshipped. The new religion, like the old, was to have its catechism,[28] its patron saints (including such figures as Newton and Galileo, but also men like Dante, Shakespeare or Julius Caesar). Women (*especially* Clothilde de Vaux) were to be deified, and the whole Church should have a chief priest, analogous to the Pope. Comte saw himself in this role.

All this may well strike one as somewhat absurd,[29] though the Chinese, for example, through their ancestor worship, have for long thought it sensible to pay homage to deceased human beings rather than transcendent metaphysical entities. Be that as it may, Comte was by no means entirely unsuccessful in his efforts, and his system was taken up by a not inconsiderable number of followers after his death.[30] It was particularly successful in France and South America, where for example, in Brazil, university curriculums are still influenced by positivist theory. The various humanist movements of today, though small in active membership, very likely represent perhaps the unspoken metaphysical position of the majority of people amongst Western industrial societies; and they have their roots in nineteenth-century positivist thought.

Thus, the desire for a positivist science-based society, free from metaphysics, though now perhaps diminishing, has certainly been a potent force since Comte's day. But, of course, Comte cannot be held personally responsible for all these matters. He was as much a spokesman for his age as a shaper of it — an effect as much as a cause. And even in the field of philosophy of science, the influential positivist movement was only in a remote sense an outcome of his teaching and writing. So what, then, was the significance of Comte for the purposes of our present inquiry? His belief that there could be a unification of the sciences in which the laws of nature could successfully be determined by empirical inquiry was, I suggest, the main consideration. And by eschewing anything more than the search for the laws according to which phenomena are related, Comte sought to wean science from any connection with matters of value (though values returned, one must suppose, in the 'religion of humanity'). Thereby, a positivist, 'scientific' society might be achieved — presumably one that would be happier than the endlessly contentious system it was supposed to supersede.

Mach

Next, instead of pursuing the various channels of subsidiary influence that flowed from Comte's work, and his somewhat bizarre 'religion of humanity', it will be better to look in a little detail at the work of the most important representative of nineteenth-century positivist philosophy of science, Ernst Mach (1838–1916), a man of high reputation as physicist, mathematician and experimental investigator of the sensory processes. In addition, he was a scholarly historian of science, and made important contributions to metascience — with a distinctive phenomenalist epistemology. Indeed, the many-sided character of Mach's work was one of its most impressive features. According to William James, who visited Mach in 1882, he had 'read everything and thought about everything'.[31]

It would be wrong to suggest that Mach was a direct disciple of Comte (though he did on one occasion refer to the theory of three stages as if it were to be taken seriously[32]). Mach's ideas were formed in the austere school of nineteenth-century German experimental physics. But he certainly regarded himself as a positivist philosopher of science, and, as we shall see, his epistemology was phenomenalist in character. In his rejection of hypotheses concerning matters that lay beyond the reach of experimental inquiry Mach certainly was in tune with Comte's positivist program; and his great interest in the history of science stood within the Comtean tradition.

The works of Mach with which we shall be concerned were *The History and Root of the Principle of Conservation of Energy*,[33] *The Science of Mechanics*,[34] *The Analysis of Sensations*,[35] *Popular Scientific Lectures*,[36] and *Knowledge and Error*.[37] In *The Analysis of Sensations*[38] Mach tells us that he first read Kant's *Prolegomena* when he was fifteen — a natural enough action for a studious German youth with philosophical ambitions — and a couple of years later, in reaction to Kant, he had 'a brilliant vision — an abrupt revelation of the constitution and nature of things'.[39] In Mach's own words:

> [T]he superfluity of the role played by "the thing in itself" abruptly dawned upon me. On a bright summer day in the open air, the world with my ego suddenly appeared to me as *one* coherent mass of sensations, only more strongly coherent in the ego.[40]

Thus, like a number of philosophers, Mach as quite a young man had a kind of vision of the essential nature of reality; and he spent much of the rest of his life using this as a guiding principle, working out the details and further implications of the idea that the world and the ego appear as 'one coherent mass of sensations'.

From reading Kant, Mach found himself introduced to the idealism of Berkeley (which Kant discussed in the Appendix to his *Prolegomena*), and, as we shall see, the philosophy of science that Mach subsequently developed can be construed as a refurbished Berkeleyian theory of knowledge. So perhaps we should see Berkeley rather than Comte as one of the chief inspirations of Mach's philosophy. (Actually, some historians of philosophy would be happy to classify Berkeley as a positivist.)

Like some of his philosophical predecessors such as Fichte, Mach rejected the Kantian domain of *noumena* – of 'things in themselves' — as a hypothetical superfluity. Like Berkeley, Mach would have been prepared to say that *'Esse* is *percipi*'; though he did not accept Berkeley's notion of God standing in the wings, so to speak, preventing the degeneration of a thoroughgoing empiricism/sensationism/phenomenalism into a useless solipsism. Mach also rejected any kind of Cartesian innate ideas or apriorism. In Mach's view, all scientific laws and principles rest on empirical foundations, and are certified in no other way than by experience. There are no Kantian mental categories. Rather, as Hume had claimed much earlier, what may appear as innate concepts of the understanding are in fact no more than habits fixed in the mind as a result of past experiences.

So what Mach tried to do was to build up a refurbished empiricist theory of knowledge, based on a consideration of sensations, and their patterns and relations. Then, this epistemology was to serve as an underpinning for a general theory of science. Let us, therefore, look more closely at Mach's sensationism, as it appeared in his *Analysis of Sensations*.

Austerely, Mach described the 'I' or the 'Ego' as: 'that complex of memories, moods, and feelings, joined to a particular body (the human body)'.[41] Also:

> Colors, sounds, temperatures, pressures, spaces, times, and so forth, are connected with one another in manifold ways and with them are associated dispositions of kind, feelings, and volitions. Out of this fabric, that which is relatively more fixed and permanent stands prominently forth, engraves itself on the memory, and expresses itself in language. Relatively greater permanency is exhibited, first by certain complexes of colors, sounds, pressures, and so forth, functionally connected in time and space, which therefore receive special names, and are called bodies.[42]

So, on this view, both the human 'Ego' and bodies are, so far as we can possibly know, nothing more than complexes of *sensations*. This is, indeed, a position of which the good Berkeley would have approved, though certainly it may appear bizarre to one approaching the matter for the first time. The point, of course, is that if we can *only* know sensations then our account of the world is (arguably) best given *only* in terms of sensations.

So Mach tried to give a closer account of complexes of individual elements of sensation (which were taken as given, and not themselves susceptible to further analysis). The elements of the complexes were regarded as simple sensations — such as a patch of blue in the visual field, of such and such an intensity, in such and such a direction, and on some particular occasion.[43] Mach gave labels to various groups of simple sensations as follows:

> *A, B, C,...*, represented a complex of elements of the type normally called 'bodies'.
> *K, L, M,...*, represented a complex of elements of the type normally called 'our own bodies'.
> α, β, γ,..., represented a complex of elements of the type normally called 'minds' (eg, volitions, memories, and so on).
> α, β, γ,..., *K, L, M,...*, is the 'Ego'.

Now, Mach maintained, '*A, B, C,...*,' and 'α, β, γ,..., *K, L, M,...*,' are not independent. For example, as I approach an object it appears to get bigger. '*A, B, C,...,*' is influenced by the location of 'α, β, γ,..., *K, L, M,...,*'. So, said Mach: '[T]he ego can be so extended as ultimately to embrace the entire world. The ego is not sharply marked off, its limits are very indefinite and arbitrarily displaceable.'[44] So really one always had to contemplate a more general complex — 'α, β, γ,..., *A, B, C,..., K, L, M,...*'. And the idea of an independent 'Ego' had to be given up. (We may note here the relationship of this view to Mach's early 'visionary' experience.)

If the foregoing theory of knowledge is accepted, then, according to Mach, scientific research should be concerned with the determination of the modes of connection, or the relations, between all the elements of sensation. This was, of course, more or less the positivist program of Comte; but Mach had now given a rather more careful epistemological grounding. It followed that physics and psychology were not distinct fields of inquiry. Mach's philosophy was essentially *monistic*, and was a very thoroughgoing form of sensationism/phenomenalism. Any Kantian realm of *noumena* was, as has been said, totally rejected, for the objective of the positivist was to purge epistemology (and science) of all metaphysical elements.

Turning, then, to a discussion of actual scientific investigations, Mach introduced what may be seen — with hindsight — to have some similarities to the doctrine of operationalism.[45] He claimed that all scientific *concepts* were ultimately traceable back to particular sensations:[46]

> I maintain that every physical concept means nothing but a certain definite kind of connexion of the sensory elements...*A B C*... The elements...are the simplest materials out of which the physical, and also the psychological, world is built up.[47]

And this meshed with the positivist program that the aim and object of scientific research should be the exact determination of the laws of nature — the regularities in the empirically determinable relationships between phenomena.

But a good deal more than this flowed from Mach's epistemology. In his *History and Root of the Principle of the Conservation of Energy*, he wrote:

> What we represent to ourselves behind the appearances exist *only* in our understanding ... [having] only the value of a *memoria technica* or formula whose form, because it is arbitrary and irrelevant, varies very easily with the standpoint of our culture.[48]

For this reason, Mach was always opposed to such important nineteenth-century ideas as the atomic theory. The chemists' atom seemed to play a role analogous to the Kantian 'thing in itself'. One could not see or touch atoms, and it was impossible to determine direct phenomenal relationships between them.[49] But the atomic theory was enormously helpful in coordinating a mass of empirical data. For example, it is much easier to remember the formulae of molecules than their chemical composition by weight. It is this latter, however, that is determined (more or less directly) by empirical means. So Mach was (perhaps reluctantly) willing to receive atoms as '*memoria technica*',[50] which might have appeal to those brought up in the Western intellectual tradition; but they were certainly not to be granted any real status. They might, to be sure, have heuristic or didactic value. But *that* didn't make them real.

Another name that can be and has been applied to Mach's philosophy of science is '*instrumentalism*'. If the atoms were 'useful' in science in one way or another, perhaps as aids to the prediction of phenomena, but were not necessarily thought to be real just because of this practical utility, then they might be regarded as 'instruments' of research — but nothing more. This philosophy of science, which accepts that certain hypotheses may be useful without necessarily being true, was, of course, by no means new in the work of Mach. Indeed, it was really equivalent to the mediaeval doctrine of 'saving the appearances', which we have already mentioned briefly in an earlier chapter.[51] If a theory accounted successfully for the observations, then that was fine; but according to the instrumentalist no ontological claims should be made as a result of this success.

One might think, therefore, that the 'instrumentalist' would be quite unable to *explain* phenomena. For example, if the atomic theory were nothing but a '*memoria technica*' then to use the theory would not *explain* why particular chemical compounds have their own particular chemical compositions by weight. These would be no more than 'brute facts' for an instrumentalist chemist — even though he might use the atomic theory for convenience, or as a memory aid. Mach, of course, willingly subscribed to this view. He held that scientific explanation is nothing more than the economical description of phenomena by means of laws.

But if this is the appropriate way to envisage scientific theories, then it is hard to see what *meaning* can be attached to them. As calculating devices, really explaining nothing, they seem to be devoid of content — quite meaningless. They say nothing about the real world, only about regularities in our subjective sensations. Perhaps surprisingly, Mach seems to have baulked at this point. At least, his thought took a rather unexpected turn. He drew on ideas derived from the popular evolutionary theory of his day — the Darwinian theory of evolution by natural selection.[53] Thoughts or scientific hypotheses, Mach supposed, might or might not be well adapted to 'facts'. Those that were so survived; those that were not were eliminated. Moreover, some thoughts might be well or ill adapted to other thoughts; and when they are well adapted we have a successful theory.[54] Thus, it was by this curious and perhaps unexpected route that Mach felt able to accept the role commonly assumed for hypotheses and theories in science.

However, I suggest that Mach was really skating round the ontological issues. In effect, it seems to me, in talking about a kind of biological adaptation of thoughts to 'facts', Mach was admitting that the world *was* made up of more than sensations and the relations between them. There were things as well as thoughts. Be this as it may, we should note this significant departure in Mach's thought towards 'biologism'. There were, in fact, as R.S.Cohen has clearly shown,[55] *two* Machs — one the austere,

hard-line phenomenalist, the other a more commonsense type of philosopher, who was willing to admit hypotheses and theories. For the
second Mach, these could be grounded in the 'biological' survival of
those ideas, thus somehow conforming with a real world lying behind the
world of sensations.[56] So by introducing his 'biologism' in this somewhat
unexpected way, Mach managed to make his description of science fit
more nearly to that which is commonly practised, and the way it is more
commonly understood.

We may now turn to look briefly at Mach's celebrated *Science of
Mechanics* (1883), and particularly his criticisms of the foundations of
Newtonian mechanics. His criticisms may be seen to be related to the
general phenomenalist epistemology that has been outlined above.

Mach's treatise was partly historical and partly critical. Here we need
only concern ourselves with the critical aspect, though it should be
emphasised that Mach's critique of Newton emerged from a very detailed
examination of the history of physics, as well as from the epistemological
position that has been described. Also, it appears that Mach's critique was
following that previously put forward by Berkeley, whom Mach had read
with care. However, this debt was not made clear in *The Science of
Mechanics*, for Mach did not wish his brand of phenomenalism to be
confused with Berkeley's idealism, which saw fit to place a strong
reliance on God as an essential feature of both its ontology and its epistemology.

It is well known that Newton defined mass, in a somewhat circular
fashion, as the quantity of matter in a body.[57] But Mach found this unsatisfactory. Newton's matter was a bit like an Aristotelian 'substance' or a
Kantian 'thing in itself'. So Mach sought to provide an understanding of
the concept of mass, wholly in terms of experienceable phenomena, just
as the positivist program would require. The approach was, in principle,
in terms of the 'observable' — acceleration. So, if two bodies, A and B,
induced in one another the mutual accelerations, $-a$ and a', then their
relative masses, said Mach, were a'/a.[58]

Figure 21

That is:

$\text{Mass}_A/\text{Mass}_B = a'/a$

If the accelerations produced were 'equal and opposite', then the two
masses were held to be the same. And if one of the objects was arbitrarily
assigned a mass of 1 unit, then the mass of the other could, in principle,
be found by determining experimentally the observed gravitationally-
induced accelerations. In all this, insisted Mach, there was supposedly no

theory involved (beyond, we might add, Newton's own stipulation — in his Second Law of Motion — that forces were measurable by accelerations). The notion of 'quantity of matter' had been excised from the problem, and a wholly empirical method was available for comparing masses. One had a *relational* definition of mass. It was an observable fact that the mass ratios of bodies did not vary with their chemical, electrical, magnetic, or whatever, state.[59]

Thus, by means of his conceptual analysis of the phenomenon of mass or inertia, Mach sought to eliminate some residual metaphysical obscurities within the Newtonian system of mechanics. It may be queried, however, whether he *really* succeeded in this aim. It is not possible in practice to set up the hypothetical situation depicted in the preceding figure, for all extraneous forces cannot be eliminated. Thus Mach was really indulging in speculation (or a 'thought experiment') as to what might be expected to occur in an ideal situation, rather than describing actual empirical possibilities. So at this point, his argument was at odds with his radical empiricism/phenomenalism. In practice, of course, masses may be compared with a beam balance, and mutually induced accelerations are not determined as such. Mach was, perhaps, justified in drawing attention to a kind of mystery at the heart of Newtonian mechanics. But if Newton regarded the characteristics of massive bodies as in some measure unexplained — like axioms in Euclidean geometry — the same might be said of forces for Mach. Were not they as problematic as Newtonian masses?

Mach's *Science of Mechanics* is also celebrated for its examination of Newton's doctrines of absolute space and absolute time, and Newton's so-called 'bucket experiment', according to which the absolute rotational motion of a liquid might be recognised by the curvature of its surface (when rotating in a 'bucket'). As we may readily imagine, Mach took exception to the notions of absolute space and time, maintaining that they were quite divorced from any kind of phenomenal experience — that is, they were quite impossible to observe or detect in any way whatsoever. And it was easy enough for him to show that Newton made no practical use of these notions in the *Principia*, the bucket experiment notwithstanding. But did Newton's 'thought experiment'[60] with the bucket really demonstrate *absolute* rotations, and hence absolute motion and the meaningfulness of absolute space?

Mach thought not. He argued that one could say nothing about what might or might not happen in a rotating bucket of water in the absence of all other objects in the universe, or if the sides of the buckets were several leagues thick.[61] In any case, the experiment as described by Newton, did not eliminate all possible empirical reference frames. For example, although Newton showed well enough that the water surface curved when it was rotating — no matter what the liquid was doing with respect to the sides of the bucket — this did not necessarily mean that it was rotating in an absolute sense. Other reference frames had not been eliminated. For example, for all Newton could show to the contrary, it might be the motion relative to the fixed stars or all the other objects in the cosmos — not any kind of absolute motion — that was responsible for the curvature of the water. Thus, arising from his positivist/ phenomenalist epistemology, Mach came to the conclusion that there was another fundamental theoretical problem right at the centre of Newtonian mechanics. But, as we have seen,[62] Berkeley had long before said something rather similar. And *in practice* Newtonian mechanics made no use of the concept of absolute space. So Mach's criticism only touched Newton's metaphysics, leaving the grand structure of Newtonian mechanics undisturbed for all practical purposes.

The discussions of absolute space and time, as initiated by Newton and taken up by Mach, involved the employment of what are now often referred to as 'thought experiments', or *'Gedankenexperimenten'*. It may be useful here to make a few remarks about Mach's ideas on these.[63] Mach was one of the first to give them serious consideration, but since his day their importance has been increasingly recognised and they have come to play an important role in science, particularly in discussions of relativity theory and quantum theory.

As a practising scientist, Mach was well aware that one doesn't just go into the laboratory and carry out an experiment in a systematic way. It is the thought that goes into the problem in advance, both in the way of conceptual analysis and clarification and in the design of a suitable experimental procedure, that is every bit as important as the actual performance of the experimental techniques. Today, when we talk of 'thought experiments' we often think of attempted conceptual clarifications sought by asking such questions as: 'What could I see if I were travelling at the speed of light?'; and so on. It is true, indeed, that conceptual clarification may sometimes be gained by asking questions of this kind, so that theoretical progress may be made as a result. Such 'experiments' are, after all, cheap to perform and may be particularly useful in pedagogy — points that Mach specifically emphasised.

But from Mach's phenomenalist/positivist perspective, 'thought experiments' were of prime, indeed essential, importance, not merely serving as short-cuts or aids to teaching. He maintained that the scientist has to form abstract and ideal conceptions out of the vast mass of information that he receives through his senses. By forming such conceptions, and asking himself mental questions about them, the investigator is then able to formulate questions, to pose to nature, so to speak. That is, by means of thought experiments he is able to have a basis for setting up his experimental arrangement. So wrote Mach, by way of example:

> Galileo asks, in his ingenious and primitive way: is v proportional to s, [or] is v proportional to t? Galileo, thus, *gropes* his way along synthetically,[64] but reaches his goal nevertheless. Systematic, routine methods are the final outcome of research, and do not stand perfectly developed at the disposal of genius in the first steps it takes.[65]

It is clear, then, that 'thought experiments' for Mach had more than a peripheral role. Without them, scientists would be unable to proceed at all, if (as Mach strongly believed) the phenomenalist epistemology were correct. So Mach's meaning of 'thought experiment' was a little wider than that which is commonly given to it today.

Some American pragmatists

I want now to make what may seem as something of a detour within this chapter, in order to bring into consideration some of the ideas on philosophy of science of the members of the so-called pragmatist movement of American philosophy.[66] Important representatives of this school from the point of view of philosophy of science were the philosophers Chauncey Wright (1830–1914), John Fiske (1842–1901) and Charles Peirce (1839–1914), the psychologist William James (1842–1910), and the educationist John Dewey (1859–1952). It is true that some pragmatist philosophers specifically chose to oppose their ideas to those of the positivist school.[67] Nevertheless, it will not be wholly inappropriate to consider the pragmatist movement at this juncture. We shall look particularly at the work of Peirce, who though he would not himself have accepted the positivist label, has been designated as such by at least one commentator.[68] And when we come to Dewey, an instrumentalist *par excellence*, we shall find that the ideas of the pragmatist movement had begun to merge again with those of a positivist character. Also, in defence of a consideration of the pragmatist movement at this juncture, one may emphasise that it was chiefly a nineteenth-century phenomenon, which needs to be treated at this stage of our exposition of nineteenth-century philosophical thought. Moreover, Peirce in particular had some most interesting things to say about methods of scientific discovery, which form a vitally important constituent of any book on the history of arch building.

The common feature of the pragmatist school was the belief that the truth and/or meaning of ideas was sufficiently given by their practical consequences. In a *very* simple way, then, one might say that the truth of certain engineering principles used in the building of bridges might be ascertained by finding whether or not the bridges constructed according to those principles stood up firmly and carried their loads, or collapsed with ignomony when first used. This is pragmatic doctrine in its simplest and crudest form, yet it may reasonably be understood as the common way of thinking of the movement as a whole.

So far as the emphasis on practical, empirical testing of ideas was concerned, the pragmatists certainly held common ground with their positivist contemporaries. But in relation to the general features of positivist thought, as given by Kolakowsky and mentioned above, we cannot claim that any of them — except perhaps a belief in a general unity of scientific method — were part of the pragmatist position. Nevertheless, as I say, we shall find a substantial amount of common ground with positivist philosophy, and in terms of chronology it is appropriate to deal with the pragmatists here.

It should perhaps be said before we go any further that the pragmatists' criterion of truth cannot be deemed adequate, even though it might seem eminently satisfactory to the practical man. Thus, in the history of astronomy, Ptolemy's geocentric model of the cosmos *worked* pretty well, and could be used successfully for navigation purposes. But, as we can see with hindsight, this did not mean that the Ptolemaic cosmology was *true*. Faulty theories may often describe phenomena with great success.

Such elementary considerations immediately suggest that hypothetico-deductivism, which we have mentioned on a considerable number of occasions in our preceding pages, cannot be regarded as the be-all and end-all of scientific method and the search for a satisfactory epistemology of science. For one may often formulate hypotheses, deduce their consequences, test these consequences empirically and find them 'confirmed'[69] — indeed go up and down our arch in a most careful and proper manner — yet still have no guarantee of the truth of the experimentally (pragmatically) confirmed hypotheses.

However, it may be that truth, certainty, or whatever, are expectations that are much too stringent. Perhaps pragmatic success, as a result of one's scientific inquiries, is all that may reasonably be asked for. And it may, in the end, be all that science, pure and applied, can produce for us. This would seem to lead us to a relativist view of truth — a view that is reinforced when we recall the extent to which different people may have different perceptions as to what is useful or successful.

So, we may perhaps have provisional doubts about pragmatist philosophy as a guide to the search for truth. Nevertheless, we find on closer examination that it was an interesting and subtle instrument, and that Peirce, in particular, made contributions of lasting significance. Particularly important for our purposes are his ideas on method — or, as we might say, on 'arch-climbing'.

Peirce's ideas on a 'logic of discovery' were evolved over a considerable period, and it is beyond the scope of this book to treat the historical development of his thought in detail.[70] We shall, therefore, merely seek to give a general view of his account of hypothesis construction, without tracing all the twists and turns of his thinking.

Throughout his philosophical work, Peirce was greatly interested in logic, and it was, therefore, within this general framework of interest that he sought to construct a 'logic of discovery'. In 1878, for example, he sought to clarify the distinctions between deduction, induction, and the process of hypothesis formulation, by means of the following simple examples:[71]

(i) *Deduction*:
     Rule — All the beans from this bag are white.
     Case[72] — These beans are from this bag.
     Therefore Result — These beans are white.

(ii) *Induction*:
     Case — These beans are from this bag.
     Result — These beans are white.
     Therefore Rule — All the beans from this bag are white.

(iii) *Hypothesis*:
     Rule — All the beans from this bag are white.
     Result — These beans are white.
     Therefore Case — These beans are from this bag.

So, by shuffling terms around in this way, Peirce could readily illustrate three distinct forms of reasoning or 'inference' that were commonly employed in science. It was the third one that particularly interested him. He wished to know what kind of 'logic' — what kind of reasoning — was involved in the process of formulating hypotheses.

The term that Peirce coined for the process which occurs when a hypothesis is formulated, in order to explain some puzzling phenomenon, was *'abduction'*.[73] This may be represented by the following simple schema:

> The surprising fact, *C*, is observed;
> But if *A* were true, *C* would be a matter of course;
> Hence, there is reason to suspect that *A* is true.[74]

In fact, Peirce believed, the 'logic' of an abductive procedure could not be isolated from the processes of testing the hypothesis that is formulated in the abductive movement.[75] For at the moment of formulating an hypothesis, one should necessarily consider how it would explain that which needed to be explained.[76] It was not just a matter of plucking any old hypothesis out of the air, so to speak. So, having (somehow) invented a hypothesis, its consequences should be *deduced*,[77] and then *tested*.[78] This last step was, perhaps unexpectedly, called *induction* by Peirce. All three steps together constituted 'scientific method'.

Now out of these three stages, the one that particularly interests us is obviously abduction. How could there be a 'logic' of scientific discovery? Could it be more than shrewd guesswork? Well, contrary perhaps to what one might expect, Peirce considered that there was a sort of ethical or moral component of logic,[79] and ultimately, perhaps, an aesthetic dimension. Logic was 'the theory of right reasoning, of what reasoning ought to be, not of what it is'.[80] Logic was 'not the science of how we *do* think; but... how we *ought* to think... in order to think what is true'.[81] The unexpected connection between hypothesis construction (abduction), practical consequences (pragmatism), and a kind of 'moral' dimension, is brought out in the following single sentence of Peirce:

> [T]he only way to discover the principles upon which anything ought to be constructed is to consider what is to be done with the constructed thing after it is constructed.[82]

So hypotheses should[83] be formulated that could explain phenomena satisfactorily, were capable of experimental testing (a point with which the positivists would have agreed), and which could be investigated as *economically* as possible. Here questions of both mental economy (as in the employment of Ockham's razor) and economy in terms of materials, time and other costs were relevant. Evidently, Peirce was giving sanction to the significance of socio-economic factors in the very cognitive structure of scientific knowledge, a position which the positivists such as Mach would have found quite unacceptable. Their concern was to purge science of all such supposedly irrelevant factors. Nevertheless, positivists and pragmatists could agree on the absolute necessity of hypotheses being capable of experimental testing.

But Peirce was more broad-minded in his acceptance of hypotheses than the orthodox positivists. We recall that Comte and Mach were only interested in hypotheses that could be tested by direct observation; and on that basis entities such as atoms and so on were disallowed. For Peirce, however, they were more than welcome if they allowed scientific explanation of phenomena, and were capable of being tested — albeit indirectly.

It is interesting to consider the pragmatic view of *truth* in all this. The truth of hypotheses was to be gauged and understood by their practical consequences; and if no distinction could be reached at that level then two competing hypotheses would have to be judged equally true. For example, in the sixteenth century the geocentric theory of Ptolemy and the heliocentric theory of Copernicus might have to be judged equally true, for they were equally efficaceous for practical purposes such as navigation, and also so far as all observational tests were able to show at that time. Today — and indeed since the invention of the telescope — the practical consequences of the two theories are not the same. So one of them is judged true and the other false. Yet is not this somewhat paradoxical? In the sixteenth century the two competing theories would — according to the pragmatic criterion of truth — have to be judged equally true. This seems strange, given that the two theories were diametrically opposed. It is not surprising, then, that pragmatism could readily find itself 'degenerating' into instrumentalism; and this actually happened in the case of Dewey (though he is commonly regarded as a member of the pragmatist tradition and in any case would not himself have regarded the process as degenerative). Also, as we have seen, there was overlap between instrumentalism and positivism. So these various 'isms' interlocked in a number of intricate and obscure ways, and often it is difficult to characterise a particular figure as positivist, pragmatist or instrumentalist in some definite and clear-cut way. For the positions themselves overlap in some degree.

But returning to Peirce and considering the question of the truth of hypotheses, it is interesting to note that, like Mach, he offered a kind of biologistic approach to the problem. Through the long process of human evolution, men had supposedly acquired a special aptitude for selecting true (or truer) hypotheses from among the infinite number that were always potentially available to explain any puzzling phenomenenon.[84] So rather like Aristotle and St Thomas Aquinas long before, though for quite different reasons, Peirce supposed that there was a kind of natural affinity between man and nature, such that one could have confidence in the procedures of induction (Aristotle) or abduction (Peirce). Unfortunately, however, it does not seem (to me) that such confidence has any adequate or satisfactory basis. The natural selection of human ideas has yielded ghosts and demons, phlogiston and the aether, just as much as the double-helix DNA model or the discovery of Neptune by the 'Method of Residues'. At best, one might have the hope that pragmatic success leads one *towards* the truth (that is, I suppose, towards a correspondence of some kind between ideas and actual states-of-affairs, or the actual natures of things). But it does not, I suggest, furnish a satisfactory criterion of truth in a certain and unambiguous manner, such as the metascientist might hope to discover, display or demonstrate.

Peirce, I think, hoped to reveal — through his examination of abduction — a *logic* of scientific inquiry, or of hypothesis formulation. But this he failed to do.[85] At best, he was able to offer some reasons as to *why* some hypotheses should be preferred over others. He was very likely giving a satisfactory statement of the *form* of argumentation used when hypotheses are propounded. But this form stubbornly refused to be deductive in character. It still seemed to require flashes of insight, wit, genius, or whatever. Sherlock Holmes had this knack; Dr Watson did not. And probably no amount of work at the metascientific level would ever give it to him! Peirce's description of abduction really threw very little light on the problem of understanding the nature of scientific creativity when new and fruitful hypotheses are formed.

Peirce's account of induction is particularly interesting, for he believed that he could give a satisfactory account of it in accordance with pragmatic doctrine. The results of inductive generalisations could, he suggested, be accepted if the practical consequences flowing from them were favourable. This was, he claimed, what was constantly happening in science, for *self-correcting* tests were used. For example, if a man is constructing a table of mortality on the basis of some census returns he will soon discover that certain false returns have been sent in, for some young men like to overestimate their age, while old men like to underestimate it. The statistician will find, for example, that the number of men who are given as exactly twenty-one is in excess of those that are exactly twenty. He must therefore correct the figures to allow for the distortion of the data. In other words, a carefully conducted inductive procedure can provide information whereby its own results may be corrected during the course of the investigation.[86] In this way, the methods of science may yield pragmatic success. Peirce, consequently, was a 'fallibilist' in relation to scientific knowledge, not a sceptic. He thought knowledge evolved, with 'well-adapted' ideas surviving. For him, induction was an 'on-going' process, not a single inference.

The developmental and instrumental character of knowledge was also emphasised by William James, who wrote:

> True ideas are those that we can assimilate, validate, corroborate and verify. False ideas are those that we can not. That is the practical difference it makes to us to have true ideas; that, therefore, is the meaning of truth for it is all that truth is known-as.... The truth of an idea is not a stagnant property inherent in it. Truth *happens* to an idea. It *becomes* true, is made true by events. Its verity is in fact an event, a process: the process namely of its verifying itself, its veri-*fication*. Its validity is its process of valid-*ation*.[87]

This is an attractive account of the pragmatic theory of truth. But while possibly meshing with the view that some scientists may hold of the nature of truth, it probably does not seem philosophically satisfactory. The standard problem for the epistemologist is knowing when one has achieved a satisfactory corroboration or verification of ideas. And the 'pragmatic' test for correspondence between ideas and states-of-affairs doesn't allow them to be lined up 'one-to-one' and viewed from a neutral, independent vantage-point. Truth and practical success are seemingly two different kinds of animal. It may be, of course, that truth is not something that science can produce in an unequivocal manner, but it would seem that the pragmatic conception of truth involves a change in meaning of the term that can scarcely be tolerated.

But this criticism would not necessarily have been thought serious to the pragmatist philosopher, certainly not to John Dewey, the 'philosopher of action',[88] who was in fact seeking a redefinition of truth, which he held to be no more than pragmatically-based 'warranted assertability' rather than correspondence between states-of-affairs and ideas about those states. We have not space here to say much about the work of Dewey, despite its enormous range and extent, and its very considerable influence in America and, to a lesser extent, elsewhere. But a few words will be said about a limited section of his work, in order to give some idea of its empiricist/pragmatic/instrumentalist/positivist character.

Dewey had no sympathy with the sensationalism of one such as Mach, with the reduction of objects to nothing more than phenomenal appearances. For Dewey, the everyday objects of common experience had to be conceded philosophical primacy. A philosopher's account of the world had to start from the assumption of the real existence of *things*, not phenomena.[89]

So thinking, for Dewey, was concerned with problems encountered by man in his day-to-day confrontation with objects. To illustrate this, he described in a well-known little work, *How We Think*, how one might be walking along a road and encounter a branching of the ways.[90] This would immediately create a problem, and hence a stimulus to thought and action. The traveller's thoughts would be aimed at the discovery of facts that would relieve his indecision, and serve his purposes, whatever they might be. Although this was a very simple case, it provided for Dewey a model of the whole way in which experience, thinking, and action were dynamically interrelated. He deplored the traditional social distinction between theory and practice, between thinkers and doers.[91]

Dewey's emphasis on action quickly led him from pragmatism to instrumentalism:

> [A] method of action, a mode of response, intended to produce a certain result... is... uncertain till tested by its results.... [N]otions, theories, systems, no matter how elaborate and self-consistent they are, must be regarded as hypotheses. To perceive this fact is to abolish rigid dogmas from the world. It is to recognize that conceptions, theories and systems of thought are always open to development through use.... They · are tools. As in the case of all tools, their value resides not in themselves but in their capacity to work shown in the consequences of their use.[92]

If, then, theories might be thought of analogically as tools for particular purposes, the 'instrumental' character of thinking was at once revealed. Thought was analogous to an instrument (or tool), which might be well or ill fitted to the function that it was to serve: it depended upon the purposes in view. For example, different systems of classification would be developed according to the particular needs and intentions of different taxonomists. So for this reason Dewey expressed agreement[93] with the nominalist view of language that had been expressed by Locke in the seventeenth century. 'Real essences' could not be revealed by linguistic definitions.

Quickly, then, Dewey was carried to the view that 'truth' was identifiable with pragmatic 'success': 'Confirmation, corroboration, verification lie in works, consequences. Handsome is that handsome does. By their fruits ye shall *know* them.'[94] Thus pragmatism immediately led Dewey to an underscoring of the social conception of knowledge. For what *counts* as success, within science, or any other human activity, is evidently determined by some form of social system — by some mechanism of social control.

So Dewey could perceive a social and moral dimension to knowledge, and he repudiated the hypocritical cant that (he believed) passed for much of moral philosophy (or moralising). He was a utilitarian. He was prepared to see economic well-being as an end in itself, rather than as the means to some end.[95] But this meant that *his* notion of what was 'good' might not coincide with yours or mine. In fact, like most Americans of the twentieth century, he saw 'growth itself... [as] the only moral 'end'.'[96] And this value judgment, according to the epistemology that we have just outlined, would provide the final arbiter (for Dewey, at any rate) of the 'truth' of scientific knowledge. It will be seen that the position to which an apparently commonsense, no-nonsense, philosophy might lead one was really rather extraordinary. We shall not pursue Dewey, his pragmatism, and his instrumentalism further at this juncture. But attention should be drawn to the enormous influence he has exerted on American life and culture in the twentieth century, through his ideas being taken up so enthusiastically by the American education system. Readers may care to reflect on the extent to which American political actions of the past fifty years have been determined by 'pragmatic' considerations, and the extent to which such considerations have furnished a satisfactory basis for political and social action. In the case of Dewey at least, we do find that philosophers can influence public affairs in no small measure.

Poincaré

Another version of nineteenth-century positivism, closely related to instrumentalism, was the position known as conventionalism. The best-known example of this is to be found in the metascientific work of the notable French mathematician, Jules Henri Poincaré (1854–1912). A professor in mathematics and physics at the University of Paris, Poincaré was a prolific writer and Member (later President) of the Académie des Sciences and the Académie Française. His writings on philosophy of science include *Science and Hypothesis* (1903),[97] *The Value of Science* (1904),[98] and *Science and Method* (1908).[99]

Poincaré's influence on the development of philosophy of science was considerable, as may be gauged by the following remark of the positivist physicist–philosopher and writer, Philipp Frank (1884–1966):

According to Mach the general principles of science are abbreviated economical descriptions of observed facts; according to Poincaré they are free creations of the human mind which do not tell anything about observed facts. The attempt to integrate the two concepts into one coherent system was the origin of what was later called logical empiricism.[100]

But on reading this one might immediately be puzzled as to how scientific principles could be 'free creations of the human mind'. Did Frank really mean that, according to Poincaré, a scientific principle was nothing but an invention of the mind, with no empirical input whatsoever? Was this, in fact, the conventionalist view as espoused by Poincaré, and if so, how could it be tenable? To answer such questions, we must examine Poincaré's views in a little detail.

Poincaré's account of science was considerably influenced by the fact that he was a mathematician, for he was greatly impressed by the discovery (or invention) of the non-Euclidean geometries of G.F.B. Riemann (1826–1866) and N.I. Lobatschewsky (1792–1856).[101] The work of these two mathematicians seemed to show that the apparent truth and necessity of Euclidean geometry could *not* be ascribed somehow to the structure of the mind, as Kant had sought to do. For it was evident that the mathematician was well able to develop new geometries, quite different from the standard version of Euclid. And the new systems could be every bit as logically coherent as the ancient system.

So Poincaré came to the conclusion that the axioms of geometry, even Euclidean geometry, were not synthetic *a priori* truths. They were not necessarily true in the philosophical sense of necessary (though with the suggestion that they were human creations Poincaré's position might loosely be described as neo-Kantian). And contrary to what Mill had asserted, it now seemed clear that geometrical axioms were not experimental or empirical truths. For the geometry of the pure mathematician deals with exact or ideal situations, such as are never realised in the world of experience. So Poincaré proposed (and this was quite plausible so far as pure geometry was concerned) that geometrical axioms were *conventions* or *definitions*, and consequently were simply not open to empirical falsification.[102] To be sure, they might (or might not) be useful in saying something about how the world actually happens to be, but in themselves they could not be said to be true or false. The axioms of Euclid *might* have been suggested in the first instance in various empirical ways.[103] And Euclidean geometry *might* offer a convenient or useful means for finding our way around the world. But this did not mean that the truths of Euclidean geometry, as a mathematical axiom system, held good because the world, or space, was Euclidean. Rather, one could arbitrarily choose particular sets of geometrical axioms, and one or other of them might happen to fit (or could be modelled by) the real world. There was apparently a freedom of choice for geometrical axioms that Kant would not have recognised.

All this might readily be granted so far as pure mathematics was concerned. After all, pure mathematics doesn't lay claim to be a science or to discover the laws of nature. But Poincaré went further, suggesting that science also, as well as mathematics, had important 'conventional' aspects. Consider, for example, Newton's three laws of motion, or his inverse square law of gravitation.[104] Newton claimed that his three laws were inductions from experience. This might have been so, historically speaking, but Poincaré did not regard the historical truth or otherwise of this suggestion as the point at issue. In his view, no matter what might once have been the case, they certainly did not have the status of empirical generalisations any longer. They had become true by *convention*.[105] Thus, no physicist or astronomer would be willing to accept empirical evidence contrary to Newton's laws. If, for example, a situation were found in which a body appeared to move in such a way that it contravened Newton's laws, it would be usual to postulate some hidden force (perhaps due to an unobserved mass) that could be held responsible for the anomaly.[106] One would *not* suppose that Newton's laws were disproven by the wayward observations. What the laws do, rather, is 'package' a large quantity of empirical information. When suitable equations, derived from the fundamental Newtonian axioms, and suitable 'boundary conditions' are given (eg Earth/Sun distance), then the motions of large numbers of objects in the heavens can be predicted and understood. Thus much information is 'contained within' (or 'classified' by) the Newtonian axioms and equations. Accordingly, Poincaré likened science to a library to which books were constantly being added. Experimenters constantly added new books to the collection. It was the function of the theoreticians to draw up the catalogue (or derive suitable equations to describe the 'behaviour' of the books).[107]

Now library catalogues can be drawn up well or ill; they may be convenient or inconvenient to use. But one is straining language to say that a method of classification, such as is employed in a library catalogue, is 'true' or 'false'.[108] By reason of such analogies, the conventionalist or instrumentalist view of science has commonly been represented as one in which scientific theories serve as useful aids to the classification ('packaging') of phenomena or data, but lack any real explanatory function, and do not purport to explain states-of-affairs in the 'real world'.

But one should not make a caricature of Poincaré's position: he did not himself wish to push his views too far. In fact, despite what one might gather from the quotation of Frank given above, Poincaré certainly did not claim that science was wholly made up of conventions.[109]

Such a view was expressed by the philosopher-mathematician Edouard Le Roy (1870–1950), but Poincaré reacted against this in an important clarification of his views, in *The Value of Science*.[110] He referred to Le Roy's doctrine by the term 'nominalism', thereby straining further this generally rather overworked word. Said Poincaré of Le Roy's position:

> Science consists only of conventions, and to this circumstance only does it owe its apparent certitude; the facts of science and, *a fortiori*, its laws are the artificial work of the scientist; science therefore can teach us nothing of the truth; it can only serve us as a rule of action.[111]

Le Roy's position, therefore, was one of 'extreme' conventionalism, and from this passage we can see how it was related in a general way to instrumentalism and pragmatism (note the word 'action'). Thus could one arrive at different brands of positivism from different starting points.

Nevertheless, despite his repudiation of the extreme conventionalism ('nominalism') of Le Roy, rejecting the view that all scientific knowledge is conventional, Poincaré does seem to have retained his stance so far as the major theoretical principles of the sciences were concerned. He imagined a situation in which some stars were found which seemed not to obey Newton's law of gravitation exactly. In such a situation, he supposed, one might retain the inverse-square principle, but look for some additional force at work, pulling the stars out of their expected lines of motion.[112] Thereby, the principle would retain its conventional character, and would not be subject to falsification. All testing would be directed towards the possible 'disturbing' causes accounting for the anomalous motions, the inverse-square law remaining sacrosanct.

In such a situation, the high-level principle (that is, the inverse square law) could no longer be regarded as either 'true' or 'false'. It would merely be convenient, simple, or economical — or the reverse. In the history of science, then, high-level principles would never be given up owing to clashes with experimental results, but merely because some more useful or convenient ones were found. This would mean that two quite contradictory theories could be employed without difficulty — for example the theories of Ptolemy and Copernicus, or the wave and corpuscular theories of light. This would be perfectly legitimate, on the conventionalist view, for according to this metascientific doctrine science does not claim that its theories represent the truth, in a realist sense.

But this was not quite the end of the matter for Poincaré. While believing that science was essentially a 'taxonomic' enterprise, as we have seen in the library simile mentioned above, he also supposed that science was a study of a system of relations. And while he did not suppose that science could penetrate to the essence of things, being 'before all a classification', he did believe that it could display the real and true relations between things. Thereby, he aligned himself with the positivism of Mach, without becoming involved in all the phenomenalist epistemology of that writer. In fact, Poincaré was not a thoroughgoing phenomenalist or sensationalist. He did not wish to suggest that there was no reality other than that of the phenomenal or relational level. There could be (or is) a sub-phenomenal level, but one could not obtain objective knowledge of it by science. So theories didn't truly explain phenomena. They were devices that enabled phenomena to be connected and predictions to be made. But they did not describe reality. Theories might change, though the relations between phenomena would stay the same.[113]

What, then, are we left with when we consider Frank's statement that the principles of science were, for Poincaré, 'free creations of the human mind which do not tell anything about observed facts'? Poincaré's own words were: 'all the scientist creates in a fact is *the language in which he enunciates it*';[114] and 'facts are facts, and *if it happens that they satisfy a prediction, this is not an effect of our free activity*'.[115] Presumably, then, stubborn 'facts' *did* play a role in science for Poincaré, and were not to be invented at the scientist's whim. Nevertheless, it cannot be denied that Poincaré held that high-level scientific *principles* were beyond attack by recalcitrant facts that might turn up in the laboratory. There was some element of caricature in Frank's presentation of the case, but it was by no means wholly a misrepresentation.

We should also notice Poincaré's interesting contributions to the literature of arch construction, though he did not, of course, express himself in such terms. As might be expected from the author of a work entitled *Science and Hypothesis*, Poincaré believed in the importance of the role of hypotheses in science, and he adopted what one may reasonably call the hypothetico-deductive methodology as a general account of the 'shape' of scientific inquiry. But he did not attempt to say much about the 'logic' of the process of hypothesis formation, as Peirce had sought to do in his account of abduction. Rather, Poincaré drew on his own personal experience as a highly distinguished mathematician to say something about the psychological processes that seem to be at work in the very creative act of hypothesis formation — of leaping 'upwards' to hypotheses, explanatory principles, or whatever.

Thus, in *Science and Method* we find the following remarkable autobiographical account of one of Poincaré's own mathematical discoveries:

> For a fortnight I had been attempting to prove that there could not be any function analogous to what I have since called Fuchsian functions. I was at that time very ignorant. Every day I sat down at my table and spent an hour or two trying a great number of combinations, and I arrived at no result. One night I took some black coffee, contrary to my custom, and was unable to sleep. A host of ideas kept surging in my head; I could almost feel them jostling one another, until two of them coalesced, so to speak, to form a stable combination. When morning came, I had established the existence of one class of Fuchsian functions, those that are derived from the hypergeometric series. I had only to verify the results, which only took a few hours.
>
> Then I wished to represent these functions by the quotient of two series. This idea was perfectly conscious and deliberate; I was guided by the analogy with elliptical functions. I asked myself what must be the properties of these series, if they existed, and I succeeded without difficulty in forming the series that I have called Theta-Fuchsian.
>
> At this moment I left Caen, where I was then living, to take part in a geological conference arranged by the School of Mines. The incidents of the journey made me forget my mathematical work. When we arrived at Coutances, we got into a break [carriage] to go for a drive, and, just as I put my foot on the step, the idea came to me, though nothing in my former thoughts seemed to have prepared me for it, that the transformations I had used to define Fuchsian functions were identical with those of non-Euclidean geometry....
>
> I then began to study arithmetical questions without any great apparent result, and without suspecting that they could have the least connexion with my previous researches. Disgusted at my want of success, I went away to spend a few days at the seaside, and thought of entirely different things. One day, as I was walking on the cliff, the idea came to me, again with the same characteristics of conciseness, suddenness, and immediate certainty, that arithmetical transformations of indefinite ternary quadratic forms are identical with those of non-Euclidean geometry.[116]

This splendid account was followed by several further interesting examples, the details of which need not concern us here.[117] What they all seemed to have in common was a period of deep and intense thought which failed to yield a solution to the problem, followed by a subsequent flash of illumination when the mind was not actively concentrating on the matter. Poincaré envisaged the work of the mathematician (and, we would add, the creative scientist) as a process in which choices were made between different possible combinations of ideas. He envisaged a 'subliminal ego' which could form such combinations and occasionally produce some that were harmonious, beautiful and useful, with appeal to the mathematician's aesthetic sensibility.[118] Once this was aroused, the harmonious combination might suddenly come to the attention of the 'conscious ego'.

Unfortunately, however, Poincaré's autobiographical remarks, and his suggestions as to how the mind might operate, do not tell us anything very useful about the processes of theory and hypothesis construction in science. To be sure, one can imagine Sherlock Holmes's mind rapidly sieving through innumerable explanatory hypotheses until one with acceptable aesthetic features suddenly drew attention to itself. Yes, Poincaré's account was compatible with Peirce's theory of abduction. But it really left the mind as a kind of 'black box', and a vitally important component of metascience was presented more or less as a mystery. However, it must be conceded that the subsequent very large literature on creativity has not yielded anything like a satisfactory account of the 'logic of scientific discovery'.

We should note further that Poincaré had a firm belief in the importance of *simplicity* and *elegance* in scientific theory. He held that the choice between competing principles could be made in terms of considerations of convenience, simplicity and utility — or 'elegance' as the mathematicians say. This cohered well with the role that he gave to the aesthetic sense in the processes of mathematical discovery, or theory construction in science.

These discussions of Poincaré on the psychology of invention and discovery are only remotely connected with positivism. Indeed, in talking about such matters as the 'subliminal ego' he was using language that would have been an anathema to strict positivists such as Mach. However, Poincaré's conventionalism can fairly be represented as a close relative of instrumentalism which, it has been suggested, lay within the domain of positivist thought. It is appropriate, therefore, to treat him within the context of an account of positivist metascience.

Duhem

Our next figure, Pierre Duhem (1861–1916), with whom we complete this chapter, was likewise by no means a 'pure' positivist and, as we shall see, it is even debatable whether he may properly be called an instrumentalist, though that has been a common interpretation. Whatever his exact taxonomic position, however, Duhem was undoubtedly a most important figure in the history of the philosophy and methodology of science, and many of his thoughts about the nature of scientific theories were to have considerable influence amongst twentieth-century philosophers of science. For example, the shape of a scientific theory, as construed by Duhem, was very much akin to the shape that the later logical positivists or logical empiricists envisaged. However, we can also see it as a variant of the long-lived 'arch of knowledge'.

Duhem was one of the most perceptive and influential metascientific writers of his day, and was distinguished also as physicist and mathematician, occupying a chair at the University of Bordeaux. His work in the philosophy and methodology of science was greatly enhanced by his professional expertise in physics and also his original studies in the history of science, particularly mediaeval mechanics and early chemistry. He was a skilled linguist and was able to work with original documents in mediaeval Latin without difficulty.

Duhem's two major contributions to the history and philosophy of science were *The Aim and Structure of Physical Theory*[119] and *To Save the Phenomena*.[120] As an historian of science, Duhem adopted an interesting historiographical position, namely that there were no great breaks in scientific thinking — no concepts for which some kind of intellectual ancestors could not be found. The Galilean/Newtonian concept of inertia, for example, could (according to Duhem) be traced back through the Middle Ages, in the discussions of impetus by Buridan and his school in Paris, the ideas of the Mertonian school in Oxford, and ultimately back to the ideas of antiquity. Duhem was thus a gradualist or evolutionist in his historiography of science. He did not see 'revolutions' or great intellectual divides at every turn, as do some historians of science.

Duhem was a devout Catholic, but wished to keep science and religion quite distinct. He thought that only religion could provide an entrée to the more fundamental aspects of reality. By contrast, he held that:

> A physical theory... [does not offer] an explanation [of reality]. It is a system of mathematical propositions, deduced from a small number of principles, which aim to represent as simply, as completely and as exactly as possible a set of experimental laws.[121]

Science could not reveal the depths of being that were only accessible to the metaphysician. Duhem was not, therefore, an essentialist in his view of the reach of scientific knowledge. His position was akin to that of the mediaeval doctrine of 'saving the appearances',[122] of which he made a specific study in his book *To Save the Phenomena*.

So Duhem's picture of a scientific theory was as follows:
1. the definition and measurement of physical magnitudes;
2. the selection of hypotheses;
3. the mathematical development of the theory;
4. the comparison of the theory with experiment.[123]

And this last step was the only one that could serve as a criterion for the truth of a theory. Even so, successful comparison of the results of an experiment with theoretical predictions was not enough to prove the theory's truth. A theory was an instrument of intellectual economy. It enabled the scientist to apprehend a great mass of empirical facts, for the 'condensing of a multitude of laws into a small number of principles... [affords] enormous relief to the human mind, which might not be able without such an artifice to store up the new wealth it acquires daily'.[124] So the scientific theory was to be regarded as a device for pigeon-holing or classifying facts — as a kind of taxonomic device. As such, it should not

be thought of as either true or false, but merely convenient or inconvenient. Poincaré's analogy of a library catalogue would also have
appealed to Duhem.

But this does not seem to have been Duhem's position exactly. There
was a certain ambivalence in his language, for he compared the 'pigeon-
holing' process of the physicist to the work of a naturalist seeking a
natural – as opposed to an *artificial*[125] — system of classification:

> The neat way in which each experimental law finds its place in the classification created
> by the physicist and the brilliant clarity imparted to this group of laws so perfectly
> ordered persuade us in an overwhelming manner that such a classification is not purely
> artificial, that such an order does not result from a purely arbitrary grouping imposed on
> laws by an ingenious organizer. Without being able to explain our conviction, but also
> without being able to get rid of it, we see in the exact ordering of this system the mark
> by which a natural classification is recognized. Without claiming to explain the reality
> hiding under the phenomena whose laws we group, we feel that the groupings
> established by our theory correspond to real affinities among the things themselves.[126]

It seems, then, that for all his positivism, and for all his seeming
instrumentalist language, Duhem did hanker after a realist theory of
knowledge. That is, he could not help the feeling that when there was a
nice coherence of scientific theories — a consilience of inductions, in
Whewell's terminology — and when it proved possible to use scientific
theory to make successful empirical predictions,[127] then the theory was
more than a mere instrument or convention, but truly said something
about the real world. However, Duhem did not come fully into the open
on this issue, but spoke of the physicist's 'suspicion' that 'the data of
observation correspond to real relations among things'.[128] Yet, having
admitted this natural propensity towards realism on the part of the
physicist, Duhem immediately sought to smother it, saying:

> The physicist cannot take account of this conviction. The method at his disposal is limited
> to the data of observation. It therefore cannot prove that the order established among
> experimental laws reflects an order transcending experience; which is all the more
> reason why his method cannot suspect the nature of the real relations corresponding to
> the relations established by theory.[129]

My belief is that Duhem's remarks about physicists' leanings towards
realism were partly autobiographical. And even today one can well see
why anyone might be uncertain on the realist/instrumentalist issue.

So we pass on from this ambivalent section of Duhem's text to
consider his description of the 'shape' of a physical theory. He stated that
a physical theory was made up of 'mathematical symbols serving to
represent the various quantities and qualities of the physical world' and
'general postulates serving as principles'.[130] And from these together a
mathematical structure could be set up which had to be free from logical
contradictions but otherwise was not subject to any particular structural
restrictions. Ultimately, when the chain of logical inferences had been
carried through sufficiently, conclusions would be reached that could be
tested experimentally:

> This comparison between the conclusions of theory and the truths of experiment is
> therefore indispensable, since only the test of facts can give physical validity[131] to a

theory. But this test by facts should bear exclusively on the conclusions of a theory, for only the latter are offered as an image of reality; the postulates serving as points of departure for the theory and the intermediary steps by which we go from the postulates to the conclusions do not have to be subject to this test.[132]

Thus the kind of model for an established scientific theory that Duhem seems to have had in mind may be represented as in Figure 22. It was like our arch, but with one leg missing! (In fact, it was rather as Plato represented the case of the Greek geometers, who started from apparently self-evident first truths without bothering to guarantee them in any way with the aid of dialectic.)

Figure 22[133]

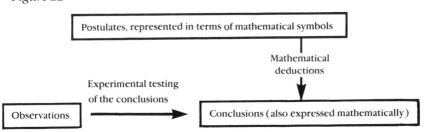

So, if the conclusions were in agreement with the experimental tests, one might say that one had an acceptable theory — one that could pigeon-hole facts satisfactorily. Whether it was thereby *true*, according to Duhem's system, was presumably something of a moot point.

So far so good. In fact, we have an interpretation of physical theory not unlike that entertained by Galileo, back in the seventeenth century, in his discussion of the phenomena of uniformly accelerated motions. (See pp.55–58.) But now we encounter a difficulty, hitherto largely unregarded. For how may one *link* the experimental tests and observations with the mathematically represented conclusions of the theory? The two are given in quite different ways. The tests involve things done in the laboratory, yielding instrument (pointer) readings and the like. The theorems — or mathematical conclusions of the theory — are expressed in strictly mathematical terms. Duhem himself drew particular attention to this problem, pointing out that there was a 'disparity between the *practical* fact, really observed, and the *theoretical* fact, the symbolic, abstract formula stated by the physicist'.[134]

There could be 'correspondence' between symbol and fact, but no 'complete parity'. Later writers, building upon Duhem's simple model of a scientific theory, introduced the term 'correspondence principles' (or 'rules') in reference to links of various kinds between 'symbol' and 'fact', in order that the theoretical conclusions drawn from a theory might be' tested experimentally. Here, in fact, we get the beginnings of that great sacred cow of logical positivism — the distinction between observational language and theoretical language, and what has come to be called the 'received view' of the structure of scientific theories.[135]

To put all this into clearer perspective, I should like to offer a simple example of the kind of thing that Duhem probably had in mind when he was writing of the structure of a scientific theory. It is, however, my example, not Duhem's. Suppose that we are interested in constructing a theoretical model for gases, in terms of the molecular and kinetic theories; and then we wish to test the model experimentally. We can imagine that a gas is made up of small molecules each of mass m, moving with a velocity c in a cubical container with sides of length l (see Figure 23). We imagine further that one-third of the molecules move backwards and forwards across the cube in the direction of the x axis, one-third in the direction of the y axis, and one-third in the direction of the z axis. All collisions of the molecules on the walls of the container are presumed to be perfectly elastic. Newton's laws are assumed to apply at each collision and during the flights of the molecules. The molecules are assumed not to collide with each other in the container(!).

Figure 23

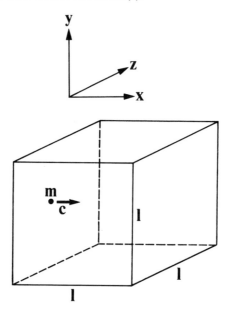

Given these (rather bizarre) assumptions, then:
Momentum of a molecule moving along the x axis = mc.
Momentum change on impact with wall = $mc - (-mc)$
$$= 2mc.$$
Number of impacts of the molecule per unit time = c/l.
So rate of change of momentum due to molecule = $2mc.c/l$
$$= 2mc^2/l.$$
Hence, the force exerted on the walls of the container by a single molecule = $2mc^2/l$ (by Newton's Second Law).
If there are n molecules in the box, then the total force exerted = $2mnc^2/l$.
But the total surface area of the container = $6l^2$.

And pressure (p) = force/area
$$= 2mnc^2/l \text{ divided by } 6l^2$$
$$= mnc^2/3v \text{ (where } v = \text{volume).}$$
So $pv = mnc^2/3$.
Also, the mass of gas in the box = mn.
Let the density of the gas (= mass/volume) = d.
Thus $p = dc^2/3$,
or $c = (3p/d)^{1/2}$.

In words: the molecular velocity is inversely proportional to the square root of the gas density, if the pressure (and temperature, for c to be constant) is constant. Or: c is proportional to $(d)^{1/2}$.

At this stage, we have carried through our reasoning to reach a conclusion, expressed in mathematical/theoretical terms, from the postulates of the theory. Now we have to test the deduction. Obviously this cannot be done directly. One cannot climb inside the container and measure the molecular velocity with a ruler and a stop-watch. So we have to introduce some so-called 'correspondence rules'. We assume that the rate of effusion or diffusion of a gas is directly proportional to the velocity of the molecules. Also, we have to have some specified method, using gas containers, balances, etc, for comparing gas densities. And we need some practical way of comparing rates of gaseous effusion under the same conditions of temperature and pressure. This can be done conveniently with an old burette, with a bit of metal foil glued to its tip and a minute pin-hole pierced in the foil (as in Figure 24).

Figure 24

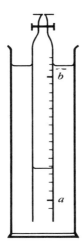

Two different gases are placed in the burette in turn. On opening the tap, one can measure the time taken in each case for the water level to rise from a to b, as the gas effuses from the pin-hole. The average times (T_1 and T_2) for the two gases can be measured with a stop-watch.

Now if c is proportional to $(d)^{-1/2}$
and T is proportional to $1/c$,
then for two different gases:

$$c_1/c_2 = T_2/T_1 = (d_2/d_1)^{1/2} \quad \text{(Equation 1)}$$

So now we have an empirical test of the postulates of the theory, and if the measurements made for T_1, T_2, d_1 and d_2 fit Equation 1, then we can say that the results of the experiment are in accordance with the postulates of the theory. If any reader cares to carry out the experiment, it will be found (in my experience) that theory and practice accord quite nicely with one another.

Incidentally, the law that the 'Rate of diffusion (or effusion) of a gas is inversely proportional to the square root of its density' is called Graham's law of diffusion, after its discoverer, the British chemist Thomas Graham (1805–1869).[136] Graham discovered his law empirically, independently of the theoretical argument from the molecular/kinetic theory, a very rough version of which has just been given. But as can be seen from the reasoning presented above, Graham's Law can readily be *explained* in terms of the molecular/kinetic theory, even with the very crude model that has been offered here. We may say, then, that with the help of the theory we have an 'economical' description of the very many facts that fall under Graham's Law. And also a theoretical explanation of these facts has been offered.

Of course, the model that has been suggested for the molecules in their box is crude to the point of absurdity. In fact, no one could imagine it to be true. Other more sophisticated models, however, lead to the same conclusion that $pv = mnc^2/3$. So, clearly, the experimental test that has been suggested is powerless to distinguish between such competing and conflicting models. This brings us face-to-face with the well-known point that *evidence in favour of a theory does not necessarily demonstrate the truth of that theory*.[137]

Now let us suppose that when the experiment with the burette was performed it was found that T_2/T_1 did not equal $(d_2/d_1)^{1/2}$, within acceptable limits of experimental error. What would this mean? It would mean, I suggest, one of the following:

1. The model was incorrect.

2. Rates of effusion were not (as was supposed) directly proportional to the velocities of the molecules in the gas.

3. Newton's mechanics were faulty.

4. A mistake was made somewhere in the reasoning set out above, or in the arithmetical computation of the results.

5. The experiments were carried out incompetently, or with inaccurate instruments, or the apparatus needed to be redesigned.

Assumptions about all these matters had to be made before the argument could be carried through successfully. Points 4 and 5 could, let us suppose for the sake of discussion, be checked without undue trouble. So we are left with Points 1, 2 and 3. The main hypothesis we may call H_1. H_2 and H_3 become 'auxiliary hypotheses', imported from other branches of sciences for the purposes of this investigation. So the situation may now be represented as in Figure 25.

Figure 25

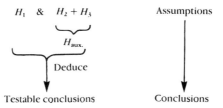

The tests, we supposed, were found to be unfavourable. So, either H_1, H_2 or H_3, or two of these, or all three, were faulty. The experiment, therefore, would not in itself be able to locate the source of difficulty with any certainty.

So far as I am aware, Duhem was the first to point out this simple but extremely important methodological problem with any clarity. It (or what some regard as a distinct but related point) has also been discussed by W.V.O.Quine,[138] and so it is today known as the Duhem–Quine thesis. It has generated a great deal of metascientific discussion. Duhem's own words on the question were:

> [T]he physicist can never subject an isolated hypothesis to experimental test, but only a whole group of hypotheses; when the experiment is in disagreement with his pre- dictions, what he learns is that at least one group of the hypotheses constituting this group is unacceptable and ought to be modified; but the experiment does not designate which one should be changed.[138]

He also gave a number of historical examples to illustrate the problem.

Duhem realised that if what has been said above concerning the question of the falsification of hypotheses were correct then it would never be possible for there to be a truly 'crucial' experiment in science, contrary to what Bacon and many others had supposed. Science was not like a watch, said Duhem, such that a watch-maker could take it apart to look for a faulty component. Science was more like a human body, functioning as a whole. Its hypotheses interlocked, one with another, and logically could not be teased apart. However, if certain elements in a system of hypotheses under test are regarded as virtually beyond question (as would have been the case for the principles of Newtonian mechanics in the example from the molecular/kinetic theory discussed above[140]) then the falsifying data could be brought to bear more directly upon a smaller number of hypothesis. But no single one could ever be isolated for individual scrutiny.

Duhem also pointed out that theories aren't often falsified in science — or regarded as being falsified. For if a system of hypotheses is found wanting it is normal to modify them with the introduction of various adjustments or limitations (today commonly known as '*ad hoc*' hypotheses), rather than permitting the collapse of the whole intellectual edifice, with nothing to put in its place. Thus Duhem well understood how some aspects of science could readily become 'conventionalised'.

Duhem emphasised that the mathematical principles of a theory operated, so to speak, in a kind of 'ideal' world. As a result, taken in isolation they had no empirical meaning, for they did not indicate how much tolerance was to be allowed them in their experimental realisation. Consequently, a physical law, expressed in its mathematical terms, *could never be tested experimentally* (which would be another reason for considering that there is a radical difference between 'theoretical language' and 'observation language'). For the signs of equality in the mathematical equations in practice had to mean 'more or less equals'. The approximate values of the experimentalist *never* yielded the exactitude of the mathematical representations of physical laws. Consequently, one could not say whether the laws, as stated mathematically, were strictly true or false. So it was pointless to ask such a question — according to Duhem. He wrote: 'A law of physics is, properly speaking, neither true nor false but approximate.'[141]

There have been two general construals of Duhem's position at this point. Some commentators, notably Karl Popper,[142] have represented Duhem as a thoroughgoing instrumentalist, holding that scientific laws and theories are not proper descriptive statements but merely 'instruments' which allow observational predictions to be made on the basis of other observation statements. But before accepting Popper's characterisation of Duhem we should note statements of Duhem himself, such as the following:

> [Physical laws] are always symbolic. Now, a symbol is not, properly speaking, either true or false; it is, rather, something more or less well selected to stand for the reality it represents, and pictures that reality in a more or less precise, or a more or less detailed manner.[143]

This suggests that there was some realist aspect to Duhem's thinking. And we should remember his ambivalence as to whether theories could serve as 'natural' or 'artifical' classifications of data. Given such considerations, some commentators have preferred not to use the label 'instrumentalist' for Duhem, but have called him a 'descriptivist', in that he supposed that scientific laws did give more or less accurate descriptions of reality.[144] Such an interpretation does seem to be warranted by the preceding quotation from Duhem's book.

There is, fortunately, less doubt about whether Duhem should be called a positivist. Indeed, one section of his text was headed: 'Our physical system is positivist in its origins.'[145] I think he may also fairly be called a conventionalist. For, through his discussions of what is now called the Duhem–Quine thesis, he showed very clearly that he recognised the ways in which certain features of the theoretical terrain can readily become conventionalised. On the question as to whether Duhem was a realist or an instrumentalist, there is, as I say, some uncertainty. On the whole, I'm inclined to think that he leaned somewhat towards realism, even though, as we said above, he thought that the ultimate depths of reality were best plumbed by religion rather than by science.[146]

There is a great deal else of interest in *The Aim and Structure of Physical Theory*, including a celebrated comparison of the characteristics of French and British physics, but unfortunately we do not have space to attend to such matters here. We have now given an indication of some of the ideas to be found in the various early forms of positivism, a difficult task, given the Protean character of the positivist movement. The movement was carried forward with renewed vigour in the twentieth century in the work of the so-called 'logical positivists'. This movement attempted a unification of science through the establishment of a unified philosophy of science, based upon a collaboration of the results of investigations in logic and a more rigorously stated empiricism. The logical positivists did not refer to Comte very often as their intellectual father, though they freely acknowledged their debt to Mach. Nevertheless, the looked-for 'unification of science' was consonant with Comte's original positivist program, and both Comte and the logical positivists looked to the methods and results of science to bring about an increase in human happiness. So, leaving the nineteenth-century positivists behind us, we shall now seek to give some account of the complexities of logical positivism.

NOTES

1   L Kolakowski *The Alienation of Reason: A History of Positivist Thought* trans N Guterman Doubleday New York 1968 pp 3–10

2   It would not be unacceptable to speak of Berkeley, for example, as a positivist, though it would be historically anachronistic.

3   This issue is currently an active focus of debate. I have noticed that it tends to be the older generation of philosophers of science that holds that facts and values *can* be separated and that scientific knowledge is value-free. Here, I think, they betray the positivist schooling to which they were exposed in their youth!

4   It may be worth noting also that philosophers of law make a distinction between what they call 'positive' law and 'natural' law. The 'positive' law is that which is actually in the statute books, or enforced in the courts. The 'natural' law is that to which one might make an intuitive appeal — on the basis of what is held to be 'naturally' right or wrong — as the basis of legislation or jurisdiction. Thus, the lawyers' distinction between 'positive' and 'natural' law is somewhat analogous to the positivists' distinction between fact and value. Of course, a great many philosophers of law deny that there *is* any such thing as 'natural law'.

5   See, for example, E A Burtt *The Metaphysical Foundations of Modern Science* 2nd ed Routledge & Kegan Paul 1932 p 282. The term 'positive science' appears to have been introduced by Madame de Stael about 1800. It was taken up by Saint-Simon, and from him by Comte. So although we usually think of Comte as the founder of positivism, this is not correct in an absolutely strict sense.

6   I A Comte *Cours de Philosophie Positive* 6 vols Bachelier Paris 1830–42

7   It is worth noting here that History of Science was first instituted as a formal academic discipline at the tertiary level at the end of the nineteenth century, when a chair in the subject was established at the Collège de France; and this occurred largely as a result of pressures generated by the positivist movement.

8   The desire to explain social phenomena in terms of biology is, of course, by no means exhausted, as the present interest in 'sociobiology' amply demonstrates.

9   S Andreski ed & M Clarke trans *The Essential Comte: Selected from Cours de Philosophie Positive by Auguste Comte First Published in Paris 1830-42* Croom Helm London and Barnes & Noble New York 1974 p 49. This is at present the most accessible version of Comte's writings available in English, though it presents only a small portion of the original course of lectures. Another version, fuller but incomplete and not entirely faithful to the original, is *The Positive Philosophy of August Comte* trans H Martineau 2 vols Chapman London 1853

10 *Ibid* (1974) p 39

11 In Comte's day, of course, Latin and Greek were commonly regarded as the ideal preliminaries to the study of almost anything; so one might readily launch forth from Tacitus to physiology!

12 Comte *op cit* (note 9 1974) p 49

13 *Ibid* p 51

14 Comte supposed that the natural place of astronomy was before physics; yet he rightly pointed out (*ibid*) that optics, for example, was needful 'to complete the account of astronomy'.

15 It may be mentioned here in parenthesis that recent attempts to 'reduce' sociology to biology (in the 'sociobiology' of E O Wilson and his admirers) have been treated with great hostility by biologists who happen to be to the left of the political spectrum. It is maintained that such attempts at reduction are deterministic and obscurantist, politically motivated, and much else besides. Such critics do not seem willing to acknowledge that if such reduction were possible a good deal of explanation would be achieved, as well as a useful economy of thought.

16 Comte *op cit* (note 9 1974) p 20

17 This hypothetical 'principle' of inflammability was commonly invoked by eighteenth-century chemists before the time of Lavoisier.

18 These hypothetical corpuscles were suggested by a number of seventeenth-century mechanical philosophers such as Robert Hooke.

19 Comte maintained (*op cit* [note 9 1974] p 53) that the more fundamental sciences contained laws of higher degree of generality — a kind of Baconian notion. This is compatible with the subsumption of laws at one level of the hierarchy under laws at another, and in that sense explanation of one level in terms of another is plausible. But the idea of (say) explaining social phenomena in terms of the inverse square law of gravity is patently ludicrous.

20 This was linked with so-called 'romantic' science, which in fact was often remarkably fruitful.

21 For example, F W Ostwald *The Principles of Inorganic Chemistry* trans A Findlay Macmillan London 1902

22 J S Mill *Auguste Comte and Positivism* Trubner London 1865

23 *Ibid* p 58

24 This view is one of the main themes of the influential work of the twentieth-century scientist–philosopher, Michael Polanyi. See his *Personal Knowledge: Towards a Post-Critical Philosophy* Routledge & Kegan Paul London 1958

25 The only branch of psychology for which Comte seems to have had much sympathy was phrenology. (This is now commonly thought a somewhat disreputable enterprise, though in its day it was a line of research that seemed promising and amenable to the usual empirical, intersubjectively testable

methods of science, rather than the private mental introspections that served as the basis of most 'psychological' inquiry in Comte's era.) Comte also paid little attention to geology and palaeontology.

26 J C Greene 'Biology and Social Theory in the Nineteenth Century: August Comte and Herbert Spencer' in M Clagett ed *Critical Problems in the History of Science* University of Wisconsin Press Madison 1959 pp 419–46

27 It appears that Comte had one unsuccessful sexual engagement with this lady; and thereafter he sublimated his desires, placing the lady of his passions on a pedestal and worshipping her in this new elevated position.

28 See *The Catechism of Positive Religion, Translated from the French of Auguste Comte, by Richard Congreve* Chapman London 1858

29 Perhaps Comte's strangest suggestion was to abandon decimal numbering in favour of a septimal system, for seven was supposedly a sacred number, and being a prime number would impress one with the weakness of the human mind and the limitations of thought!

30 See W M Simon *European Positivism in the Nineteenth Century: An Essay in Intellectual History* Cornell University Press Ithaca 1963

31 H James ed *The Letters of William James* 2 vols Longmans Boston 1920 Vol 1 p 212

32 E Mach *Knowledge and Error: Sketches on the Psychology of Enquiry* trans T J McCormack & P Foulkes Reidel Dordrecht & Boston 1976 p 72 (1st German ed Leipzig 1905)

33 E Mach *The History and Root of the Principle of Conservation of Energy* trans R E B Jourdain Open Court Chicago 1911 (1st German ed Prague 1872)

34 E Mach *The Science of Mechanics: A Critical and Historical Account of its Development* trans T J McCormack Open Court La Salle 1960 (1st German ed Leipzig 1883)

35 E Mach *The Analysis of Sensations and the Relation of the Physical to the Psychical* trans C M Williams Open Court Chicago & London 1914 (1st German ed Jena 1886)

36 E Mach *Popular Scientific Lectures* trans T J McCormack Open Court Chicago 1895 (1st German ed Leipzig 1896)

37 Mach *op cit* (note 32)

38 Mach *op cit* (note 35 1914) p 30

39 C B Weinberg *Mach's Empirio-Pragmatism in Physical Science* Albee Press New York 1937 p 1

40 Mach *op cit* (note 35 1914) p 30

41 *Ibid* p 3

42 *Ibid* p 2

43 Mach's simple sensations (or their approximate equivalents) were called 'sense data' by subsequent philosophers such as Bertrand Russell.

44 Mach *op cit* (note 35 1914) p 13

45 See below, p 276

46 This position was essentially that of Hume. See above, p 111.

47 Mach *op cit* (note 35 1914) p 42

48 Mach *op cit* (note 33 1911) p 49

49 The chemists did, of course, measure atomic weights — the ratios of the weights of particular kinds of atoms to those of hydrogen. But such weight ratios could only be determined indirectly, and with the invocation of a number of somewhat insecure hypotheses. For a long time, then, the Machian positivists held out against atoms. Eventually, most of them succumbed as a result of the work of Jean Perrin on colloids, which provided a kind of bridge between the macro-and micro-worlds. But Mach continued to reject the atomic theory until the end of his life. (See M J Nye *Molecular Reality: A Perspective on the Scientific Work of Jean Perrin* Macdonald London 1972)

50 Technical memory aids

51 See above, pp 48 & 93

52 Mach *op cit* (note 36 1895) pp 194–7 and *passim*; *op cit* (note 34 1960) pp 7–8, 577, 582 and *passim*

53 For my exposition of this, see my *Darwinian Impacts: An Introduction to the Darwinian Revolution* New South Wales University Press Sydney 1980

54 E Mach 'On Transformation and Adaptation in Scientific Thought' in his *Popular Scientific Lectures* (note 36) pp 214–35

55 R S Cohen 'Ernst Mach: Physics, Perception and the Philosophy of Science' in R S Cohen & R J Seeger eds *Ernst Mach: Physicist and Philosopher* Reidel Dordrecht 1970 pp 126–64

56 A good number of other writers besides Mach have sought to develop 'evolutionary epistemologies'. For a general critique of efforts of this kind, see M Ruse 'Darwin and Philosophy Today' in D R Oldroyd & I G Langham eds *The Wider Domain of Evolutionary Thought* Reidel Dordrecht 1983 pp 133–58

57 F Cajori ed *Sir Isaac Newton's Mathematical Principles of Natural Philosophy and his System of the World* University of California Press Berkeley 1934 p 1

58 Mach *op cit* (note 34 1960) pp 266–7

59 Mach was, of course, writing before the coming of Einstein's relativity theory, which envisaged a variation of mass with velocity. See below, p 273

60 Newton himself claimed that he had carried out the experiment, and it is indeed perfectly easy to

carry it out for oneself. It is, however, the interpretation that is to be placed upon it that is important.

61 Mach *op cit* (note 34 1960) p 284. With this hint at some possible agent 'causing' the curvature, Mach, was not strictly keeping within the bounds laid down by his own phenomenalism.

62 See above, pp 108–109

63 For further information, see E Hiebert 'Mach's Conception of Thought Experiments in the Natural Sciences' in Y Elkana ed *The Interaction Between Science and Philosophy* Humanities Press Atlantic Highlands 1974 pp 339–48. See also E Mach 'On Thought Experiments' in his *Knowledge and Error* (note 32) pp 134–47

64 One may wonder what Mach meant by the word 'synthetically' here. I take it he was referring to induction rather than deduction. We are led to believe that experimental scientists 'grope', whereas logicians and mathematicians stride along, confident in their deductive pathways!

65 Mach *op cit* (note 34 1960) pp 161–2 (cf p 54 above)

66 On the history of pragmatism, see, for example: A J Ayer *The Origins of Pragmatism: Studies in the Philosophy of C.S.Peirce and W.James* Macmillan London 1974; or C W Morris *The Pragmatic Movement in American Philosophy* Braziller New York 1970. The Greek work *pragma* meant 'deed' or 'action'. We find descendants in words like 'practice' and 'practical' as well as 'pragmatism'. In his *Critique of Pure Reason* Kant had described the situation in which a doctor was uncertain as to the cause of a patient's illness, but prescribed the best medicine that he knew. If the patient got better, the doctor would know that the medicine had been appropriate, though he wouldn't know precisely *why*. In such a situation, said Kant, the doctor would have a pragmatic or practical — but not a theoretical — understanding of the problem.

67 For example, J Fiske *Outlines of Cosmic Philosophy, Based on the Doctrine of Evolution, with Criticisms on the Positive Philosophy* 2 vols Macmillan London 1874.

68 I Murphree 'The Theme of Positivism in Peirce's Pragmatism' in E C Moore & R S Robin eds *Studies in the Philosophy of Charles Sanders Peirce* University of Massachusetts Press Amherst 1964 pp 226–41

69 There is a large philosophical literature on the 'logic' of confirmation, it being commonly claimed that confirmation in a strict sense cannot be attained by the experimental testing of hypotheses. I do not, however, wish to engage in this debate here. All I mean is that 'positive' results are presumed to be found in the experimental tests.

70 For a useful account which traces the development of Peirce's ideas on this matter, see, for example, K T Fann *Peirce's Theory of Abduction* Nijhoff The Hague 1970. See also A W Burks 'Peirce's Theory of Abduction' *Philosophy of Science* 1964 Vol 13 pp 301–6

71 *Collected Papers of Charles Sanders Peirce* eds C Harshorne & P Weiss 6 vols Harvard University Press Cambridge Mass 1931–35 Vol 2 p 374. Two further volumes (7 & 8) of Peirce's *Collected Papers* were edited by A W Burks and published by the same publishers in 1958.

72 Or 'instance' or 'state of affairs'

73 See, for example, Peirce *op cit* (note 71) Vol 6 p 358 & Vol 7 p 122 (Peirce also, on occasions, called the process 'retroduction' or sometimes 'presumption'.)

74 *Ibid* Vol 5 p 117

75 For a discussion of the analogy between the methods of Peirce and Holmes, see J A Sebeock & J Uniker Sebeock *You Know My Method: A Juxtaposition of Charles S.Peirce & Sherlock Holmes* Gaslight Publications Indianapolis 1979

76 To use once again our somewhat jaded analogy, in finding a way up a mountain one is, necessarily, finding a possible way down at the same time.

77 Peirce *op cit* (note 71) Vol 7 p 122

78 *Ibid* p 124

79 *Ibid* Vol 1 pp 79–80; 315–17

80 *Ibid* Vol 2 p 5. Such a view would certainly please modern sociologists of knowledge, who claim that what does or does not pass as knowledge at any given time is that which the community sanctions. And what the community sanctions is, of course, intimately connected with questions of value. One might, however, point out that there is a pun involved in talking of 'right' reasoning. 'Right' may mean either correct or morally appropriate. But the ambivalence of our language is perhaps an indication of the way in which even logic has an element of value lurking in the wings.

81 *Ibid* p 29

82 *Ibid* Vol 7 p 138

83 The reader should note the normative component to the argument at this juncture.

84 Peirce *op cit* (note 71) Vol 5 p 421; Vol 7 p 30

85 His efforts have had some followers, notably N R Hanson in his *Patterns of Discovery: An Inquiry into the Conceptual Foundations of Science* Cambridge University Press Cambridge 1958. But generally the opinion today is that it is a forlorn hope to give a logic of abduction. See L Laudan 'Why was the Logic of Discovery Abandoned?' in T Nickles ed *Scientific Discovery, Logic and Rationality* Reidel Dordrecht 1980 pp 173–83

86 Peirce *op cit* (note 71) Vol 5 pp 400–1. Peirce gave many other instances in his writings of his belief in the self-correcting character of science (for example, Vol 2 pp 478 481 491 496; Vol 5 pp 90 218 402 405; Vol 7 pp 66–7). Science was believed to be self-correcting by iteration.

87 W James *Pragmatism: A New Name for some Old Ways of Thinking* Longmans Green & Co London

1907 p 201 (italics in original)

88 For a good discussion of Dewey's philosophy of science, see H J Kannegiesser *Knowledge and Science* Macmillan Melbourne, 1977. In treating Dewey we are here forced to extend our discussion of nineteenth-century positivism well into the twentieth century. But Dewey lived a very long time, and many of his basic views were formed in the nineteenth century.

89 J Dewey *Experience and Nature* Allen & Unwin London 1929

90 J Dewey *How We Think* Heath Boston New York & Chicago 1910 pp 10–11

91 J Dewey *Reconstruction in Philosophy* lst ed 1920 enlarged ed Beacon Press Boston 1948 p 140

92 *Ibid* pp 144–5

93 *Ibid* p 152

94 Dewey *op cit* (note 90) p 156

95 *Ibid* p 171

96 *Ibid* p 177

97 J H Poincaré *La Science et l'Hypothèse* Bibliothèque de Philosophie Scientifique Paris 1903; *Science and Hypothesis* trans W J G Walter Scott Publishing Co London & Newcastle-on-Tyne 1905

98 J H Poincaré *La Valeur de la Science* Flammarion Paris 1904; *The Value of Science* trans G B Halsted Dover New York 1958

99 J H Poincaré *Science et Méthode* Bibliothèque de Philosophie Scientifique Paris 1908; *Science and Method* trans F Maitland Nelson London 1914

100 P Frank *Modern Science and its Philosophy* Harvard University Press Cambridge Mass 1949 pp 11–12

101 According to Euclidean geometry, only one straight line can pass through two points, the shortest distance between two points is a straight line, and only one line can be drawn parallel to another line through any given point. But according to the geometrical system proposed by Lobatschewsky, several parallels can be drawn through a point to a given straight line. In Riemann's geometry, no parallels can be drawn through a given point to a straight line. With different axioms, the three geometries naturally yield three different sets of theorems. For example, in Euclidean geometry the angle sum of a triangle is two right angles. For Lobatschewsky's, it is less than two right angles; and for Riemann's it is greater.

102 Poincaré *Science and Hypothesis* (note 97) 'Non-Euclidean Geometries' pp 35–50 esp p 50

103 *Ibid* p 70

104 See above, p 78

105 Poincaré *op cit* (note 97) p 110

106 This happened when the planet Neptune was discovered on which see M Grosser *The Discovery of Neptune* Harvard University Press Cambridge Mass 1962. More recently, the close-up photographs sent back from the space-ship *Voyager* of the rings of Saturn show a 'braided' structure for some of them, which seems to be very hard to explain in terms of orthodox mechanics. But this hasn't led to the assumption that Newton's laws should be waived in the special case of the rings of Saturn.

107 Poincaré *op cit* (note 97) p 144

108 Of course, there can be books in the library for which there are no entries in the catalogue, and *vice versa*. Or all may coincide exactly. But the classification, as such, cannot be said to be true or false.

109 Frank, it may be noted, spoke of the principles of science, not science as a whole. It is a nice question as to whether science as a whole must be conventional if its principles are so.

110 Poincaré *op cit* (note 98 1958) pp 112–28

111 *Ibid* p 112

112 *Ibid* pp 124–5. This is essentially the same as the case of the discovery of Neptune through recognition of perturbations in the orbit of Uranus, mentioned in note 106.

113 One might think that phosphorus burns in oxygen, and gains weight in the same proportion, no matter whether one adopts the phlogiston theory or the oxygen theory. This would have been Poincaré's view. He wrote, however, without knowledge of the large literature that has been developed since his day on the theory-ladenness of observations. What one sees may often be influenced to an extraordinary degree by what one expects to see.

114 Poincaré *op cit* (note 98 1958) p 121

115 *Ibid* p 122

116 Poincaré *op cit* (note 99 1914) pp 52–4

117 I confess I have only the haziest notion of what a Fuchsian function is, let alone a theta-Fuchsian! This may seem a gross inadequacy to the mathematician, but some general points can usefully be made *about* mathematics, even by non-mathematicians.

118 Poincaré *op cit* (note 99 1914) pp 57–60

119 P Duhem *La Théorie Physique: Son Objet, Sa Structure* Riviere Paris 1906; *The Aim and Structure of Physical Theory* trans P P Wiener Atheneum New York 1977

120 P Duhem *'SOZEIN TA PHAINOMENA'. Essai sur la Notion de Théorie Physique de Platon à Galilée* Hermann Paris 1908; *To Save the Phenomena: An Essay on the Idea of Physical Theory from Plato to Galileo* trans E Doland & C Maschler Chicago University Press Chicago 1969

121 Duhem *op cit* (note 119 1977) p 19

122 See above, pp 48 & 93

123 Duhem *op cit* (note 119 1977) p 21

124 *Ibid*

125 If the Darwinian theory of evolution is correct, there are species in nature which have come about through natural selection, divergence, geographical isolation, etc. It is the naturalist's task to frame a classification corresponding to these natural biological divisions. However, nature's divisions are by no means exact and clear-cut, and 'higher' groupings such as genera, orders and classes are sometimes even more difficult to determine than species. In desperation, then, many naturalists have proposed 'artificial' systems of classification, where groupings can be established with certainty, but which may bear only a very limited similarity to the natural groupings. Linnaeus's system for plant classification, based chiefly on the counting of the numbers of stamens and carpels in the flower, is the classic example of an artificial system. But some artificial systems — for example, those based on economic considerations — may be very useful for certain purposes.

126 Duhem *op cit* (note 119 1977) pp 25–6

127 *Ibid* pp 27–30

128 *Ibid* pp 26–7

129 *Ibid* p 27

130 *Ibid* p 205

131 There is, surely, a hint of realism here.

132 Duhem *op cit* (note 119 1977) p 206

133 Duhem actually spoke of the logical structure reaching 'its highest point' (p 206). I have, however, inverted his metaphor, so to speak, in order to maintain analogy with our general theme of arch building.

134 Duhem *op cit* (note 119 1977) p 151

135 H Putnam 'What Theories Are Not' in E Nagel P Suppes & A Tarski eds *Logic Methodology, and Philosophy of Science: Proceedings of the 1960 International Congress* Stanford University Press Stanford 1962 pp 240–51

136 T Graham 'On the law of Diffusion of Gases' *Philosophical Magazine* 1833 Vol 2 pp 175–90 269–76 & 351–8

137 Treated as a point in logic, this had been realised in Antiquity. See above, pp 26 & 45

138 W V O Quine 'Two Dogmas of Empiricism' in *From a Logical Point of View: 9 Logico-Philosophical Essays* Harper New York 1953 pp 20–46 (at p 43). The Duhem–Quine thesis has become associated in the literature with the notion of the 'underdetermination' of theories — that is that in principle there is always an indefinite number of theories capable of accounting for some observed facts, and that any of the theories can be maintained no matter how much contrary evidence there may be, if sufficiently radical adjustments are made somewhere in the auxiliary hypotheses, not in the hypothesis directly under test at any given moment. Such a view has been found congenial by proponents of sociology of knowledge doctrines. (See Chapter 9.) But, as has been pointed out by Roger Ariew, the conflation of the ideas of Duhem and Quine is not really warranted on the basis of Duhem's own texts. See R Ariew 'The Duhem Thesis' *British Journal for the Philosophy of Science* 1984 Vol 35 313-25

139 Duhem *op cit* (note 119 1977) p 187

140 One could, of course, feel uncertain whether the Newtonian mechanics of the macro-world could be applied in the micro-world.

141 Duhem *op cit* (note 119 1977) p 168

142 K R Popper 'Three Views Concerning Human Knowledge' in his *Conjectures and Refutations* Routledge & Kegan Paul London 1963 pp 97–120 (at p 99)

143 Duhem *op cit* (note 119 1977) p 168

144 On descriptivism, see E Nagel *The Structure of Science: Problems in the Logic of Scientific Explanation* Routledge & Kegan Paul London 1961 pp 117–29

145 Duhem *op cit* (note 119 1977) p 275

146 For a good discussion of whether Duhem was or was not an instrumentalist, see J Giedymin 'Instrumentalism and its Critique: A Reappraisal' in R S Cohen *et al* eds *Essays in Memory of Imre Lakatos* Reidel Dordrecht & Boston 1976 pp 179–201. Giedymin points out that many of the 'instrumentalists' claimed by critics of instrumentalism such as Popper tend to be straw men.

CHAPTER 6

LOGIC AND LOGICAL EMPIRICISM

One of the characteristic features of philosophy of science in the first half of the twentieth century was its preoccupation with logic — hardly surprising considering the fact that many of its leading practitioners were themselves logicians. In this chapter, then, we shall seek to give a general account of the way in which philosophy of science developed, in the late nineteenth century and into the twentieth century, from a union of interest between logicians and philosophers of an empiricist persuasion. At the time, this union appeared to be both natural and fruitful for the consideration of an enterprise such as science, which claimed to be logical in its approach and was concerned with the observable 'facts' of the external world. Today, the union of 'logicism'[1] and 'empiricism' may seem less feasible. However, this need not interfere with our historiographical enterprise. The key figures that we shall consider are Gottlob Frege (1848–1925), Bertrand Russell (1872–1970), Ludwig Wittgenstein (1889–1951), Rudolf Carnap (1891–1970), and Hans Reichenbach (1891–1953). The work of Carnap will be taken as illustrative of that of the well-known group of philosophers and scientists who formed the so-called Vienna Circle.

Frege
F.L.G.Frege came from the small town of Wismar in Pomerania, son of a schoolmaster. He studied mathematics at the University of Jena, and also philosophy, physics and chemistry. His doctorate in mathematics was taken at the University of Göttingen, and his professional life was spent as a teacher at the University of Jena. His thinking was particularly influenced by the work of Leibniz and Kant, and also the Neo-Kantian German philosopher R.H. Lotze (1817–1881).

Frege sought to confirm Kant's apriorism, holding that truth is not just a question of subjective psychology, but is objectively grounded. However, differing from Kant, Frege considered that there was an important distinction between geometry and arithmetic. He supposed that the axioms of geometry were derived from 'intuitions' (that is, they were empirical in origin); but the laws of arithmetic were considered to be purely mental constructs.[2] From Lotze, Frege derived the idea of logic as being concerned with the investigation of the forms of human thought, so that it could be, as it were, a 'pure' *a priori* science.[3] Logic was considered to be fundamental for *all* human knowledge — even more fundamental than mathematics. So there arose the program — partly carried out by Frege, and subsequently undertaken more fully by Russell in conjunction with A.N.Whitehead (1861–1947) — of 'reducing' arithmetic and algebra to logic. All this accorded with the much earlier unfulfilled plans of Leibniz to develop a 'universal characteristic',[4] representing, as it were, the 'alphabet of human thought'. Each single character in the language would represent a single concept, rather as each sign in arithmetic represents a definite number, or a particular arithmetical operation.

Kant, as we have seen, had said that 'thoughts without content are empty, intuitions without concepts are blind'.[5] But he thought we could have *a priori* knowledge of some 'synthetic' propositions. Likewise, Frege believed that there could be *a priori* knowledge that was not merely trivial: 'Propositions which extend our knowledge can have analytic judgments for their content.'[6] So, for Frege, the laws of arithmetic could be known by thought alone — without any recourse to the counting of pebbles or anything of that kind. And so there seemed to open up the prospect of a marvellous clarification of knowledge if important elements of it, such as mathematics, could be seen to be a product of logical principles. But for this a reformulation of logic itself was required.

In Chapter 1, we made some brief remarks about the logical work of Aristotle in his *Prior Analytics*. What is remarkable is that the subject of logic remained in essence the same throughout the long period between the Greeks and the nineteenth century. There were some steps towards symbolic logic in the work of Lewis Carroll, for example, but it was only with the work of Frege that a radically new approach was taken. The syllogism was no longer regarded as the be-all-and-end-all of logic, and was seen to be quite inappropriate for the analysis of arithmetical inferences.

In his epoch-making work of 1879 entitled *Begriffsschrift* or *Conceptual Notation*,[7] Frege sought to develop a new method for the symbolic representation of the contents of thought. In this new system, various propositions were written down separately, each on a line of its own, and the logical connections between them were displayed by various connecting lines drawn to the left of the propositions.[8] This two-dimensional method of symbolic representation has not been preserved in subsequent works; it is, rather, Frege's *method* of logical analysis that has been retained, and it is the essence of this that we must try to explain.

The whole system was based on the introduction (by an analogy drawn from mathematics) of the notion of a *function*. In algebra, we may have a variable, x, and a function of x such as $3x^2 + 5$, or in general $f(x)$. If the value of the variable, x, is 2, then the function has the value of 17; if $x = 3$, the function has the value 32; and so on. The number substituted for the value, x, is called the 'argument' of the function.

Now suppose we have a proposition such as 'My neck is hurting'. For this, one can think of a function, 'x is hurting', into which various terms may be substituted for x — such as neck, head, leg, heart, and so on. Such terms supply possible 'arguments' for the function. Some will yield true propositions, while for others false propositions will be obtained. So Frege's method allowed the separation of judgments into two components: that which was able to be judged; and the process of actually making the judgment, or recognising it to be true. Thus, as he later wrote:

> One can express a thought without asserting it. But in ordinary language there exists no word or sign whose exclusive job it is to assert. That is why even logic books confuse predicating with judging... To think is to grasp thoughts. After a thought has been grasped, one can recognize it as true (judging) and express that recognition (asserting).[9]

As we shall see, this distinction between the statement of a proposition and the assertion of its truth was to be of the highest importance in the work of Russell and Wittgenstein.

The ideas in Frege's *Conceptual Notation* were poorly received by his contemporaries, the six published reviews being generally unsympathetic.[10] Frege had made a major advance in logic by substituting the notions of function and argument for the Aristotelian subject and predicate; but the significance of this was not apparent to the reviewers. Nevertheless, Frege persevered with his work, his next major effort being an attempt to show how the method of logical analysis developed in *Conceptual Notation* might be employed for displaying the logical basis of arithmetic. The results of this project were published as *Die Grundlagen der Arithmetik* (*The Foundations of Arithmetic*) in 1884.[11] In this, however, Frege did not use the two-dimensional symbolism of *Conceptual Notation*, which had proved so unpopular with his readers and reviewers.

Leibniz had attempted to establish the logical basis of number by defining the notions of '0', '1', and 'following after'. So, if $2 = 1 + 1, 3 = 2 + 1, 4 = 3 + 1$, then:

$$2 + 2 = (1 + 1) + (1 + 1)$$
$$= (2 + 1) + 1$$
$$= 3 + 1$$
$$= 4.$$

And other arithmetical operations might be treated similarly. But this told one very little about the 'nature' of number. Frege's innovation (the exposition of which was based upon an analogy with the geometrical notion of parallelism) was as follows. If you have two groups of objects, and if it is possible (somehow) to correlate the members of each group one-to-one, then the constituents of the two groups of objects are held to be equal in number; that is, they are said to be '*equinumerate*' (see Figure 26, which is not one offered by Frege himself).

Figure 26

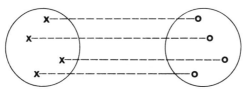

It will be recognised that this process of correlation, and establishment of equinumeracy, does not entail the actual *counting* of the numbers of objects in the two groups; so the assumption of the meaning of some particular number ('4' in Figure 26) is not involved in the process. So extending the foregoing kind of argument, Frege wrote: 'the Number which belongs to the concept *F* is the extension of the concept "equal to the concept *F*" [or "equinumerate with the concept *F*"]'.[12]

The notion of 'extension of a concept' is not, perhaps, as clear as one might wish. For example, did Frege mean that the number '4' is the *class* of all things that can be correlated, one-to-one, with a set such as (* * * *)? If this were the correct interpretation, it would seem that he was, in fact, giving arithmetic an empirical basis, as had John Stuart Mill. Yet Frege explicitly argued against the empiricist theory of mathematics, and so one may doubt that the foregoing interpretation is satisfactory. Even so, if by 'extension of a concept' he meant the class of things to which the concept applies, there remains, it might seem, the empirical problem of knowing what the extension of any particluar concept actually is. Frege himself seemed to evade the issue by writing: 'I assume that it is known what the extension of a concept is'.[13]

It may be, then, that Frege was not wholly successful in expunging all empirical elements from number theory by appeal to the notions of 'equinumeracy' and 'extension'. But what he had in mind, it seems, was to pull arithmetic up by its bootstraps, so to speak, beginning from a non-empirically-based notion of 'zero'. Thus he suggested that the number 'zero' was the number belonging to a concept with no extension; that is, the extension of the concept that is equinumerate with the extension of the concept 'not equal to itself'. 'One' was the extension of the concept that is equinumerate with the extension of the concept 'identical with zero'. Subsequent numbers in the number series were defined on this 'logically-characterised' basis, with the help of a carefully defined notion of 'following after' in the series. We can see that Frege's method was leading in the directions of set theory and the logic of classes through his notions of 'equinumeracy' and 'extension', and the analogy of parallelism. It sought to explicate the notion of number without appeal to the counting procedure.

However, Frege felt that he had still not fully succeeded in demonstrating his 'logistic' thesis that all arithmetical propositions were strictly deducible from purely logical laws, and he continued to work on his project with the preparation of a manuscript on *The Basic Laws of Arithmetic*, which was eventually published in two volumes in 1893 and 1903.[14] There was a long delay between the appearance of the two volumes, for following the poor financial success of his earlier works, and their chilly reception by the reviewers, Frege had considerable difficulty in finding a publisher for his *Basic Laws*. Eventually, the Jena publisher Hermann Pohle undertook the work, but he only agreed to publish Volume 1 in the first instance; the issue of the second volume was to be dependent on the financial success of the first. In fact, Volume 1 received only two reviews, and both were unfavourable. So Frege's second volume was greatly delayed, and he had to meet the publication expenses himself. But the long delay had some interesting historical consequences, as we shall see.

While Frege was at work trying to display the logical foundations of mathematics, Georg Cantor (1845–1918) was busy with the investigation of set theory, and in fact the work of the two was beginning to overlap. For Frege's notion of one-to-one correlation in equinumeracy was essentially the same as that underlying Cantor's notion of the equivalence of classes. So using the language of set theory, we may summarise Frege's fundamental idea for the logical foundation of arithmetic as follows:

1. The cardinal[15] number of a set is the set of all sets that are equivalent to it.
2. For the natural numbers 0, 1, 2, 3, etc., which are to be thought of as sets of sets of entities, one takes as exemplars particular sets that are defined by a process in which one builds upwards from a definition of the number '0'. Thus, the number '0' is defined as the set of entities such that they are not identical with themselves. (There are, naturally, no such entities!) Then the number '1' is the set whose only member is the number '0', there being only one such set. '2' is the set whose members are the sets '0' and '1'. '3' is the set whose members are '0', '1' and '2'. And so on...[16]

These give what appear to have been the equivalents of the essential features of Frege's number theory, even though they are not to be found stated explicitly in this way in his own writings. The first statement gives the definition or conception of number that was subsequently used by Whitehead and Russell.[17] For example, as Russell suggested, the number '2' is the class of all couples; the number '3' is the class of all triads; and so on.

This definition of number was implicit in Frege's *Foundations* of 1884, and offered a rejection of Kant's view that arithmetical propositions were 'synthetic', somehow involving a reference to time.[18] It was also, of course, fundamentally opposed to Mill's empiricist account of mathematics. Nevertheless, despite the emendations found in Frege's *Basic Laws*, and its merging with set theory, there was a fundamental problem lurking within the structure of the work. His fifth 'basic law' may be stated as: 'The set of *F*s is identical with the set of *G*s if and only if all *F*s are *G*s and all *G*s are *F*s.'[19] It was here that an unexpected antinomy was later to emerge.

The first volume of Frege's *Basic Laws* was reviewed by the Italian logician Giuseppe Peano (1858–1932) in 1895. The review was, as we have said, not favourable, and Peano seems to have been at pains to show that his own system of logic was superior to that of Frege. However, there followed a fruitful correspondence between Peano and Frege, and it was through Peano that Frege's work was drawn to the attention of Bertrand Russell, who was also searching for a suitable way to 'reduce' mathematics to logic — or demonstrating the logical foundations of mathematics.

By 1902, Frege was ready to publish the second volume of his *Basic Laws of Arithmetic* and it was in fact already in press when he was shocked to receive a letter from Russell which suggested that there was a serious flaw — an antinomy — in Frege's analysis of arithmetic in logical terms.[20] This was the famous 'Russell Paradox', which arose from the consideration of certain problems in the theory of classes.[21] Faced with such an unwanted and unforeseen situation, which seemed to threaten — indeed destroy — the whole basis of Frege's system with its hoped-for 'reduction' of mathematics to logic, Frege worked desperately to find a way out of the difficulty. He devised an amendment of his Basic Law 5, which was intended to prevent the generation of Russell's Paradox within his logical system. The revision was given in an appendix to the work and Frege expressed the hope that he had successfully coped with the difficulty and rescued the logistic foundations of mathematics.[22] However, subsequent commentators have concluded that Frege did *not* successfully find a 'way out' of his difficulty, and Frege himself seems to have lost confidence in his proposal as the years went by.[23] We shall not, therefore, follow Frege in the rather poignant efforts of his later years, but will turn now to see what Russell was doing in the philosophy of mathematics, and how he sought to cope with the difficulty that he had himself revealed. We shall, however, have occasion to make some further reference to Frege's work.

Russell

Bertrand Arthur William Russell[24] was a member of the British aristocracy (son of Lord and Lady Amberley), though by no means a typical one, as his outspoken comments and writings on social affairs bear ample witness. Though educated privately his precocious talent was early in evidence. He proceeded to Trinity College, Cambridge, where he read mathematics, later turning to philosophy. Russell's examiner for the scholarship examination for college entrance was A.N.Whitehead (1861–1947), who was highly impressed with the student's abilities, and the two men subsequently collaborated to produce their masterpiece, *Principia Mathematica*, published in three volumes just before the First World War.[25]

Russell's first major work was his *Essay on the Foundations of Geometry*,[26] which secured him a fellowship at Trinity College in 1895. In his early phase as a philosopher, he was influenced by the so-called Anglo-Hegelian movement and the idealist philosophy of F.H.Bradley (1846–1924). Partly in reaction against the scientific naturalism associated with Darwinism and evolutionary biology, the Anglo-Hegelians adopted doctrines from the German idealist philosopher G.W.F.Hegel (1770–1831), which offered a system that was a descendant of Plato's doctrine of Ideas. Hegel supposed that there was some mysterious transcendent entity — the 'Absolute' — which displayed its activity through the actions of human history.[27] Bradley adopted and adapted the notion of a transcendent 'Absolute', and the doctrine achieved considerable popularity in British philosophical circles in the late nineteenth century, being concerned with much more than the philosophy of history.[28] In Russell's case, he was attracted to a 'realist' view of mathematics, and at one stage he went so far as to suppose that he could imagine all the numbers having a state of 'timeless being', rather like Platonic Ideas.[29] But this phase of Russell's philosophical development was not to last long.

In 1900, Russell and Whitehead attended an International Congress of Philosophy in Paris. There Russell met the logician Peano, was introduced to his system of mathematical logic, and became interested in the general problem of 'reducing' mathematics to logic. In addition, Russell's attention was drawn by Peano to the work of Frege, who, as we have seen, was also deeply concerned with the relationship between mathematics and logic. Subsequently, it was through Russell's advocacy that Frege's work was gradually made known to mathematicians and philosophers, and its merits came to be appreciated.

Returning to England, Russell set to work on a manuscript which was to emerge in 1903 as *The Principles of Mathematics*.[30] In addition, he was at that time giving close attention to the work of Leibniz, and this resulted in the publication of *A Critical Exposition of the Philosophy of Leibniz*, also in 1903.[31] Like Frege, Russell could see in the work of Leibniz some anticipation of attempts to display the logical basis of mathematics.

The essence of Russell's work in *The Principles of Mathematics*, and subsequently with Whitehead in *Principia Mathematica*, was somewhat as follows. Numbers were to be defined as classes of classes: 'the number of a class [is] the class of all [those] classes similar to the given class'.[32] For example, the number '2' was the class of all couples or pairs; '3' was the class of all triads or trios; and so on. This approach, which was essentially the same as that previously adopted by Frege (though it was worked out independently by Russell), meant that a number was really the property that a set of similar classes (numerically speaking) had in common.

But the difficulty once more seems to present itself: how can one know that two classes are equinumerate, or contain the same number of entities, without empirically counting them and thereby deploying the notion of number? Again, Russell's procedure was like that of Frege, with the use of the concept of the 'one-to-one' relation:

A relation is said to be 'one-one' when, if x has the relation in question to y, no other term x' has the same relation to y, and x does not have the same relation to any term y' other than y.[33]

Pictorially, we may represent the matter as in Figure 27.

Figure 27

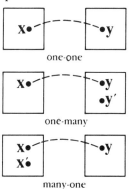

one-one

one-many

many-one

This gave Russell a definition of equinumeracy without necessitating the use of the number '1' in the definition. He could then proceed as follows:

> (1) O is the class of classes whose only member is the null-class. (2) A number is the class of all classes similar to any one of themselves. (3) 1 is the class of all classes which are not null and are such that, if x belongs to the class, the class without x is the null-class; or such that if x and y belong to the class, then x and y are identical. (4) Having shown that if two classes be similar, and a class of one term be added to each, the sums are similar, we define that, if n be a number, $n + 1$ is the number resulting from adding a unit to a class of n terms. (5) Finite numbers are those belonging to every class s to which belongs 0, and to which $n + 1$ belongs if n belongs.[34]

'This', concluded Russell, 'completes the definition of finite numbers'. As we can see, his procedure was rather similar to that of Frege.

So, all might seem well. Without pursuing matters further to consider problems of infinity, complex numbers, irrational numbers, and so on, one may ask the reader to allow that at least a *prima facie* case had been made out by Russell that arithmetic — and perhaps mathematics as a whole — could be grounded in logic, as Frege had intended. But it will be noted that the whole argument was dependent on a satisfactory under-standing of the notion of classes; and in particular it must be possible to speak of classes of classes, or indeed of classes of classes of classes. Classes had to be able to be members of other classes.

Now Russell noted[35] that there appeared to be two kinds of classes: those that *were* members of themselves (such as the class of things that can be counted, or indeed a class of all classes); and those that *were not* members of themselves (such as the class of all writers on the history and philosophy of science — which is not *itself* a writer on H.P.S.). It is when we consider this second kind of class that we encounter Russell's Paradox — the antinomy that had caused so much pain and con-sternation to Frege. For one may ask whether the class of classes that are *not* members of themselves is a member of itself. Suppose it is a member of itself: then it is not — for the entities within that class are by definition not members of themselves. Suppose it is not a member of itself: then it is — for it should belong in the class of things that are not members of themselves, by the very definition of that class.

This antinomy is analogous to the well-known paradox of the liar, in that it involves the notion of self-reference.[36] It can be seen that if taken seriously it would throw a large spanner in the works of the theory of classes, and would cast doubt on the whole logical basis of mathematics that Frege, Peano, Russell and others were trying to construct. Certainly Frege and Russell both took the matter very seriously. Russell discussed the problem in Chapter 10 of his *Principles of Mathematics*, and made some attempt to deal with it in an appendix to the volume, entitled 'The Doctrine of Types'.[37] The problem was worked out in greater detail in a paper published in 1908[38] and more fully in *Principia Mathematica*. But before we seek to attend to this, we should say a few words about Russell's treatment of the philosophy of mathematics through symbolic logic and Frege's apparatus of 'function' and 'argument'.

Like Frege, Russell found it useful to make a distinction between 'function' and 'argument'. Logic should be treated in terms of 'prop-ositional functions', into which various 'arguments' might be substituted as appropriate. With this plan in view, the idea was to show that logic provided the essential basis of mathematics. But certain difficulties had to be overcome before this program could be carried out successfully. When Russell wrote *The Principles of Mathematics*, he supposed that the meaning of a name corresponded with the thing it denoted. But this simple 'correspondence theory' presented great difficulties when one considered non-existent entities like 'unicorns', 'the present King of France', and things of that kind.

Russell sought to deal with such problems in a famous paper of 1905, entitled 'On Denoting'.[39] He made a clear distinction between the *meaning* of a phrase and its *denotation*. Contrary to his earlier views (which had it that if a phrase or term failed to denote something, it was meaningless), he now held that meaning and denotation were not to be regarded as one and the same. So on this view, a phrase which is a definite description (such as 'the present King of France') doesn't actually have to denote something in order to have meaning. Russell, then, would deal with problems such as those generated by the sentence 'The present King of France is bald' by rewriting them as follows:

1. The propositional function 'x is presently King of France' is true for at least one value of x.
2. The propositional function 'if x is presently King of France and y is presently King of France, then x is identical with y' is true for all values of x and y.
3. The propositional function 'if x is presently King of France, x is bald' is true for all values of the variable x.

It can be seen that each of these three propositional functions might be true or false, according to the circumstances. The third one would be false if there were a present King of France who was not bald; the second would be false if there were two present Kings of France; and the first would be false if there were no present King of France.

It will be further noted that what Russell did here was to present an *analysis* of the original proposition in terms of three propositional *functions*. In doing so, he helped to clarify the status of names, meaning and denotation, and prevented the world from being populated with mysterious entities such as 'the present King of France', 'round squares', 'golden mountains', etc., which could no longer be conjured into existence merely by virtue of being used in a meaningful way in various propositions. Russell's method was derived, in part, from his mentor, the philosopher G.E.Moore (1873–1958), with whom he had become acquainted at Cambridge. Moore taught his students that philosophical problems could be clarified (and hopefully dissipated) by breaking prop-ositions down (analysing[40] them) into their 'components', in order to provide a clarification of meaning.[41] It is clear that Russell's method for dealing with his problem of 'denoting' utilised this approach. Thus Russell was a leading member of the tradition of *analytical* philosophy. But he was also, it should be noted, using the notions of 'propositional function' and 'argument'.

In their *Principia Mathematica*, Whitehead and Russell seemingly succeeded in effecting a reduction of mathematics to logic (although today it is not universally accepted that they were successful). Moreover, they used a form of symbolism that has lasted, whereas Frege's two-dimensional representations are now quite obsolete. The logical axioms deployed by Whitehead and Russell were as follows:

1. $(p$ or $p)$ implies p
2. q implies $(p$ or $q)$
3. $(p$ or $q)$ implies $(q$ or $p)$
4. $[p$ or $(q$ or $r)]$ implies $[q$ or $(p$ or $r)]$
5. $(q$ or $r)$ implies $[(p$ or $q)$ implies $(p$ or $r)]$.[42]

These five axioms (which were written symbolically in the original text) may be 'translated' as follows:

1. If either p is true or p is true, then p is true.
2. If q is true, then 'p or q' is true.
3. 'p or q' implies 'q or p'.
4. If either p is true, or 'q or r' is true, then either q is true, or 'p or r' is true.
5. If q implies r, then 'p or q' implies 'p or r'.

It was subsequently shown by Bernays[43] that the fourth axiom could be derived from the others, and was therefore superfluous. And, as was shown by Nicod in 1917,[44] the whole system could be reduced to the use of a single logical 'constant', the so-called Sheffer stroke, $p|q$, meaning 'neither p nor q'.[45] But such a 'simplification' is rather illusory. Theorems written out by means of this symbolism become grotesquely cumbersome, and modern texts of symbolic logic do not normally employ the Sheffer system.

We shall not seek to pursue these matters further here. Let us take it for granted, for the present purposes, that Whitehead and Russell did successfully write out a large number of the fundamental theorems of mathematics, deriving them from the basic axioms of their system of symbolic logic. But we have, so far, failed to give an account of Russell's treatment of the problem of self-referent sentences (like the paradox of the liar). As we have said above (see note 38), Russell dealt with the difficulty by what is called his 'theory of types'. This involved dealing with the problem that had arisen in the logic of classes by treating numbers and classes as propositions of functions; so his procedure drew from ideas to be found in the work of Frege.

Russell defined a type as: 'The range of significance of a propositional function, ie, as the collection of arguments for which the said function has values'.[46] He then sought to avoid the antinomies arising in relation to self-referent or reflexive statements by arranging matters so that no 'totality' (class) contained members defined in terms of itself. This could be achieved by arranging propositional functions in a hierarchy. At the lowest level or first order would be propositional functions ranging over individuals, which individuals constituted the first logical type. Then there would be second-order functions which ranged over functions of the first-order propositions (which were of the second logical type). Next would come third-order functions dealing with functions of the second order. And so on, *ad infinitum*, should one wish to carry on with the game indefinitely. The particular point, then, was that there had to be no conflation of types for good logical hygiene to be preserved. It was, according to Russell's doctrine, *meaningless* to have propositions of order n containing propositions of order *n* within their range. Only propositions of order $n-1$ or less were permitted. Thus it was literally meaningless for the Cretan Epimenides to say 'All Cretans are liars'. But the difficulty could be evaded and meaningfulness restored by a suitable invocation of the theory of types. In Russell's words:

> [W]hen a man says 'I am lying', we must interpret him as meaning: 'There is a proposition of order n, which I affirm, and which is false'. This is a proposition of order $n + 1$; hence the man is not affirming any proposition of order n;...[47]

By means of this general procedure, Russell and Whitehead sought to evade problems similar to those that emerged in Frege's work, in relation to the program of reducing mathematics to logic. The procedure has subsequently met with general approval. We may note also that the doctrine of 'metalanguages' is analogous to Russell's theory of types. For example, at the beginning of this book we spoke of philosophy of science as being a 'metadiscipline' with respect to science itself. In philosophy of science one 'talks *about*' science. Thus if we envisage science as consisting of propositions of order n, then the language of philosophy of science contains propositions of order $n + 1$; and keeping this distinction clearly in mind allows good philosophical (or historiographical) hygiene to be preserved.[48] However, the notion of metalanguages is more closely associated with the work of Alfred Tarski (see p.247) than Russell.

In relation to Russell's philosophy of mathematics, we should mention one further point. It will be recalled that according to Russell's doctrine the number '2' meant the class of all couples. And obviously there are very many pairs of couples of things in the universe. But the number series should extend to infinity. Are there groups of (say) 10^{53} in the universe? Or 10^{1000}? There ought to be, if Russell's doctrine were to be satisfactory. To ensure himself against such doubts, Russell introduced the assumption — dignified with the name 'axiom of infinity' — that there was an infinite number of things in the universe.[49] But this seemed to introduce a kind of empirical element into mathematics — which Russell had sought to show was strictly deducible from logical principles. It is not, therefore, altogether surprising that the axiom of infinity has given rise to much subsequent discussion, and has generally been regarded as a weakness in Russell's system. Wittgenstein, for example, took exception to it in his *Tractatus* (para. 5.551), referred to below, note 57.

All these matters may seem somewhat recondite and of only marginal relevance to philosophy of science proper. But they were of considerable interest to Russell's most famous pupil at Cambridge, Ludwig Wittgenstein. Wittgenstein's work, in its turn, exerted a profound influence on the members of the Vienna Circle. And it was they who really formed the mainstream of work in philosophy of science in the first half of the twentieth century. However, before we proceed to a consideration of these matters, we must say something about Russell's theory of knowledge and his doctrine of logical atomism.

Russell's most lasting contribution to philosophy was, it is generally agreed, his philosophy of mathematics, and in particular the extraordinary achievement of *Principia Mathematica*. Perhaps it was this great intellectual *tour de force* which led Russell to suppose that the structure of the physical world is essentially the same as that of logic or of a logically perfect language. Thus the structure of the world was envisaged as being analogous to the logical structure of *Principia Mathematica*![50]

Russell worked on the project of displaying a 'logical structure of the world' for a number of years, having deep discussions on the matter with Wittgenstein when he was with Russell at Cambridge in 1912–13. The doctrine was first put forward publicly in detail in a series of eight lectures given in London in 1918, and in the introduction to the published version of these Russell stated that they contained 'certain ideas which I have learnt from my friend and former pupil, Ludwig Wittgenstein'.[51] It is believed, however, that this may have been somewhat over-generous to Wittgenstein, so far as Russell knew the state of his philosophy in 1918.

In describing his method, Russell explained that he called it 'logical atomism' because he was interested in carrying through the process of logical analysis to the very simplest kinds of notions — 'logical atoms'. Some of these would be 'particulars' — 'such things as little patches of colour or sounds, momentary things';[52] others would be predicates, relations, logical connectives, and so on. The terms 'logical analysis' and 'logical atoms' were presumably selected for their analogy with chemical analysis and chemical atomism; but the two kinds of analysis were obviously quite distinct. We have already noted that following Moore, Russell supposed that the analysis of concepts into their constituents was particularly conducive to the clarification of thought. And his supposition that there were logical atoms or 'simples' because there are logical 'compounds' is reminiscent of the opening section of Leibniz's *Monadology*. Thus it seems likely that Russell's early interest in the philosophy of Leibniz had a very considerable impact on the subsequent development of his thought and the establishment of logical atomism.

It will be noticed that Russell offered a patch of colour as an example of a logical atom. He also, on other occasions, referred to such perceptual appearances as 'sense data', a term he is thought to have derived from G.E.Moore. Russell supposed that the direct materials constituting our sensory experience were these 'sense data' — patches of colour, simple sounds, tastes, smells, and such like. And it was on the basis of these that people were thought to have built up their knowledge of the world. Russell, then, was very much part of the tradition of British empiricism, as well as that of Leibnizian rationalism. Since Russell sought to show how his 'logical atoms' might be treated according to the rules of a well-constructed logical language (like that of *Principia Mathematica*), he became one of the chief founding-fathers of the movement known as logical empiricism.

In his lectures of 1918, Russell then maintained that 'the world contains *facts*', a fact being 'the kind of thing that makes a proposition true or false'.[53] Suppose one says 'It is raining'; then it is a *fact* that makes this *proposition* either true or false. A fact is 'the sort of thing that is expressed by a whole sentence'; and 'facts belong to the objective world' — a statement expressive of Russell's realist position. It is propositions, not facts, that are either true or false. But propositions are not names for facts. (Russell acknowledged here his indebtedness to Wittgenstein for this point.) Indeed facts cannot be named. But the meaning of symbols in propositions is to be had, Russell informed his listeners, through direct acquaintance. One can only understand the meaning of a term such as 'red' by seeing red things. (The taste of a pineapple was, we remember, Hume's example.) However, a term like 'red' may be described, as it is in dictionaries or with the help of optical theory. Russell referred to 'atomic facts' and also to 'atomic propositions' such as 'This is white'. 'Particulars' were defined as 'terms of relations in atomic facts'.[54]

Also in his 1918 lectures Russell went on to show how one could seek to deal with 'molecular' (or composite) propositions, by considering the truth values ('true' or 'false') for their constituent 'atomic propositions'. That is, one could seek to combine the 'atomic propositions' in various ways, and then determine whether the 'molecular' combinations were true or false, given the truth or falsity of their 'atomic' constituents. For example, suggested Russell, for the molecular proposition 'p or q', using 'TT' for 'p and q both true', 'TF' for 'p true and q false', etc, one could say:

T T	T F	F T	F F
T	T	T	F

where the lower line stated the truth or falsehood of the molecular proposition 'p or q'.[55] Here we have a forerunner of the well-known 'truth-tables' of Wittgenstein, discussed below, p. 225.

Russell's logical atomism was originally developed within the context of his studies in the philosophy (or logic) of mathematics and the work of Leibniz, but it appeared that it might offer possibilities for philosophical advance within the broader terrain of epistemology. The truth or falsity of complex statements might, it seemed, be assessed by consideration of the truth or falsity of their most simple (empirical) atomic constituents. Thus the empiricist program, started so long before by Locke and carried forward by such distinguished philosophers as Hume, Mill and Mach, might at last be coming to fruition with the aid of the new methods of mathematical logic. Of course, there remained the difficulty of being really certain about the truth or falsity of the absolute bedrock of empirical knowledge — the 'sense data'. Did they genuinely represent reality? Well, dismissing such worries for the moment, we may, perhaps, see opening up before us the prospect of a grand union of logic and empiricism — logical-empiricism — which would provide a suitable framework for the philosophical analysis of science.

Yet, on reflection, one may find the whole enterprise rather suspect. For if one is going to start from the bedrock of empirical information (sense data), then it is clear that one must be seeking to *ascend* the 'arch of knowledge', a process that we have come to associate with *induction* and logical insecurity. Is it the case, then, that sense data can have certainty attached to them, so that they may be used for the purpose of constructing an empirically-grounded 'logical' picture of the world? This is implausible, for the sense data are in fact obtained as the last terms of analysis of compound or 'molecular' knowledge. But if the last terms of analysis are *empirical* elements, the 'arch of knowledge' has become inverted, so to speak. The empirical 'elements' (sense data) serve as the 'starting points' for a logical construction, but only back to 'molecular' knowledge of empirical matters, whence we started. It is not clear, then, that much is to be gained by all this — and certainly not theoretical explanation of empirical matters. The philosopher might claim, perhaps, that he could show in principle how to reach and display the empirical bedrock of science, offering philosophical clarification thereby. But the reader may be inclined to think that we are really likely to be more confused by all this than when we started.

Let us not, however prejudge the matter. Let us, rather, seek to follow the historical trail a little further. This requires us to look immediately at the work of Wittgenstein, and then at Carnap in whose work we find an attempt to construct a theory of knowledge which does move from 'molecular' knowledge to 'bed-rock' empirical simples, and thence (deducing all the way) to knowledge of things, and higher constructs still, such as psychological and cultural objects. (See Figure 32.)

Wittgenstein

Wittgenstein was a man of rare and extraordinary genius, yet there seems to have been something unbalanced in his personality, and he dropped out of the academic world for quite long periods of his life, though no doubt this was sometimes with the intention of thinking problems through with undivided attention.[56] Coming from a wealthy Viennese family (with a tendency towards suicide), he studied at the *Technische Hochschule* in Berlin, whence he moved to Manchester at the age of nineteen to study aeronautics. Then, as we have said, he studied under Russell at Cambridge in 1912–1913, having been advised to do so by Frege. Wittgenstein fought as a volunteer in the Austrian army during the Great War, apparently with great valour. Most men would not have spent such a time puzzling with problems in mathematical philosophy and logic. But Wittgenstein did, and by the end of the war he had ready for publication a short work that has subsequently become notorious for its difficulty, but which is recognised as being of fundamental importance in the history of philosophy — the *Tractatus Logico-Philosophicus*.[57]

One of Wittgenstein's most fundamental tenets was that philosophy is not a science; it does not tell one anything about actual states-of-affairs in the physical world.[58] Philosophy, in Wittgenstein's view, aimed at the clarification of thoughts. It was an *activity* rather than a body of doctrine.

Most modern philosophers would agree with this.

The *Tractatus* was written in the form of a series of aphorisms, each of which was numbered according to a special code. Each main section was numbered (1–7); then comments followed on the first aphorism in each section; then comments on the comments; comments on the comments on the comments; and so on, as the needs of the author's argument required. So the numbering of the aphorisms in the first section, for example, is: 1, 1.1, 1.11, 1.12, 1.13,1.2, 1.21.[59]

Perhaps the most important feature of the *Tractatus*[60] was its treatment of the problem of the relationship between ideas expressed in language and actual states-of-affairs in the world. This is one of the most fundamental problems in all philosophy, and it is still a puzzle today, in that there are many different views on the question. Wittgenstein sought to deal with the problem by means of his celebrated 'picture theory' of language.

As we have seen,[61] Russell's program of logical atomism had supposed that it was reasonable to treat the physical world as if it were a kind of 'logical object'. There was, he supposed, a form of correspondence between the logical operations of our mind, the logic of mathematics, and the way things actually are and behave in the world. But for Wittgenstein this position was not quite correct. In his view, logic *made possible* the representation of the world in thought, the thoughts being expressed through the symbolism of language. But the propositions of logic did not *themselves* represent the world. On the other hand, logic did reveal which states of affairs were *possible*. That is why he wrote (para. 1.13): 'The facts in logical space are the world'. Logic did not, of course, determine how the world was structured, like some external causal agent. It was, however, said Wittgenstein (para. 6.13), a 'mirror-image of the world'.

To illustrate these insights, Wittgenstein compared propositions to pictures. A picture, by depicting some scene, can represent some state-of-affairs. Language can do likewise, but by means of different symbolic conventions. Now, considering a picture, it bears some relationship to the physical state of affairs that it depicts. For example, if it is a picture of a house the chimney may be on the roof on both the physical thing being depicted *and* in the picture of the house.[62] In Picasso it might be very differently represented. But whatever the actual relationship between physical objects and their pictorial representations, the picture does not depict those *relations*; it depicts the physical world directly. Of course, one might try to elucidate the relationship between object and picture: for example, by means of some perspective construction lines. But this attempted clarification only introduces another picture, which itself will require further elucidation. Ultimately, the 'sense' of a picture has to be grasped directly; otherwise one gets into an infinite pictorial regress.

So in Wittgenstein's 'picture theory' of language, then, propositions making up a language in use are regarded as analogous to, or equivalent to, a series of pictures. But there is a kind of logical structure in language, reflecting the 'logical structure' of the world, for it is supposed that the world and language used to describe the world have the same logical form. Language 'pictures' *possible* states-of- affairs in the world. Its sense lies in this possibility. Propositions are *meaningful* when they possess the possibility of correlation with the world. Actual examination of the world will reveal whether they are true or false. Thus it is meaningful (and true) to say: 'There are nine figures in Botticelli's *Primavera*'. It is simply meaningless to say: 'Nine are *Primavera* Botticelli's there figures in'. Different logical (grammatical) rules could in principle be developed whereby such a sentence might be meaningful. But as it stands it has no 'logical structure' at all.

In considering the particular propositions of logic, Wittgenstein made use of ideas that were similar to those previously published by Russell[63] and which the two philosophers presumably discussed together when Wittgenstein was at Cambridge. For example, for the molecular proposition 'p or q', one could construct a so-called 'truth-table' as shown in Figure 28.

Figure 28

p	q	p or q
T	T	T
F	T	T
T	F	T
F	F	F

It follows, therefore, that the proposition 'p or q' could be rewritten as '$(TTTF)(p,q)$', if one so desired. Thus, we can see that the *sense* of a proposition is revealed by its truth possibilities.

Further, if all complex molecular propositions could in principle be analysed to their simple atomic constituents (rather as Russell did in his paper 'On Denoting', which we discussed above, p. 217), then there seemed to be the possibility of deploying a more-or-less mechanical method for establishing first the sense of molecular propositions, and then their truth or falsity — provided there was a means of knowing the truth-values of the atomic constituents. All this meshed with Wittgenstein's picture-theory of language. Language was, in its logical essence, truth-functionally structured, and there was (supposedly) a parallelism between the logical structure of language and of the world. But a suitable correspondence theory of truth — required at the level of the atomic propositions for the empirical basis of science to be secured — was by no means satisfactorily established. For the sceptic might still worry about how the truth-values of the atomic constituents might be securely determined.

However, at this stage, Wittgenstein was not so much concerned to establish the truth or falsity, as such, of scientific propositions. So proceeding to a less ambitious (or less ill-advised) task, he constructed various 'truth- functions' for some simple molecular propositions (para. 5. 101), as in Figure 29.

Here, 'truth' and 'falsity' might be regarded as 'arguments' in prop- ositional functions, in a manner analogous to that in which Frege had used arguments and functions.

Figure 29

pq	TT	FT	TF	FF
1. Tautology (If *p* then *p*, and if *q* then *q*)	T	T	T	T
2. Not both *p* and *q*	F	T	T	T
3. If *q* then *p*	T	F	T	T
4. If *p* then *q*	T	T	F	T
5. *p* or *q*	T	T	T	F
6. Not *q*	F	F	T	T
7. Not *p*	F	T	F	T
8. *p* or *q* but not both	F	T	T	F
9. If *p* then *q*, and if *q* then *p*	T	F	F	T
10. *p*	T	F	T	F
11. *q*	T	T	F	F
12. Neither *p* nor *q*	F	F	F	T
13. *p* and not *q*	F	F	T	F
14. *q* and not *p*	F	T	F	F
15. *q* and *p*	T	F	F	F
16. Contradiction (*p* and not *p*, and *q* and not *q*)	F	F	F	F

Now, considering particularly the cases of tautology and contradiction, Wittgenstein claimed (rightly) that they 'said' nothing at all about the world; so they 'lacked sense' (para. 4.461). On the other hand, they were not nonsensical (para. 4.4611). So, he suggested, they 'showed' something of the nature of logic — for one could construct a truth-table for them.

Wittgenstein further argued (para. 6.1) that all the reasoning procedures of logic were tautologies — saying nothing. Nevertheless, they displayed their logical form, and could be understood. On the other hand, one could not 'prove' the validity of logic. If I say:

If *p* then *q*
p
———————
Therefore q

there is no point in my trying to 'prove' this. You either see it or you don't. If I sought to prove it, or justify it in terms of some other logical principle, then that principle also would be in need of justification; and so one would have an infinite regress on one's hands. This was the point lying behind Wittgenstein's fundamental distinction between saying and showing. In his view, logic could only be shown, not stated. The same would hold for the sentences of everyday life, which *show* their internal relational structure, but do not *say* it.

Having thus sketched some of the features of Wittgenstein's doctrine, we may consider how he applied it to such matters as causal propositions and the laws of science — the realm of the contingent, rather than that of logical necessity. Here his position was quite like that of Hume. 'There is', he said, 'no possible way of making an inference from the existence of one situation to the existence of another, entirely different situation' (para. 5.135); for there is 'no causal nexus to justify such an inference' (para. 5.136). There is never any *logical* connection between states-of-affairs. And the procedure of induction is no more than an assumption of a principle of simplicity, compatible with our experiences (para. 6.363).

It would appear, then, that on this view the laws of science were not law-like in a *logical* sense. They could be broken, so far as logic was able to show to the contrary; they were simply summaries of experience, and did not allow true explanation. To explain laws in terms of other laws once again got one into a regress; so the Ancients (suggested Wittgenstein) were perhaps sensible to place the ultimate burden of explanation on God or Fate — some ultimate entity, deemed not in itself to be in need of explanation (para. 6. 372). Modern scientists do just this, of course. At any stage in the development of a science, some entity (eg mass or electric charge) is taken as fundamental and not itself capable of being explained.   Wittgenstein's approach to the problem of causality, which had troubled so many philosophers, was not merely that of Hume restated. 'The law of causality', Wittgenstein said, 'is not a law but the form of a law' (para. 6.32). By this he would appear to have meant, for example, that the principle 'Every event has a cause' was not empirical — not descriptive of nature at all. For two similar but separate events *always* differ in *some* respect from one another. Therefore, we can *always* ascribe differences between those events to some causal factor or factors. Consequently, in thinking about, or representing, the world as if the principle of causality held, we are merely saying something about the kind of rule we use for representing facts. It does not follow that the physical world actually *is* causal. In putting forward this view, Wittgenstein's position had clear affinity to that of Kant.

All this may, however, seem very puzzling when one considers that a scientific system such as Newtonian mechanics purports to give a mathematical description of the world, and tautologies are employed in the mathematical reasoning process. As we shall see in Chapter 7, one way of dealing with the problem is to think of the mathematical description as abstract and theoretical, consistent or (maybe) inconsistent, but neither true nor false. Connection with the world is achieved through so-called 'correspondence rules'. (In our discussion of Duhem, we saw nineteenth-century presentiments of this description of science.) As for Wittgenstein, his view of the matter (para. 6.341) was that Newtonian mechanics provided a means for description of the world — but there were other possibilities (Aristotelian, Einsteinian, we might suggest). The Newtonian description deployed causal connections in the description — but that did not mean that they 'existed' in the world itself. Newtonian mechanics offered *a way of talking about* the world, but not more. Nevertheless, Wittgenstein suggested, we could *show* that there are laws of nature by finding them.

So Wittgenstein asked his readers (para. 6.341) to think of Newtonian mechanics by means of the following analogy:

> Let us imagine a white surface with irregular black spots on it. We then say that whatever kind of picture these make, I can always approximate as closely as I wish to the description of it by covering the surface with a sufficiently fine square mesh, and then saying of every square whether it is black or white. In this way I shall then have imposed a unified form on the description of the surface.

But of course there can be different meshes of different gauges, and with holes of different shapes, corresponding to different forms of theoretical description. One could, I suppose, look at the world through a 'causal' mesh, or an 'acausal' one, and get different views accordingly. But this would not mean that the system could be *wholly* conventional. The actual figures on the white surface would certainly have a bearing on what is seen through the mesh. On the other hand, some meshes would be more convenient to use than others, and might be preferred accordingly. But one should always recognise that there are two components to knowledge: that which is analogous to the marks on the paper, and that which is analogous to the net — the objective and the subjective; or the *a posteriori* and the *a priori* in Kantian language. One can never eliminate the net and see the marks on the paper as they are 'in themselves'. There is, however, the possibility that we can really know more about the world itself by finding which kind of net (or scientific theory) allows it to be described most simply (para. 6.342).

We can relate all this back to Wittgenstein's discussions of language. Obviously, since we use our language to talk about the world, then particular features or limitations of our language will place limitations on the ways in which we can think about the world. As a trivial example, if a language has only three numbers — 'one', 'two', 'three', and then 'many' — a person using that language must think that a table has 'many' legs. Similarly the theoretical language, with the help of which one seeks to describe and make sense of the world, must necessarily shape the way in which one thinks about the world. In the *Tractatus*, Wittgenstein wrote: 'The *limits of my language* mean the limits of my world' (para. 5.6). And 'We cannot think what we cannot think; so what we cannot think we cannot *say* either'(para. 5.61).

In introducing the *Tractatus* to English readers, Russell maintained that the most fundamental of Wittgenstein's theses was that there must be something in common between the structure of a sentence and the structure of the fact that it asserted. (This, we may recall, was the main claim of the 'picture theory' of language that Wittgenstein espoused.) But that which was in common between sentence and fact could not be said in terms of the language being used for the formulation of the sentence. It could only be 'shown', not 'said'.

But what could be 'shown', not 'said', formed the domain of philosophy. Said Russell:

> Mr.Wittgenstein maintains that everything properly philosophical belongs to what can only be shown, to what is in common between a fact and its logical picture. It results from this view that nothing correct can be said in philosophy. Every philosophical proposition is bad grammar, and the best that we can hope to achieve by philosophical discussion is to lead people to see that philosophical discussion is a mistake... . The object of philosophy is the logical clarification of thoughts. Philosophy is not a theory but an activity... . The result of philosophy is not a number of "philosophical propositions", but to make propositions clear... . In accordance with this principle the things that have to be said in leading the reader to understand Mr Wittgenstein's theory are all of them things which that theory itself condemns as meaningless.[64]

Wittgenstein himself put the matter more metaphorically in the penultimate section of the *Tractatus* (para. 6.54):

> My propositions serve as elucidations in the following way: anyone who understands me eventually recognizes them as nonsensical, when he has used them — as steps — to

climb up beyond them. (He must, so to speak, throw away the ladder after he has climbed up it.)

He must transcend these propositions, and then he will see the world aright.

And then he concluded his extraordinary book (para. 7) with what has become one of the most famous pronouncements in the whole of philosophy: 'What we cannot speak about we must pass over in silence'.

This last was a reference, it is sometimes thought, to Wittgenstein's contention about the ineffability of mystical intuitions and ethical understandings. (So often morality can best be taught by example — by showing, not saying.) However, there was also, I suggest, an epistemological aspect to para. 7, summing up so much of the *Tractatus*'s earlier arguments. There is the world. We try to describe it in some language, scientific or otherwise. But there is a problem as to whether what we say about the world corresponds with the way the world actually is. We would like to know the true nature of that relationship of correspondence. But we can only indicate the relationship in an uncertain way, for we have to use language itself (of one form or another) to try to express the relationship. Wittgenstein tried to do this as best he could with the help of the picture theory of language. But he could only seek to *show* the relationship with the help of picture analogy — not state it explicitly in words. In seeking to make all this clear to his readers, he was, of course, engaged in philosophical activity. To be sure, he might seek to give an account of his argument in language; but the regress would begin, and nowhere could one come to an end. There would be descriptions of descriptions of descriptions ... *ad infinitum*. Nevertheless, he could try to bring his readers to the point where they too understood what Wittgenstein was seeking to show. But the whole thing could never be made exactly explicit. That is why he said that the philosophical analysis he was giving was ultimately without sense — for what needed to be expressed was inherently inexpressible. The sense of the relationship between the world and its description in language could not be expressed in language. *So*: 'What we cannot speak about we must pass over in silence'.

There is, I think, something extraordinarily intriguing about this remarkable conclusion that Wittgenstein reached, presumably some time while he was fighting the war. It is not surprising, therefore, that the *Tractatus* provided the subject for intense discussion amongst the members of the Vienna Circle, offering an enormous stimulus to their efforts to forge a coherent philosophy of science from a union of logic and empiricist epistemology.

After the war, Wittgenstein worked for a few years as a teacher in village schools in Austria. A fascinating account of this period of his life has been pieced together by W.W.Bartley III, chiefly by visiting Wittgenstein's former village acquaintances and pupils. It appears that he taught some of them the niceties of the paradox of the liar![65] But despite the genius that he displayed as teacher, Wittgenstein was eventually more or less run out of the village where he worked, very likely because of his homosexuality and partly because of his inability to relate to the peasant families in the villages where he worked, though he had excellent rapport with the children. He therefore returned to Vienna, where he did some work as an architect, and had discussions with selected members of the Vienna Circle, though he never became a full member of the group. He returned to Cambridge in 1929, and eventually occupied a chair there. At Cambridge, he exerted very considerable influence as a teacher, though publishing almost nothing. He wrote voluminously, nevertheless, and most of the philosophical material that he put together in this second period at Cambridge is now in print. Of these posthumous publications, the *Philosophical Investigations*, published in 1953,[66] were the most important, and particularly interesting since in them Wittgenstein revealed a remarkable reversal of many of the opinions earlier expressed in the *Tractatus*.

It is not necessary to examine Wittgenstein's later writings in detail at this juncture, but I cannot forbear from giving just a hint of the most important changes from the earlier position. We have seen that in the *Tractatus* there was a somewhat ambivalent attitude towards the relationship between logic and the world. The principles of logic supposedly represented the 'scaffolding of the world' (para. 6.124). They didn't represent the world itself, but the possibilities according to which it might be fashioned. But *why* should the principles of logic be 'seen' by someone to be true, even if it was not possible to put the reason in words (because of the distinction between showing and saying that we have previously discussed)? Is it because there is some kind of underlying logical structure either to the world, or to our thought systems, that somehow can be held responsible for the apparent self-evidence of the propositions of logic? Or is there a kind of Platonic 'form' for logic? (This would seem most unlikely!)

Partly as a result of such considerations, Wittgenstein became generally dissatisfied with his picture theory of language, and the notion that there was a correspondence relationship between states of affairs in the world and the propositions asserting or denying those states. This relationship seemed obscure if one put forth propositions about non-existent states of affairs, and the picture theory seemed ill-equipped to handle beliefs, hopes, fears, commands, expectations, and so on, as opposed to descriptions, assertions or denials. The problem came to a head for Wittgenstein when he asked himself *what made a portrait a portrait of some particular person*.[67] The answer, it appeared, was not some truth-functional relationship between states of affairs in the world and propositions asserting or denying the existence of such states, but the *intention* in the mind of the painter. So we have what is called in philosophy 'the problem of intentionality'.

After long reflection on such matters, Wittgenstein came to the conclusion that there could be no underlying logical structure to the world to which our minds must adhere, or *vice versa*. In the last analysis, the propositions of logic appear to us to be valid simply because of the processes of our education and upbringing. The propositions of logic reflect the rules of language, and these are known to us by our use of language in everyday life and linguistic experience. It is also through the use of language that we come to know, tacitly, which sentences in language are well-formed and meaningful, having a proper logical structure, and which are improperly constructed gibberish. Since Wittgenstein expressed these ideas even in relation to logic, let alone scientific hypotheses and theories, he was in fact putting forward views that were forerunners of those held by modern sociology-of-knowledge theorists, whose ideas I shall refer to again in Chapter 9. That is, Wittgenstein was giving expression in *Philosophical Investigations* to the view that meaning, and thereby knowledge, reside in the last analysis in social practices, particularly with respect to language.

The work of the later Wittgenstein was also influential in leading to the important 'concepts-influence-percepts' thesis that is now widely accepted by philosophers of science. A particularly persuasive exponent of this thesis has been the philosopher Norwood Russell Hanson (1924–1967) in his several entertaining publications, notably his *Patterns of Discovery* (1958).[68] Citing Wittgenstein extensively, Hanson produced a wealth of argument — psychological, historical and philosophical — to convince readers that we never just 'see' the world, passively absorbing impressions and then interpreting them, as the positivist sense-data theorists might lead one to believe. Observation, Hanson argued, is always an active process that is shaped by one's theoretical expectations, cultural assumptions, and linguistic background and attributes. A scientist doesn't just 'see' when he makes his observations: he 'sees that' or 'sees as'. Observing is, therefore, a 'conceptual' process. If this point be granted, it has considerable implications for one's understanding of the 'structure' of scientific theories. As we shall see below, in the heyday of logical positivism, theories came to be regarded as axiom systems, stated in terms of some 'theoretical language'. The 'theorems' arrived at by means of these deductive calculi were supposedly testable by observations, expressed in terms of some 'observation language'. Suitable translation rules ('correspondence rules') would serve as a form of 'interpretation' or 'connection' between the logically distinct theoretical and observational languages. But, if observations and theory are inextricably interconnected, as Hanson's arguments might lead one to conclude, then the whole structure of scientific theories, as envisaged by positivists, would be under threat. Thus while the 'early' Wittgenstein's philosophy furnished a major theoretical framework for the establishment of the logical positivists' view of science, the 'later' Wittgenstein produced arguments that were eventually to undermine the grand meta-scientific structure that the positivists created.

But we must not pass too swiftly to the post-positivist era. First, we must try to unravel the intricacies of the logical positivists' (or logical empiricists') account of science, as they appear in the voluminous writings of the Vienna Circle and the Berlin School, considering chiefly the works of Rudolf Carnap and Hans Reichenbach as exemplars.

The Vienna Circle

It will be recalled that Ernst Mach held a chair at Vienna in the philosophy of the inductive sciences from 1895 to 1901. In 1922, the chair was filled by Moritz Schlick (1882–1936), and it was he that established the famous series of (Thursday evening) meetings that ultimately gave rise to the movement known as the Vienna Circle,[69] espousing the philosophies of logical empiricism (or logical positivism) and physicalism[70] — also the so-called 'unified science' movement. Some of the most prominent members of the Circle were:

Alfred Ayer *philosopher*
Gustav Bergmann *mathematician*
Rudolf Carnap *philosopher and logician*
Herbert Feigl *philosopher*
Philipp Frank *physicist*
Kurt Gödel *mathematician and logician*
Hans Hahn *mathematician*
Felix Kaufmann *lawyer*
Victor Kraft *historian and philosopher*
Karl Menger *mathematician*
Otto Neurath *sociologist*
Kurt Reidemeister *mathematician*
Moritz Schlick *theoretical physicist and philosopher*
Friedrich Waismann *philosopher*
Edgar Zilsel *philosopher*

Karl Popper and Wittgenstein, both of whom were living in Vienna in the 1920s, were not members of the Circle, and did not attend its meetings, but they had discussions with some of the members, and Wittgenstein at any rate had a considerable influence on the movement's thought. They were greatly interested in the *Tractatus*. In particular, the members of the Circle maintained that their celebrated 'verification principle' (about which we shall say more below) was derived from Wittgenstein's work. The name 'logical positivism' was coined in 1931 by A.E.Blumberg and Herbert Feigl[71] to describe the general philosophical position of the members of the Circle. The name is in common use today, but its approximate equivalent, 'logical empiricism', is really more apt since it conveys directly the main intention of the group, namely to form a satisfactory philosophy of science from the union of logic and the epistemology of empiricism — thereby establishing a unified science, from which all metaphysics would be eliminated. All was to be achieved by the clarification of the meaning of language with the help of logical analysis. In fact, the logical positivists held that they were not doing philosophy at all — or if they were, it was a new form of 'scientific philosophy'. This claim does not seem very plausible today, but one can see why it was made. After all, Wittgenstein believed that the propositions of philosophy were unverifiable in experience, and so (in his book) nonsensical. Thus a 'scientific' philosophy, free from this grievous defect, could naturally have its attractions.

Earlier in this chapter, looking at the work of Frege, Russell and Wittgenstein, we have already examined what was perhaps the most important component in the historical background to the work of the Vienna Circle. But there were a good many other strands that really should be taken into consideration in a fuller account. Some of them have been given by Joergensen, whose list we may reproduce here:

1. Positivism and empiricism (especially the work of Hume, the Enlightenment philosophers, Comte, Mill, Avenarius and Mach).
2. The basis, aims and methods of the empirical sciences (Helmholtz, Riemann, Mach, Poincaré, Duhem, Boltzmann, Einstein).
3. Logistics and its application to reality (Leibniz, Peano, Frege, Russell, Whitehead, Wittgenstein).
4. Axiomatics (Peano, Hilbert).
5. Eudaemonism (ie, ethical and social systems directed to the attempted maximisation of happiness) and positivistic sociology (Epicurus, Hume, Bentham, Mill, Comte, Marx, Feuerbach, Spencer, Popper-Lynkaeus, Menger).[72]

Space forbids me to try to deal with all these figures, but readers will recognise that a number of them have already been accorded some discussion.

The Vienna Circle gradually broke up during the 1930s, partly through political pressure exerted by the Nazis — for some of the Circle's members were Jewish and the group tended to be left-wing in its political affiliations — and partly because of gradually widening disagreements within the group. Several members took university posts in America, and it was really as a result of the break-up of the Circle that its views came to be so influential in the English-speaking world, beyond the confines of central Europe. The logical positivists published a well-known journal, *Erkenntnis*, and later there appeared the important series of monographs called the *International Encyclopedia of Unified Science*, the title of which was an indication of one of the major features of the logical-positivist program — the achievement of a unification of the various sciences. To this endeavour, Auguste Comte, the founder of positivism, would surely have wished to have given his blessing.

In talking about the work of the Vienna Circle, it is probably convenient to begin with the discussions of 'meaning' and the well-known 'principle of verifiability', which stated that the meaning of a proposition is equivalent to its method of verification. (A proposition is that which remains true or false, as the case may be, throughout a variety of related sentences or ways of stating the proposition.) This principle of verifiability was given as a major tenet of the Circle by Waismann in 1930.[73] Further, only two kinds of cognitively meaningful propositions were recognised by the logical positivists: propositions that were analytic (in a Kantian sense[74]) and those that were empirically verifiable. This division, it may be noted, was very similar to that which one finds in 'Hume's fork', suggested back in the eighteenth century.[75]

This severe limitation on the range of meaningful propositions was regarded as a source of intellectual strength, not a weakness. Besides eliminating from the realm of meaningful discourse such sentences as 'the good is round and smooth' or 'the cosmos is infinite', it also excluded statements of a theological character, such as 'God exists'. For the notion 'exists' is not 'contained within' the concept of God[76] and one cannot explore the question of God's existence empirically: faith is required, not keen eyesight. Indeed, if applied rigorously, the verification principle would, it seems, exclude all metaphysics from the realm of the meaningful. But this in no way alarmed the logical positivists. On the contrary, it was what they earnestly desired, as part of their program of establishing a unified science and a 'scientific' philosophy.

We have noted that much of the early stimulus to the members of the Vienna Circle was provided by Wittgenstein's *Tractatus*. But one doesn't find the verificationist doctrine as such stated in that work. Perhaps the nearest one gets to it is: 'To understand a proposition means to know what is the case if it is true' (para. 4.024). On the other hand, the *Tractatus* was certainly interested in meaning, and the elimination of meaningless statements from philosophical discourse. So in this, Wittgenstein doubtless pointed the attention of the members of the Vienna Circle towards the verification principle. However, it was really their own creation, though looking back to the writings of Hume.

There has been a great deal of discussion in the philosophical literature as to the adequacy of the verification principle. Initially, there was uncertainty as to whether the principle was intended to refer to the *meaning* of a sentence or its *meaningfulness*. Schlick, for example, was concerned with meaning, and sought to indentify verification with some kind of process of ostensive definition. Thus the meaning of 'yellow' could be established by *showing* a yellow object.[77] Thereby he hoped to get rid of a regress — defining the meaning of words with other words, and so on *ad infinitum*, as in a dictionary. But meaning cannot be conveyed *wholly* by showing or pointing. If one merely shows something, or points to it, one cannot know which feature amongst the many possibilities is the one which is actually intended. There are further difficulties if one takes the trouble to inquire exactly what is meant by the process of verification. It may involve a good deal more than just looking. It may involve that, but also some kind of practical investigation. So when we identify the meaning of a proposition with its verification there can be considerable ambiguity as to what is actually entailed. For this reason, the verification principle came to be associated with the *meaningfulness* of propositions, not their meaning. This approach was taken, for example by A.J.Ayer, one of the early British representatives of the Vienna Circle movement, and responsible for making logical positivism well-known to English-speaking readers through his widely read *Language, Truth and Logic*.[78]

Even so, very serious difficulties remained. It is true that it is not easy to set down exactly how one might verify the metaphysical statement that 'God is love', or something of that kind. But there is the problem that verificationism would also appear to rule out as meaningless all ethical propositions, which have to be construed as no more than the man-ifestations of human beliefs or urges. Actually, some members of the Unified Science movement were perfectly willing to accept this, and they sought a behaviourist or sociological understanding of morality. But the 'emotive' theory of ethics, espoused by logical positivists, does not seem very plausible, and most people do not regard ethical propositions as either meaningless or merely behavioural.

At one time, it was commonly held against the verificationist principle that it excluded itself — for it was neither analytic nor empirically verifiable. But this objection is not, perhaps, particularly well founded. The principle can be regarded as belonging to a different logical 'type' (in Russell's sense) to the propositions about which it speaks. So, regarding it as a 'metastatement', there is no need for it to indulge in self-reference, with all the antinomies and paradoxes that are thereby involved.

Much more serious is the objection that general empirical statements cannot be verified, because of the problem of induction. For example, one cannot verify the statement that 'All iron has a coefficient of expansion of 0.000012', and it is clearly not analytic or tautologous. There are likewise difficulties about verifying statements about past events, not to mention some of the propositions in modern physics, as for example in quantum theory. How can one verify the *existence* of a 'wave-function'? For such reasons, therefore, the verification principle fails to make a clear-cut distinction between meaningful and meaningless propositions, thereby serving as a criterion of meaningfulness.[79] Consequently, it is rarely or never invoked for this purpose these days.

Carnap

There was, however, a good deal more to logical positivism than the verification principle; for logical positivism sought to give a general account of the structure of scientific theories, supposedly grounded in empirical basic statements with the help of rigorous logic. To get an idea of the model that came to be envisaged, we shall examine the work of Rudolf Carnap (1891–1970), who gave perhaps the most thoroughly developed account of the matter. It was his plan to establish an 'empiricist language', into which scientific laws and theories — but not metaphysical propositions — might be translated.

Carnap, who 'came from a family of poor weavers' in northwest Germany,[80] studied mathematics and philosophy at Jena, working under Frege. He began a doctorate in physics, but this was interrupted by the First World War. He served in the German army at the front in the early years of the fighting, and then was moved back to Berlin to work on wireless telegraphy. After the war he continued his studies in philosophy, obtaining a doctorate concerned with an analysis of the concept of space at Jena in 1921.

In 1926 Carnap moved to Vienna, where he joined the Vienna Circle and taught philosophy at the University. He taught at Prague from 1931 to 1935 and in 1936 he moved to America, where he occupied a chair at the University of Chicago. Carnap states in his autobiography that he was particularly interested in the work of Frege and Russell.[81]

Following Russell's work, *Our Knowledge of the External World*,[82] which he read in 1921, Carnap regarded experience as the fundamental groundwork on which all scientific knowledge must rest. His program was therefore very much within the empiricist tradition. But now Carnap felt better able to approach the problem of giving a satisfactory account of an empiricist theory of knowledge, and the empirical basis of science, with the help of the recent developments in symbolic logic, and the apparent spectacular success of the demonstration of the logical foundations of mathematics by Whitehead and Russell.

Carnap worked vigorously on this project in the early 1920s, and by 1925 he had finished a preliminary version of his first major book which was published in 1928: *Der Logische Aufbau der Welt (The Logical Structure of the World)*.[83] This can, I suggest, be seen as the culmination of the program, previously envisaged by Russell, which would give a rock-solid account of our knowledge of the world, starting from the simplest elements of sensation (whatever they might be) and building upwards logically to give a full account of all our thoughts. Thereby one could hope to give a thoroughgoing account of the empirical basis of scientific knowledge. This program had, of course, a good deal in common with the earlier phenomenalism of Mach.

Now Mach had taken his sensations and thoughts as givens, and had sought to analyse them to their origins in their constituent sensations, and then offer an account of how such sensations might account for our thoughts. It was, if you like, a latter-day version of the ancient procedures of analysis and synthesis, applied to epistemology. But Carnap did not suppose that his enterprise had much to do with actual psychological processes. It was, rather, a process of *'rational reconstruction'* — that is, 'a schematized description of an imaginary procedure', consisting of rationally prescribed steps, which would lead to essentially the same results as the actual psychological process'.[84] However, he recorded[85] that he changed his plans somewhat when he became familiar with the *Gestalt* psychologists, who showed that we never see things as the mere sum of very many discrete sensations, but as 'wholes'. So, he decided, 'sense data' are not the true ultimate constituents of sensation. Carnap determined, rather, to use total instantaneous experiences as the elements of sensation, rather than discrete sense data.

To proceed, Carnap had to assume that one could recognise, and recollect, certain similarities and differences between elements of sensation within the total 'stream of experience'. These 'recollection[s] of similarity', rather than elementary experiences, were chosen as the bedrock of the whole construction system.[86] Thus *relations* rather than (say) elements of sensation formed the foundation level. But Carnap immediately began to move in a logical circle, for the basic elements of sensation (elementary experiences) were then defined[87] in terms of the primitive relation: recollection of similarity. The system did not, therefore, get off to an auspicious start!

Two or more elementary experiences could, then, supposedly be recognised as similar; and consequently so-called 'similarity circles' could be discerned by recognition of partial similarities between the elements of experience.[88] I hazard a schematic representation of a 'similarity circle' as in Figure 30. In this, all the elementary experiences, which together constitute a similarity circle, must each of them wholly or partly overlap a given 'quality sphere' (for example, a portion of the artists' 'colour sphere', of arbitrarily stipulated size). There could, of course, be innumerable such 'similarity circles'; and relating to hearing, smell, etc., as well as sight.

But the similarity circles would themselves overlap; and this allowed Carnap to entertain the notion of 'quality class'. His definition was:

A class k of elementary experiences is called a quality class if k is totally contained in each similarity circle which contains at least half of it, and if, for each elementary experience x which does not belong to k, there is a similarity circle in which k is contained, but to which x does not belong.

Figure 30

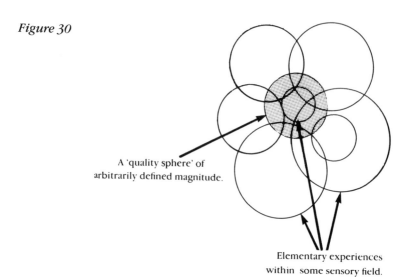

A 'quality sphere' of
arbitrarily defined magnitude.

Elementary experiences
within some sensory field.

By means of this cumbrous definition,[90] Carnap was seeking to give exact expression to the thought that although similarity circles may overlap they need not do so in a totally inchoate fashion, but in such a manner as to permit recognisable qualities to be discerned. So representing some overlapping 'similarity circles' now by broken curves, we might pick out a 'quality class' (k) in accordance with Carnap's definition, as in Figure 31.

Figure 31

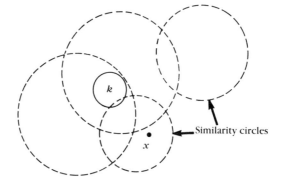

Similarity circles

The 'quality classes', said Carnap, should be the largest possible classes conformable with the definition. This definition was intended to provide that the 'quality classes' were not divided by essential overlappings of similarity classes.

The quality class k, then, might represent some particular shade of red — abstracted from the vast number of visual experiences. One could then proceed further to the recognition of similarities between quality classes, thence to sense classes, the visual sense, visual field places, colours, space and time order, the visual things, one's body, one's other autopsychological objects, physical objects, other persons, heteropsychological objects (other minds), and cultural objects (eg, a tribe).[91] The logical connections between the upper levels of Carnap's 'construction' scheme were not given as rigorously as those at the lower level. However, the same approach was intended to be applicable throughout, so that complex constructs could be regarded as logically connected with the base-level of sensations. This being so, the empiricist program of *logically* reducing all knowledge to sensations was ostensibly carried through or made possible.

The reader would not, I fear, thank me for dragging him or her through all this. I hope that enough has been said, however, to give at least the gist of what Carnap was trying to do. It was a thoroughgoing exercise in logical empiricism, seeking to show how a rigorous account could be given of our understanding of the world and its constituents (both physical and mental), working upwards from the most fundamental levels of empirical experience, using the logic of classes and their relations all the way. Also, since all the moves from the 'ground level' up were supposed to be (in principle) deductive, it followed that complex concepts could (in principle) be logically analysed down to their simple constituent empirical elements. Thus would the empiricist program be carried through.

For sheer painstaking effort, combined with a high degree of philosophical ingenuity, Carnap's system certainly has claims upon our attention and admiration. *The Logical Construction of the World* was a philosophical *tour de force*. It was also, I suggest, altogether rather pointless. It was, as Carnap himself openly acknowledged, a specimen of 'rational reconstruction', no doubt interesting as an intellectual exercise, but having absolutely nothing to do with the way we *actually* think, either about ourselves or the external world. To find out about that sort of thing, we require the science of psychology. Also, Carnap's enterprise was of negligible interest or use to practising scientists.

Carnap, however, would not have thought such criticism relevant to his enterprise. He was trying to show how, *in principle*, knowledge of complex phenomena, social, psychological and physical, could all be accounted for in terms of the most simple elements of sensation. Also, he sought to show that there was a strictly logical connection between the levels of the simple and the complex. In this, his program was not really so very different, in essence, from that of Locke and Hume. Locke, it will be recalled, had to assume that the mind could compare and contrast simple ideas. Likewise, Carnap had to assume the possibility of the mind being able to recognise similar features in the most elementary instantaneous experiences. The great difference was the vast logical apparatus he brought to bear on the problem.

But as I say, one must be sceptical of Carnap's program. For a start, it virtually ignored the active role of the mind in making sense of the world, such as we discussed in our chapter on Kant. One must grant Kant's point that the mind *does* bring something to the process of cognition, even if one is dissatisfied with his particular account of how it does so. Whether this '*a priori*' component arises within the 'ego', the neuronal synapses, the social structure, or whatever, is not here the point at issue. The thing is that it is not to be ignored. For all the magnificence of his logical apparatus, Carnap's account of cognition and the 'logical construction of the world' was still essentially based on the assumption of a 'white paper' (or a *tabula rasa*) theory of the mind. The only active power that Carnap allowed to the mind was the capacity to notice and remember similarities and differences between elements of the sensory field.

One may also have scepticism, on general grounds, towards the whole logical program, as deployed in the *Aufbau*. It starts off with the assumption that the mind can identify similarities occurring within the flux of sensory experience. It is on the basis of these perceived similarities that Carnap's whole construction proceeded. Yet, if one follows this route, one immediately runs up against the problem of induction, as do all full-blooded empiricists. Personally, I am not overly distressed about the problem of induction, which seems to me to be something of a pseudo-problem. But I would most certainly be concerned if I were trying to mount an empiricist epistemology on a rigorously logical footing. (Admittedly, however, Carnap's later work was to a large degree directed towards the problem of inductive inferences in science. My complaint here is specifically directed against the early

Figure 32

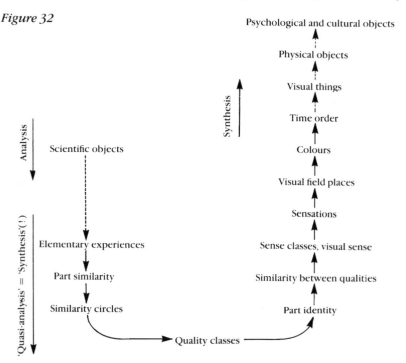

Aufbau – which Carnap himself came to regard as a less than fully successful undertaking.)

There is a curious section of the *Aufbau*, entitled 'About Analysis and Synthesis',[92] which may remind us of some of the features of the convoluted history of this doctrine, and which throws some interesting light on the relationship of Carnap's doctrine to the 'arch of knowledge'. Carnap, we have seen, having taken due note of the findings of *Gestalt* psychology, chose the stream of 'elementary experiences' as the starting point for his theory. Nevertheless, he blithely went 'below' the level of these experiences to consider 'partial similarities' and thence he built up the 'similarity circles'. But since the 'elementary experiences' were supposed to be fundamental, they should have been unanalysable. So Carnap called the process whereby the unanalysable elementary experiences were 'analysed', 'quasi-analysis'. His model, then, would seem to have been somewhat as in Figure 32.[93]

But 'quasi-analysis' could be regarded as a 'synthesis', since it involved the build-up of classes of elements, shown by partial similarities between basic experiences. The whole exercise was by this stage becoming quite bizarre, but we can see some vague resemblance to the ancient doctrine of analysis and synthesis when Carnap wrote:

> Analysis is possible only if, and to the extent in which, synthesis has preceded; it is nothing but a retracing of the path of synthesis from the final structure to intermediate entities and finally — if the analysis is 'complete' in the sense of construction theory — to the basic elements.[94]

We may note that for Carnap (using his ponderous logical apparatus) the movement of synthesis was deductive — as it was for the ancients, and for Newton. But by a kind of sleight of hand he seemingly managed to turn the arch of knowledge right over and build deductively from observational experiences. Indeed, this was what (he supposed) was required of a philosopher of science. Scientific objects should be analysed into their constituent elementary components and then 'reconstructed'. Once this had been achieved, then the process of verification could be undertaken with proper rigour, for, wrote Carnap: 'verification means testing on the basis of experience'.[95] It would appear that he was aiming for, and had gone some way towards achieving, the kind of system we foreshadowed at the conclusion of our discussion of Russell, but for which we did not consider the augeries to be very favourable.

In fact, few people nowadays seem to have much to say in favour of Carnap's *Aufbau*, and it has had few followers,[96] the task of logically 'constructing' physical objects out of phenomena (elements of *sensation*) seeming quite unrealisable. This despite the fact that phenomenalism has a very respectable intellectual history, being an essential component, for example, of Mach's positivism. In fact, Carnap himself soon repudiated his own efforts,[97] coming to prefer a 'physicalist', rather than a 'phenomenalist' basis for his 'construction' program. That is to say, observable *things*, rather than phenomenal *experiences* or *sensations* were to serve as the bedrock for giving an account of the objects of science. But this disavowal of his earlier labours may not have cost him so very much pain. In his autobiography he specifically stated that he was prepared to use a variety of different philosophical languages and approaches.[98] Even when he wrote the *Aufbau*, he envisaged alternatives to phenomenalism.

As we have previously indicated, one of the continuing preoccupations of the members of the Vienna Circle and logical positivism was the demonstration of the unity of science. This continued a tradition stemming from Comte, the founder of positivism. Carnap's *Aufbau*, had it been an unqualified success, might have served as a basis of such a unification, in that the elementary experiences were supposed to serve as the basis for the 'construction' of all 'scientific objects', and could be obtained from them by analysis. But having abandoned the phenomenalism of the *Aufbau*, some alternative basis for unification was obviously called for. This was provided in his *Unity of Science*, referred to above.[99] Here, Carnap utilised the notion of so-called 'protocol statements' or 'observation statements', instead of the elementary experiences of the *Aufbau* or the 'atomic propositions' of Russell and Wittgenstein. The word 'protocol' literally means 'first glue', and is derived from legal terminology. When a lawyer used to look after documents on a case, he would have them glued into a book for safe keeping. The one first glued in was the 'protocol'. By analogy, then, Carnap and his fellow logical positivists looked for some 'first statements' which could serve as the fundamental basis for the construction of scientific knowledge.

But these 'protocol statements' were supposed by Carnap to belong to the language of what he called the 'formal mode' (not the 'material mode'). That is, they belonged to the language of theoretical science, and hence could be incorporated as 'arguments' (in Frege's sense) into the logical structure of the reconstructed theoretical language of science. Therefore, they were not to claim to describe directly perceived experiences or phenomena, but were to be statements about language; *then* they could be fitted into the theoretical structure making up a science. When one considers this proposal, one might well think that contact with experience — with the real world of things — would have been broken. Actually, the examples suggested by Carnap might allay one's fears on that score. He suggested, for example: 'Joy now'; 'Here now blue'; or 'A red cube is on the table'.[100] Or more plausibly (for a reconstruction in philosophy of science), the protocols might be records of direct observations in a laboratory notebook.[101]

Yet Carnap's examples hardly seem to be statements about language. Otto Neurath took this requirement more seriously, and suggested examples such as: 'Otto's protocol at 3.17 o'clock: [At 3.16 o'clock Otto said to himself: (at 3.15 o'clock there was a table in the room perceived by Otto).'[102] But even this parade of precision (in the formal mode) could not eliminate all the awkward questions that the trouble-maker might wish to ask.

What day did Otto observe his table? Which room was he in? ...? So protocols, even those stated with such precision as those of the good Otto, could not be incorporated into the required logical structures of the 'formal mode' of language so that there would be no uncertainty whatsoever as to *exactly* what happened, as required for a formally reconstructed science. Or, insofar as the protocols *could* be said to belong wholly to a *formal* system, the looked-for connection with the world of experience was necessarily broken. So the never-ending problem for the formalist of trying to incorporate empirical elements into a logical system was not solved satisfactorily. In any case, it might appear that Neurath's protocols were phenomenalist in character — a feature that had supposedly been given up by the logical positivists after the *Aufbau*. It might also appear that they could not be intersubjective: they talked about Neurath's own private, personal experiences. Yet the logical positivists' reconstructed formalism was supposed to be of universal applicability and generality. Neurath believed that this difficulty could be circumvented with the help of a 'coherence theory of truth', referred to further below on p. 247. But as we shall see, this was by no means entirely satisfactory. And even more fundamentally, it was difficult to know what was and what was not a protocol sentence – the positivists being unable to find agreement amongst themselves on this issue.

In *The Unity of Science*, Carnap worked hard to argue for the intersubjectivity of protocols. People, he claimed, could check each others' observations, by different methods if necessary. Also, he claimed:

> All statements whether of the protocol, or of the scientific system consisting of a system of hypotheses related to the protocol, can be translated into the physical language. The physical language is therefore a universal language and since no other is known, [it is] the language of all science.[103]

Nevertheless, despite the apparent confidence of this assertion, with all science ostensibly reducible to physics, the search for a satisfactory basis for the unification of science continued, and in 1938 the publication of the *International Encyclopedia of Unified Science* offered a somewhat different approach. In the first volume, Carnap contributed a useful essay setting out the intended program of unification, and how it was to be achieved.[104] It involved the reduction of all the sciences (including psychology and sociology) to physics, in the following way.

'Physical language' was defined as the sublanguage of science that contained physical terms — together with logical and mathematical terms.[105] It would contain such terms as temperature, electron, pressure and so on. Then there was the 'physical thing-language' (also called merely the '*thing language*') which was made up of the common parts of physical language and everyday prescientific language.[106] It would contain terms referring to visible objects such as tables — neither too small (such as atoms) nor too large (such as nebulae). Also, the 'thing language' would contain terms such as 'hot' (which was common to both prescientific and physical languages), but not ones such as 'temperature'. Special scientific instruments were needed to measure temperatures, so this term did not belong to the vocabulary of prescientific language. Now terms from the 'thing language' such as 'heavy', 'red' or 'soluble', were examples of what Carnap called '*observable thing-predicates*'.[107] It was the thesis of physicalism, therefore, that all scientific statements could (if necessary, and with a good deal of circumlocution) be reduced to, or given in terms of, the 'thing language' or in terms of 'observable thing-predicates'. This, then, was to be the basis of the program for unified science. In effect, 'observable thing-predicates' were to play the same role as the elementary experience of sensation of the *Aufbau* or the protocol sentences of *The Unity of Science*.

To effect the 'reductions' to the level of the 'thing language', Carnap used a special device (called a 'reduction sentence'), which he had described in an important earlier paper: 'Testability and Meaning'.[108] Suppose, for example, one wished to 'reduce' the statement 'The body x has an electric charge at the time t' to a sentence operating at the level of the so-called 'thing language'. To do this, one might choose to say: 'If a light body y is placed near x at t, then: x has an electric charge at t [is equivalent to] y is attracted by x at t'.[109] Putting these sentences in the form of symbols, we might write:

The body x has an electric charge at the time t

Q_3

y is attracted by x at t

Q_2

A light body y is placed near x at t

Q_1

(Physical language)

(Thing-language sentences)

So Q_3 is the physical (theoretical) property (of being electrified) that is to be reduced to an experimental situation (Q_1) and an experimental result (Q_2). Thus by substituting Q_1 and Q_2 in place of Q_3, one could replace the sentence of the physical language (Q_3) by 'thing-language' sentences, Q_2 and Q_1. And in principle, Carnap supposed, whole chains of reductions might be carried out in a similar manner, so that theoretical physics, in all its complexity, could be restated in terms of observational sentences at the 'thing level'. In 'Testability and Meaning', Carnap gave a short-hand representation of a reduction sentence (such as might be required in accordance with his program) as follows:[110]

If Q_1 then (if Q_2 then Q_3)

This could be translated as:

'If a light body y is placed near the body (x) at time t'

then

('If y is attracted to x at t') then ('x is electrified at t')

So, by the use of this kind of procedure, Carnap sought to give a general account of how the physicalist program of reduction to 'thing-predicates' might be carried through. Thereby, the empirical basis of all science could in principle be displayed, for the meaning of all sentences in the theoretical language of science could supposedly be 'cashed out' (via physics) in terms of observational sentences at the 'thing level'.

In 'Testability and Meaning', Carnap was already beginning to retreat from the hard-line 'verificationism' of the early Vienna Circle. It was recognised that because of the problem of induction a complete and full procedure of verification was inherently impossible to achieve. So Carnap was settling for something 'weaker', namely 'confirmation'.

During the past forty years or so there has been a very large philosophical literature developed round the theory of confirmation, but we shall not seek to attend to this topic here.[111] Let us simply note that the chief problem that Carnap was seeking to deal with in 'Testability and Meaning' was the relationship between the theoretical language, in which the deductive structure of the theory was represented, and the way in which that might be related to the empirical basis of science — that is, to statements about objects at the 'thing level' or, as it was called the 'observation language'. How might theory and observation be logically knitted together? That was the most pressing question.

Carnap was by no means the first person to be concerned with this problem. In fact, it was present in embryo in the description of the structure of scientific theories given by Duhem which we discussed in the previous chapter. And there were many others in Carnap's day who were seeking to deal with the same problem. But since we are taking Carnap as our exemplar of the work of logical positivism in full flower (so to speak), we shall restrict our consideration to the work of this particular philosopher.

I may remind the reader that in principle it is possible for the mathematician or the logician to set out a deductive system, from axioms to theorems, with complete rigour — that is, in accordance with an agreed system of consistent deductive procedures. (In fact, work of this kind had been going on at least since the time of Euclid, though without complete rigour in early geometry). This generated the descending deductive leg of the 'arch of knowledge'. In the case of physics, this would be the work of the theoretical physicist, whereas the task of the experimental physicist was to give an interpretation of the theoretical terms and test the predictions made in the theory by providing empirical input and testing the theoretical output. To show what could happen to a very simple piece of physics when a logician like Carnap got to work on it, seeking to reveal its full logical structure, I should like to take an example from one of his publications, first issued in 1939.[112] It may give some indication of the kind of knots in which the logical positivists could become tangled when they sought to put their logical-empiricist program into practice.

To explain the physics first: if the length of a metal rod is measured, and if the rod is heated through a measured temperature increase, it will expand and the new length may be measured. Hence the expansion may be calculated by difference; and one may also readily calculate the increase in length of each unit-length of rod for each unit-rise of temperature. This is called the linear coefficient of thermal expansion, and may be symbolised by α. For iron it has the value of about 0.000012 cm/cm/deg C — that is, 0.000012 deg C^{-1}. This means that a rod of iron 1 cm in length becomes 1.000012 cm long when its temperature is raised by 1 deg C. The equation for thermal expansion may also be written as:

$$l_{T_2} = l_{T_1}[1 + \alpha(T_2 - T_1)].$$

When Carnap had axiomatised all this, it came out as follows.[113]

If:

x = some object

t = time
l = length of object
T = temperature
α = coefficient of thermal expansion
Sol = solid
Fe = iron
te = temperature of ...
lg = length of ...
th = coefficient of ...
c = some particular object

Then:	1. c is a Sol.
	2. c is a Fe
Premisses	3. te($/c$,O) = 300
	4. te(c,600) = 350
	5. lg(c,0) = 1000.
Axiom A_1	6. For every x, t_1, t_2, l_1, l_2, T_1, T_2, α {if [x is a Sol and lg(x,t_1) = l_1 and lg(x,t_2) = l_2 and te(x,t_1) = T_1 and te(x,t_2) = T_2 and th(x) = α] then $l_2 = l_1(1 + \alpha(T_2 - T_1))$}.
Proved mathem. theorem:	7. For every l_1, l_2, T_1, T_2, α [$l_2 - l_1 = l_1\alpha(T_2 - T_1)$ if and only if $l_2 = l_1(1 + \alpha(T_2 - T_1))$].
(6)(7)	8. For every x, t_1, ... (as in [6]) ... [if [- - -] then $l_2 - l_1 = l_1\alpha(T_2 - T_1)$].
(1)(3)(4)(8)	9. For every l_1, l_2, α {if [th(c) = α and lg(c,0) = l_1 and lg(c,600) = l_2] then $l_2 - l_1 = l_1\alpha(350-300)$}.
Axiom A_2	10. For every x, if [x is a Sol and x is a Fe] then th(x) = 0.000012.
(1)(2)(10)	11. th(c) = 0.000012.
(9)(11)(5)	12. For every l_1, l_2 [if [lg(c,0) = l_1, and lg(c,600) = l_2] then [$l_2 - l_1$ = 1000 × 0.000012 × (350-300)].
Proved mathem. theorem:	13. 1000 × 0.000012 × (350 − 300) = 0.6.
(12)(13)*Conclusion*	14. lg(c,600) − lg(c,0) = 0.6.

Now, one would hardly want to deny that this establishes the required conclusion with full logical rigour. The axiom system relevant to the expansion of iron rods has seemingly been given and used successfully. An empirically testable conclusion has been derived from the axioms. It may not seem a very attractive kind of enterprise, and given the complexity of even the very simple example just considered, it is obvious that it would be a horrendous task to attempt to reconstruct the whole of physics in the manner that has just been illustrated. Admittedly, however, Carnap was not necessarily concerned to do this. His task was fulfilled if he could show how the problem might be tackled in principle.

In discussing his example, Carnap described it as follows:

The calculus is first constructed floating in the air, so to speak; the construction begins at the top and then adds lower and lower levels. Finally, by the semantical rules [i.e., those which give meaning to the theoretical terms, or symbols, in the calculus], the lowest level is anchored at the solid ground of the observable facts.[114]

We see, very obviously, that we have here something analogous to the descending side of Plato's arch, left with the axioms unsecured, just as were the axioms of the Greek geometers, to which Plato took such exception. In so far as Carnap's arch was secured, this was done at the lowest level, where an empirically testable consequence emerged from the thicket of the logical calculus.

But we should notice several important points. The first is that the rationally reconstructed theory does not consider the problems of experimental errors, which are important in considering the validity of the formal structure reproduced above. For example, the 'premisses' should really be such as $te(c,0) = (300 \pm e)$, rather than $te(c,0) = 300$, where e represents the experimental error. So, if errors are taken into consideration, the whole situation becomes even more complicated than that given by Carnap — which was surely sufficiently complex in itself! A second point is that the whole rational reconstruction undertaken by Carnap is purely '*static*'. It says nothing about the experimental or theoretical procedures that might lead one to a knowledge of the physics of thermal expansion; it merely polishes up the logic of the theory *after* it has been developed. So one leg of the arch is missing entirely. Nevertheless, it might seem to have some kind of magical existence, for we notice that an empirical element is simply brought in as an *axiom* (A_2). And there is an empirical generalisation (law) in axiom A_1. But this is not surprising: concrete empirical predictions can only be got out of the system if empirical elements have been put in. And since the same kinds of experiments on expansion would have to be performed in order to determine the coefficient of expansion of iron (0.000012 deg C^{-1}) as are envisaged in the testing of the conclusions drawn from the logical calculus, the whole exercise is like a grand argument in a circle. It is like deducing the wonderful result that 'Socrates is mortal' from the premise that 'All men are mortal'. So one may be doubtful whether Carnap's effort revealed anything of great interest about the relationship between theoretical and empirical knowledge!

However, the paper of Carnap that we have just been discussing by no means represented his last word on the topic. Turning to a paper he published in 1963,[115] we see the problem of the relationship between the language of theory and of observation considerably clarified. Here he made a firm distinction between the language of theory (L_T) and that of observation (L_O), and he sought to give a more definite account of the relationship between the two in terms of so-called correspondence rules. It was this view of the structure of a scientific theory that Carnap seems to have retained till the end of his career. One of his clearest expositions of the matter is to be found in his late publication, *Philosophical Foundations of Physics* (1966).[116] Here he gave as specific examples of correspondence rules:

(i) If there is an electromagnetic oscillation of a specified frequency,

then there is a visible greenish-blue colour of a certain colour.
(ii) The temperature (measured by a thermometer and, therefore, an observable in the wider sense...) of a gas is proportional to the mean kinetic energy of its molecules.[117]

So the general picture of a scientific theory that Carnap espoused at that time was of the form shown in Figure 33.

Figure 33

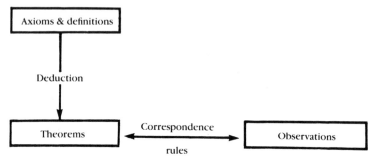

But when put in this (admittedly simplified) manner, we can see how very little further forward the whole question of the structure of scientific theories had been carried since the time of Duhem.[118] Moreover, there is a somewhat arbitrary and subjective decision as to where the 'correspondence rules' are to be placed in the system. They are, after all, postulates, or assumptions, made somewhere within the body of the theory. And it would, in principle, be possible to relocate them in the axioms. Indeed, this was, in effect, what was done by the so-called 'operationalists', whom we shall discuss in the next chapter. In other words, there seems to be the possibility of a good deal of freedom of choice as to the way one might seek to provide a rational recon-struction of a complex scientific theory. So it might become a novel kind of art form as much as a strictly logical enterprise! But worse than all this is the way in which the use of correspondence rules simply begs the whole question of the problem of the relationship between theoretical and empirical knowledge. The logical positivist, in his rational recon-struction, simply lays it down, as some kind of arbitrary decision that theory and observation are to be connected in the manner specified by the correspondence rules of the theory. But with this assumption, the whole tangled problem of the theory-ladenness of observations, or of the relative certainty that can be ascribed to observational and theoretical knowledge, is thereby swept under the carpet, ignoring the fact that cor-respondence rules may, on different occasions, function as theoretical postulates, empirical laws, theorems or definitions. In any case, one can argue that the 'gap' they are intended to bridge is really an illusory one, so far as the *meaning of terms* is concerned.

The problem of the relationship between statements about the world and actual states-of-affairs in the world has been perennial in philosophy. In this chapter, for example, we have seen how greatly it troubled Wittgenstein. What was needed was a satisfactory correspondence

theory of truth — truth consisting in some form of known correspondence between beliefs and facts. Wittgenstein's picture theory of
language was actually an example of such a theory. Another example,
which appealed particularly to Karl Popper, was developed by the Polish
logician, Alfred Tarski, in 1931.[119] Tarski utilised the notion of
metalanguages, so that a sentence asserting the truth of a sentence in
some language must not be regarded as belonging to that language, but to
an appropriate metalanguage. The usual example is: '"Snow is white" is a
true sentence if and only if it is the case that snow is white.' To put it in
this simplistic way is to do but little justice to the subtleties of Tarski's
argument. But I think the reader may be prepared to grant that an
approach through the notion of metalanguages can hardly be expected to
help the scientist to know whether his or her theoretical statements *do*
'correspond' with actual states of affairs. The scientist talks of electrons.
Are they *'real'*? Do the electrons of electrical theory refer to and/or
correspond with actual physical entities? It seems to me that such
questions will not be answered by logical analysis, but rather with the
help of some kind of pragmatic criterion of truth. Or maybe, as we shall
see suggested in Chapter 9, truth is a relative context that should be
understood within the context of the social formations within which
ideas of truth or falsity have been generated. As for the correspondence
theories of truth, chiefly developed by logicians in the first half of the
twentieth century, they have now, for the most part, been abandoned.

Another version of truth doctrine that was developed by some logical
positivists was the so-called 'coherence theory', of which Whewell's
notion of consilience of inductions was a forerunner.[120] Somewhat as
with Bacon's methodology of science, which would allow one to know
which hypotheses were false but not which were true,[121] Neurath
supposed that if the theoretical system was working well, and protocol
sentences were being fed into it without any logical contradictions
becoming apparent, then one could be confident that the sentences were
true ones. Neurath was, of course, well aware that truth was not
demonstrated with certainty by this means. One could know that
something was *wrong* within the system. It would 'cause a bell to ring if a
contradiction ensues [as a result of feeding in a particular protocol
sentence].'[122] But knowing that a particular theoretical system is not
wrong is not the same as saying that it is right. The classic case in the
history of science was perhaps provided by Ptolemaic astronomy. It
'saved the appearances' very well, and the component doctrines seemed
to cohere (at least for a time). Yet this did not mean that Ptolemaic
astronomy was *true*. Coherence may be an indication that there are no
logical contradictions within the theoretical system. But it does not serve
as a watertight criterion of truth. Even so, Neurath was satisfied that
coherence provided an *adequate indication* of the truth of a theoretical
system — even if there was considerable difficulty in knowing what a
protocol sentence actually is.

Carnap was, for a time at least, a supporter of Neurath's coherence
doctrine. Of course, if he had been able to carry through the program of
the *Aufbau* successfully he would have needed neither a coherence nor a

correspondence theory of truth. All would have been welded firmly together from the bedrock of phenomena upwards. But as we have argued above, the *Aufbau* failed: one cannot deduce one's way up the 'arch of knowledge'. The subsequent downward deductivist approach in which Carnap chiefly engaged is, I believe, now also largely superseded as a description of actual scientific practice. But it has certainly played a very important part in the history of twentieth-century metascience.

I have here dealt with only a tiny fragment of the whole history of the movement of logical positivism — and the enormous intellectual effort that was made therein to produce a synthesis of logical formalism and empiricist epistemology. I have said just a little about Carnap's views on the structure of scientific theories, but only a very little of all that might be said. The general picture of scientific theories as being made up of definitions and axioms, deductions, theorems, and correspondence rules allowing the checking of the theory against empirical observations, with a clear distinction between observation language and theoretical language, is today usually called the '*received view*' of scientific theories. The history of this doctrine, with all its many choppings and changings, has been well described by Frederick Suppe, and the interested reader is invited to examine his detailed account.[123]

Reichenbach

I have suggested above that the terms 'logical positivism' and 'logical empiricism' are approximate equivalents, and this is indeed the case, so much so that some commentators treat them more or less as synonyms.[124] It is, however, possible to identify certain distinctions if we examine the work of the German philosopher, Hans Reichenbach (1891–1953), the leading member of the so-called 'Berlin School', which paralleled the activities of the 'Vienna Circle'. Logical empiricism may be regarded as an attempted union of logic(ism) and empiricism. And as such, it was the general program of the members of the Vienna Circle. Yet they liked to call themselves 'logical positivists'. Reichenbach's work did not display *quite* such an emphasis on logic as did the writings of Carnap, and in a sense the label of 'logical empiricist' is better suited to Carnap than to Reichenbach. Yet it was Reichenbach who claimed the label of 'logical empiricist'; while Carnap, the leading Vienna Circle theorist, was the chief 'logical positivist'. The work of the two men was concerned with rather similar problems. There were, nonetheless, distinct differences and to that extent, then, it is perhaps appropriate to accept a distinction between 'logical positivism' and 'logical empiricism', the former referring to the work of the Vienna Circle and its successors, and the latter to the work of the Berlin School and its descendants. The distinction is not vital, however, and the broad agreement between the Circle and the School should be acknowledged. Reichenbach and Carnap cooperated in the editorial work for *Erkenntnis*.[125]

Reichenbach[126] came from a prosperous Hamburg family of Jewish origins. He studied engineering at the *Technische Hochschule* in Stuttgart and also mathematics, philosophy and physics under leading scholars of the day in Berlin, Göttingen and Munich. Like Wittgenstein, he fought in the Great War, and in 1920 he obtained a teaching post in Stuttgart. In 1926, with Einstein's support, he was called to a chair in philosophy of physics at Berlin. Subsequently, like many other scholars of Jewish origins, Reichenbach was dismissed by the Hitler regime and he left his country, teaching first in Turkey and from 1938 in the United States where he held a chair at the University of California.

Reichenbach wrote very extensively on the philosophical problems associated with relativity, quantum theory and time, but we shall not consider these difficult matters here. Rather, we shall examine his epistemological views and his work on induction; also his well-known distinction between the 'context of discovery' and the 'context of justification' — which can usefully be related to the history of the 'arch of knowledge'.

At the beginning of the nineteenth century, the French astronomer–mathematician–philosopher, Pierre-Simon de Laplace (1749–1827), envisaging the cosmos as an entirely deterministic, causal, machine-like entity, supposed that if the movements and configurations of all the matter in the universe were known at some point in time, and all the laws of mechanics were also known, then the whole future history of the world could in principle be calculated.[127] Such a view of the nature of things is often called Laplacian determinism. It was never accepted by those who believed in the freedom of the human will, and the nineteenth-century work in thermodynamics gave little comfort to the Laplacian view. It was further undermined with the coming of the quantum theory at the beginning of the twentieth century. But for many physicists and philosophers it seemed reasonable to accept physical determinism at the level of the world of macro-objects, even though it had to be rejected for the micro-world. Laplace, of course, recognised that we cannot predict the future with certainty, so that the best we could hope for was probabilities; but he regarded this as due to human imperfections rather than to the essentially probabilistic character of the way things are. This was the 'subjective' theory of probability.

It was, however, repudiated by Reichenbach who claimed that all knowledge must be, in essence, probabilistic. For physical predictions are *never* exact; for *all* relevant factors can never be incorporated into the calculations. So only probabilistic calculations can be made, even for events at the macro-level. This is not an inadequacy in us, as physicists; it is the way the world necessarily is in relation to our observations. What one does is make probabilistic statements that future events will occur within certain limits of exactness. It is not the case that there are exact and strictly determined events which we can only know approximately. It is, rather, that one deals with *sequences* of events that lie, with certain degrees of probability, within particular ranges. Such sequences are, according to Reichenbach, the empirical phenomena with which the physicist has to deal.[128] And the positivist philosopher should realise this and develop his metascience accordingly, not search after some Laplacian ideal. So this was the starting-point for Reichenbach's analysis of science and his attempt to deal with the problem of induction on a probabilistic basis.

To examine Reichenbach's views in a little more detail, we shall say something about his book *Experience and Prediction*,[129] which provided the chief exposition of his philosophy of science and general epistemology. Reichenbach regarded it as a corrective to the usual logical-positivist approach to knowledge and to science, which (in his view) neglected the probabilistic character of the relations between our ideas and events in the external world.

Reichenbach regarded epistemology as a 'critical', not a descriptive enterprise. Having looked at science as actually practised, the epistemologist should seek to provide a 'rational reconstruction' of the ways of thinking of the 'ideal scientist'. That is, one should 'replace actual thinking by such operations as are justifiable... [or] can be demonstrated as valid.'[130] Hence the thought-processes of actual scientists may be appraised in terms of the philosophically constructed ideal. Thus epistemology, as practised by Reichenbach, was intended to have a kind of therapeutic value: it would weed out shoddy thinking and epistemologists would have a kind of 'advisory task'.[131] With its interest in rational reconstruction, Reichenbach's work meshed well with that of many of his positivist peers; yet as we have said, it maintained its distinctive character, particularly through its emphasis on probabilistic reasoning.

It will be recalled that one of the leading concerns of the members of the Vienna Circle was meaning, with the verification principle used as a criterion of the meaningfulness of propositions. And as we have seen, this work grew from the work of Wittgenstein. For the logical positivists, a proposition was meaningful if it was verifiable *in principle*, even though at any given point in time verification might not be technically feasible. Like the positivists, Reichenbach was also interested in meaning but he sought to treat verification in a more pragmatic sense. Verification, 'in principle', might often involve future technological developments. But Reichenbach, accepting Hume's critique of induction, maintained that one must think in probabilistic terms. It this was done, although nothing could be said with certainty about the future, something useful could be said nevertheless. So, he proposed: '*a proposition has meaning if it is possible to determine a weight, i.e., a degree of probability, for the proposition*'.[132] Past experiences will give one expectations as to future events and one will place certain reliance on statements about the future; that is, one will give them certain 'weight'.[133] We see, then, that Reichenbach's discussion of meaning, via verification, can be linked to his consideration of probabilistic inferences and the problem of induction. The notion of 'weight' was of special importance. It allowed Reichenbach, as he said, to 'construct a bridge from the known [past, verified propositions] to the unknown [future, as-yet unverified propositions]'.[134] As we shall see, it was based upon his theory of probability.

We have seen that in his *Aufbau*, Carnap attempted to produce a 'logical construction of the world', starting from the epistemological bedrock of relations between elements of sensation. While denying that Carnap had carried through this program successfully, Reichenbach attempted something similar, and the fourth chapter of his *Experience and Prediction* is entitled 'The projective construction of the world on the concreta basis' — concreta being tangible entities such as 'houses, furniture, streets, other people, etc.'[135] Evidently, concreta corresponded to the inhabitants of Carnap's 'thing level' in his *Unity of Science*. They provided the 'primitive' basis for Reichenbach's theory of knowledge.

The actual argument for Reichenbach's probability-based theory of induction proceeded somewhat as follows. Events were to be regarded as members of classes. They should be, to a greater or lesser degree, repeatable — like observations in a scientific experiment. But even singular events could be thought of in this way. For example, a doctor might say that a particular patient probably died of tuberculosis, basing his assessment on his knowledge of similar cases.[136] So Reichenbach gave a 'frequency interpretation' of probability. Also, he said:

> *[T]he meaning of probability statements is to be determined in such a way that our behaviour in utilizing them for action can be justified.*[137]

Thus his system involved elements drawn from positivist philosophy (shown by the interest in meaning) and from pragmatism (shown by the fact that meaning is to be assessed in terms of practical action). Reichenbach's approach to induction was that it provided the procedure or behaviour that, on the whole, was the most favourable, or conducive to practical success.

A key notion within Reichenbach's epistemology and his theory of induction was that of the 'posit',[138] namely a statement that is *treated* as true, at least temporarily, even though it is not *known* to be true. Normally we posit those events that are thought to possess the highest degree of probability and can be assigned the greatest weight accordingly; for this is the rational thing to do. Any statements about the future are best made as if one were a person making a wager, giving, as far as possible, due weight to all the relevant factors. If we wish to achieve pragmatic success in life, we are obliged to behave in this fashion; it is the best policy — the rational thing to do. And in fact we do it all the time. We want to go to the beach. We know it involves some risks: sharks, sun-burn, jelly-fish, traffic accidents, and so on. We also can contemplate the possibility of certain pleasures: dolphins, sun-tan, good surfing, visions of beautiful bodies and so on. We assess the likelihoods of these various possible circumstances and make our decisions accordingly. So all of us engage in a kind of informal probability logic every day of our lives.

Logicians have, in fact, worked out a system known as the 'probability calculus', which is analogous to other systems of formal logic, except that it deals with the calculus of probabilities, rather than the drawing of firm conclusions (deductively) from given premises.[139] Reichenbach made extensive use of the probability calculus in his treatment of the problem of inductive inferences, but we shall seek to give a picture of his views without resorting to formalism and symbolic representation.

The aim of induction, said Reichenbach, '*is to find [a] series of events whose frequency of occurrence converges towards a limit*'.[140] We can, for example, imagine ourselves taking repeated readings with some piece of apparatus. They appear to converge to a limit and we posit that they do in fact so converge if the readings are continued indefinitely. But to do this involves, in Reichenbach's terminology, a *'blind posit'*[141] since although it is the best posit we can make, we don't know how good it actually is. But as further data are obtained, the initial blind posit may be corrected in the light of the newly acquired information. So *if* there is a limit, this method should find it.

But suppose there are other methods, such as, for example, clairvoyance or crystal gazing, that find the limit more quickly and successfully than does induction. In such a case, however, where we find the clairvoyant's predictions are subsequently confirmed by events, we then have new information (about the clairvoyant's successes) which, *by induction*, leads us to accept his or her predictions as reliable. So, concluded Reichenbach,[142] *if* there is any method which leads to a frequency limit the inductive method will do it also. Induction is at least sufficient, if not necessary, for this purpose. Moreover, if, perchance, clairvoyance or some other method for predicting future events were successful, this success would have to be judged by inductive means.

All this does not mean, however, that induction leads to 'the truth'. Said Reichenbach:

> What we obtain is a wager; and it is the best wager we can lay because it corresponds to a procedure the applicability of which is the necessary condition of the possibility of predictions. To fulfil the conditions sufficient for the attainment of true knowledge does not lie in our power; [but] let us be glad that we are able to fulfil at least the conditions necessary for the realization of this intrinsic claim of science.[143]

Thus, to work inductively gives us the best posit, though all empirical knowledge is probabilistic in character.

It should be noted that Reichenbach's claimed pragmatic justification of induction is a rule of rational inference. As such, it is not a statement that is either true or false. It is a kind of tool or instrument for research. So here a methodology, rather than a theory is regarded as instrumental in character.

Reichenbach also discussed what he called 'concatenated inductions'. Let us consider his own example.[144] In his day, chemists had been able to find the melting points of all the common elements that are solid at room temperature except carbon. So denoting 'substance ... is molten' by A, and 'substance ... is not molten' by $\overline{A}$, we can write:

Copper:
$\overline{A}\,\overline{A}\,A\,A\,A\,A\,A\,A \ldots$

Iron:
$\overline{A}\,\overline{A}\,\overline{A}\,A\,A\,A\,A\,A \ldots$

. . . .

. . . .

Carbon: $\overline{A}\,\overline{A}\,\overline{A}\,\overline{A}\,\overline{A}\,\overline{A}\,\overline{A} \ldots$

thereby representing the known data with respect to the melting points of the various elements. Such a pattern Reichenbach called a 'probability lattice'; and with it one could make inductive inferences both 'horizontally' and 'vertically'. The first two rows of data would allow one to make the blind posit that copper and iron have two particular melting points. Then, thinking vertically, we can make the posit that although carbon has no known melting point it will melt if heated to a sufficient temperature. Each successive element investigated adds weight to the posit that copper has a specific melting point. Now on the evidence aviable, the posit for carbon, if investigated in isolation, would be that it does *not* melt. But consideration of the data available for the other elements would lead one to the posit that carbon *does* have a melting point. By considering the rows collectively, the original blind posits for each row can be given 'appraised weights'.[145] And the posits thus transformed can serve as the basis for a (blind) posit made with respect to carbon. So as research proceeds, each blind posit may be converted into a posit with appraised weight, and in the process new blind posits are introduced. This process emphasises the fact that scientific reasoning is not concerned with isolated inductions, but with ones that 'intertwine', like the threads in a piece of woven cloth.

Likewise, 'cross-inductions' occur when we find consiliences, in Whewell's sense. Galileo's work led him to a formula (law) for the acceleration due to gravity, and a blind posit, supported by empirical evidence, could be made as to the truth of this law. Similarly, Kepler made a blind posit concerning the elliptical motions of the planets. The laws of Galileo and Kepler were eventually subsumed under Newton's gravitational law. When this occurred the generalisations (posits) of Galileo and Kepler could be given appraised weight and thereby lose the status of blind posits. But for this a new blind posit (Newton's law) had to be proposed.

This schema has, I suggest, a good deal to commend it, but only as a *rational* reconstruction of scientific knowledge — as an attempt to show that the empirically-grounded inductive generalisations of science 'make sense' rather than being irrational. It must be emphasised that Reichenbach's representation of the structure of inductive inference was not historically based. He made no claim that the investigations of the melting points of copper, iron, carbon, etc, had actually proceeded in the history of science in the manner indicated by his probability lattice. And he didn't suggest that the laws of Galileo, Kepler and Newton were actually discovered in the way required by his rational reconstruction of the appropriate reasoning processes. It might be suggested, then, that Reichenbach's metascience served a kind of 'apologetic' role for science.[146] It was intended to show that there was a form of rationality at work in science — even in the procedures of induction. Or, even if the actual historical thought processes were not strictly rational, they could be reconstructed in such a way that their rational structure might be displayed and given due reliance accordingly. It was for this reason, then, that Reichenbach made his distinction between the 'context of discovery' and the 'context of justification'.[147] He was chiefly concerned with the latter, which (he correctly asserted) often corresponded with the way in which scientists present the results of their investigations to their reading public.[148] The reasoning is then made to appear compact and coherent, no matter how irrational, arational or generally woolly it may have been in the first instance, in the laboratory, field, or study.

Reichenbach's distinction between the 'context of discovery' and the 'context of justification' is of considerable interest to students of the 'arch of knowledge' and it has led to a good deal of discussion in recent metascientific literature. Many commentators, such as P.K.Feyerabend (see below, pp.333–342), regard the distinction as spurious and deny that it can be made satisfactorily.[149] They regard it as yet another ill-conceived notion stemming from the positivist era. We will not be so dismissive, but will consider the relationship between discovery and justification and the 'upward' and 'downward' limbs of the traditional 'arch of knowledge'. If we think of the arch in its simplest form, an upward inductive movement followed by a downward deductive progress, it is clear that Reichenbach's 'discovery' and 'justification' do not fit this schema at all well, though they might seem to do so on first considering the matter. For while 'discovery' might seem to correspond to an upward inductive move, 'justification', as discussed by Reichenbach, involved the reconstruction of a pattern of argument such that the original speculative guesses would be made cogent and plausible. And this reconstructed pattern of reasoning would be essentially inductive, not deductive. Even so, we must recall our simile of the mountain. To find one's way up is to find one's way down at one and the same time. So if Reichenbach's (reconstructed) 'justification' is inductive, it is, in a sense, concerned with arch-climbing (to new scientific laws, principles, or whatever), even though it might follow long after the first speculative inductive leap to such principles. I suggest, therefore, that Reichenbach's model of science might be compared to the work of a rock climber making an ascent that is hazardous at the first trial, with a rope thrown up, perhaps, to catch on to some projecting bit of rock.[150] Only later is the whole made more secure, banging in pitons and the like. And then other climbers may readily climb the path thus secured (by rational reconstruction). Of course, once the ropes, ladders, steps, etc, are made secure, then one can descend as readily as one can ascend.

All this is, of course, rather fanciful, but I make these remarks in order to show that the new ideas can only with some difficulty be accommodated to the old; and the old model of a simple inductive/deductive arch (or more recent hypothetico–deductivism) is not by any means a satisfactory solution to all the problems of metascience. As we shall see in Chapter 9, some modern metascientists find themselves entirely dissatisfied with hypothetico–deductivism, the 'context of discovery/context of justification' distinction, or any sort of arch-like structure for science. So the old order changeth.

But we will consider these recent views more fully in due time. Here I may say that Reichenbach's approach to a pragmatic justification of induction seems to me to have a good deal to commend it. Recognising that science does not offer certainty in the way of knowledge of truth, we are allowed to escape from the vice-like grip of logic, and epistemological demands that are unreasonably stringent. Reichenbach suggests why scientific knowledge should be accorded some 'respect'; yet it also needs to be regarded critically. On the other hand, he tells us very little about how science is *actually* practised, either internally with respect to the thought processes of scientists engaged in creative work, or externally with respect to the social milieu of science and the influences of social factors on scientific inquiry. Also, later writers have found that the actual presentation of Reichenbach's arguments for the vindication of induction need clarification and tightening. Thus Reichenbach's former student, Wesley Salmon, for example, has done much in the way of strengthening Reichenbach's account of induction.[151] And the whole topic is, even now, an active area of metascientific research.

One of the obvious deficiencies of the logical positivist/empiricist philosophy of science that we have been discussing in this chapter was its 'static' character. Even if a wholly satisfactory representation of the logical structure of a scientific theory were given, and an exact description of the relationships obtaining between theory, language and observational experience, or a satisfactory rational reconstruction of the processes of scientific justification, we still would not have anything like a complete meta-account of science. For we need to know also about the *dynamics* of science: how hypotheses and theories are formulated, tested and modified, and how the social context influences the way in which scientific ideas are generated and in which scientific change occurs. To describe science as a frozen iceberg, so to speak, rather than a flowing river, is grotesquely inadequate. We shall therefore, in Chapter 9 (and to a lesser extent in Chapter 8), try to describe some of the work of more recent years on the dynamics of science.

But first we must go back to the end of the nineteenth century to correct an impression that has probably been created in the present chapter — namely that twentieth-century philosophy of science has been largely a product of the interaction of logic and epistemology. This was by no means the whole of the case. Developments in experimental and theoretical science were also of the highest importance. In the next chapter, therefore, we shall retrace our steps somewhat, to look at some of the remarkable innovations in physics in the early twentieth century and their impact on philosophy of science. To some extent we shall be having another look at the history of logical positivism, but this time we shall give attention to the influence of ideas drawn from science itself rather than logic. We shall also have a look at some of the ideas about models in science, which were particularly influenced by work done in the strange new field of quantum theory.

NOTES

1   This is the philosophical position that assigns primacy tó logic in all forms of human knowledge.
2   F L G Frege (ed I Angelelli) *Kleine Schriften* Wissenschaftliche Buchgesellschaft Darmstadt 1967 p 50
3   See H D Sluga *Gottlob Frege* Routledge & Kegan Paul London Boston & Henley 1980 pp 52–8
4   See above, p 104
5   See above, p 120
6   F L G Frege *The Foundations of Arithmetic: A Logico-Mathematical Enquiry into the Concept of Number* trans J L Austin *Die Grundlagen der Arithmetik: Eine Logische-Mathematische Untersuchung über den Begriff der Zahl* Koebner Breslau 1884; Blackwell Oxford 1950 p 104
7   F L G Frege *Begriffsschrift, eine der Arithmetischen Nachgebildete Formelsprache des Reinen Denkens* Nebert Halle 1879; *Conceptual Notation [A Formula Language of Pure Thought Modelled upon the Formula Language of Arithmetic] and Related Articles* translated & edited with a biography & introduction by Terrell Ward Bynum Clarendon Press Oxford 1972
8   For example, the arithmetical proposition that two plus three equals five was represented by Frege as:

$$\text{———— } 2 + 3 = 5.$$

To express the judgment that this proposition is true, Frege wrote:

$$\vdash\text{———— } 2 + 3 = 5.$$

(The horizontal line represented the 'content stroke'; the vertical line the 'judgment stroke'.)
To express the negation of a proposition, Frege wrote, for example:

$$\text{————}_\top\text{ } 4 + 2 = 7.$$

And to express the judgement that four plus two does not equal seven, he wrote:

$$\vdash\text{————}_\top\text{ } 4 + 2 = 7.$$

To express the judgment 'not (not A and B)', Frege wrote:

$$\vdash\begin{array}{l}\text{———— } A \\ \text{———— } B\end{array}$$

For example:

$$\vdash\begin{array}{l}\text{———— } x^2 = 4 \\ \text{———— } x + 2 = 4.\end{array}$$

This meant: 'It is not the case that, if $x + 2 = 4$, x^2 does not equal 4'. This amounts to saying: 'If $x + 2 = 4$, $x^2 = 4$'.
So Frege's symbol:

$$\vdash\begin{array}{l}\text{———— } A \\ \text{———— } B\end{array}$$

was equivalent to 'B implies A'; or 'If B then A'. Thus his symbolism was designed to accommodate the logic of the *modus ponens* (see above, p 45). Frege gave a set of eight logical terms by means of his symbolism as follows:

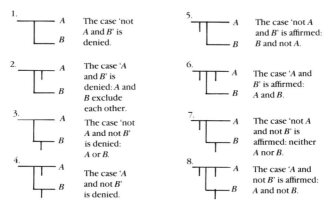

To indicate 'generality', he used a Gothic 'a' in a 'dip' in the content stroke. For example:

$$\begin{array}{l}\text{———— } x = 0 \\ \text{——}_{\mathfrak{a}}\text{— } a = x \\ \text{———— } a^2 = x\end{array}$$

meant: 'If each square root of x is x itself, then $x = 0$'. So to express the judgment that a function, $\varnothing(A)$, is true for all values of A, Frege wrote:

$$\vdash\text{——}_{\mathfrak{a}}\text{———— } \varnothing(\mathfrak{a}).$$

The components of an Aristotelian syllogism were thus handled by Frege as follows:

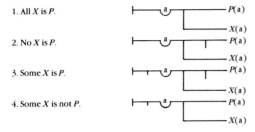

1. All *X* is *P*.

2. No *X* is *P*.

3. Some *X* is *P*.

4. Some *X* is not *P*.

Thus the functions of Frege's system could be predicates of relational expressions, not merely mathematical expressions. We shall not seek to give a further account of Frege's system here. For a more detailed exposition see Bynum *op cit* (note 7) pp 55–80

9   F L G Frege (ed H Hermes *et al*) *Nachgelassene Schriften* Meiner Verlag Hamburg 1970 p 201

10   For English translations of the six reviews that the book received, see Bynum *op cit* (note 7) pp 209–35

11   Frege *op cit* (note 6)

12   Frege *op cit* (note 6) pp 79–80. Frege regarded equinumeracy as analogous to parallelism in geometry.

13   Frege *op cit* (note 6) p 800. It should be noted that Frege's account of arithmetical numbers is seemingly grounded in logic, for the notion of there being nothing that is not identical with itself is a logical principle.

14   F L G Frege *Grundgesetze der Arithmetick: Begriffsschriftlich Abgeleitet* 2 vols Pohle Jena 1893–1903; *The Basic Laws of Arithmetic: Exposition of the System* trans & ed M Furth University of California Press Berkeley & Los Angeles 1964

15   These are the numbers 'one', 'two', 'three', etc., as opposed to the ordinal numbers 'first', 'second', 'third', etc.

16   For further discussion of Frege's number theory, see C Parson 'Frege's Theory of Numbers' M Black ed *Philosophy in America* Cornell University Press Ithaca 1965 pp 180–203; and G Currie *Frege: An Introduction to his Philosophy* Harvester Press Sussex and Barnes & Noble New York 1982 Chapter 3

17   In Russell's words: 'The number of a class is the class of all those classes that are similar to it.' (B Russell *Introduction to Mathematical Philosophy* Allen & Unwin London and Macmillan New York 2nd ed 1920 p 18)

18   See above p 127

19   Frege *op cit* (note 14 1964) p 72; as restated by Bynum (*op cit* note 7) p 37. It may be noted that even Frege himself, long before Russell's paradox was made known to him, had expressed some reservations about his Law 5 (*op cit* [note 14 1964] pp 3–4).

20   See 'A Letter from Bertrand Russell to Gottlob Frege' trans B Woodward in J van Heijenoort ed *Source Book in Mathematical Logic* Harvard University Press Cambridge Mass 1967 pp 124–5

21   See p 216

22   Frege *op cit* (note 14 1964) pp 127–43. In Frege's modified version of Law 5, he suggested that two concepts should be said to have the same extension if every object which fell under the first, but which was not itself the extension of the first, fell also under the second — and vice versa.

23   See Sluga *op cit* (note 3) pp 162–70

24   Russell wrote an excellent *Autobiography* (3 vols Allen & Unwin London 1967–69). For introductions to his work, see for example: A J Ayer *Russell* Fontana/Collins London 1972; E R Eames *Bertrand Russell's Theory of Knowledge* Allen & Unwin London 1969. More technical are: P A Schilpp ed *The Philosophy of Bertrand Russell* Northwestern University Press Evanston & Chicago 1944; D F Pears *Bertrand Russell and the British Tradition in Philosophy* Fontana/Collins London & Glasgow 1967. A valuable collection of Russell's major papers is B Russell (ed R C Marsh) *Logic and Knowledge: Essays 1901–1950* Allen & Unwin London and Macmillan New York.

25   A N Whitehead & B Russell *Principia Mathematica* 3 vols Cambridge University Press Cambridge 1910–13; 2nd ed 1927

26   B Russell *Essay on the Foundations of Geometry* Cambridge University Press Cambridge 1897

27   See G W F Hegel *The Philosophy of History* trans J Sibree Dover New York 1956

28   See F H Bradley *Appearance and Reality: A Metaphysical Essay* Clarendon Press Oxford 1893

29   B Russell 'My mental development' in P A Schilpp ed *The Philosophy of Bertrand Russell* Tudor New York 3rd ed 1951 pp 1–20 (at p 13)

30   B Russell *The Principles of Mathematics* Cambridge University Press Cambridge 1903; 2nd ed (with important new introduction) Allen & Unwin London 1937

31   B Russell *A Critical Exposition of the Philosophy of Leibniz* Cambridge University Press Cambridge 1903

32   B Russell *op cit* (note 30) p 115
33   B Russell *Introduction to Mathematical Philosophy* Allen & Unwin London and Macmillan New York 1919 p 15 (This *Introduction*, written by Russell when he was in gaol for his anti-war work, provided a popular exposition of his philosophy of mathematics.)
34   B Russell *op cit* (note 30) p 128
35   *Ibid* p 102
36   The paradox of the liar is also sometimes known as the paradox of Epimenides the Cretan, who is supposed to have said: 'All Cretans are liars.' An analogous example is the sentence that reads:

> 'Every statement within this rectangle is false.'

If one assumes such sentences to be true, then what they actually say requires one to conclude them to be false; and if one assumes that they are false then immediately it seems that they are true! The paradox of the liar and the set-theory paradox discovered by Russell belong to the same 'family', though they are generally regarded as distinct by modern logicians.
37   There was also an appendix giving an appreciative exposition of the work of Frege.
38   B Russell 'Mathematical Logic as based on the Theory of Types' *American Journal of Mathematics* 1908 Vol 30 pp 222–62; reprinted in Marsh ed *op cit* (note 24) pp 59–102
39   B Russell 'On Denoting' *Mind* 1905 Vol 15 pp 528–33; republished in Marsh ed *op cit* (note 24) pp 41–56
40   This has rather little to do with the ancient traditions of mathematical and methodological analysis or Kant's concept of analyticity that we have discussed earlier in this book. There is some analogy with the ancient methodological analysis.
41   For example, suggested Moore, '*x* is a brother' could be replaced by '*x* is a male sibling'. The meaning of more difficult sentences such as '*x* is the cause of *y*' could supposedly be achieved by analysis. (In fact, Hume had undertaken just this task.) Ultimately, of course, as in chemical analysis, one would come to an unanalysable residue. Moore recognised this, and in his *Principia Ethica* (1903) he stated that the concept of 'good' was unanalysable and could only be known by direct acquaintance.
42   Whitehead & Russell *op cit* (note 25 1927) Vol 1 pp 96–7. The actual symbolism used in the original text was somewhat less easy to grasp, since it used dots instead of brackets:

   1. $\vdash .p \lor p. \supset .p$ (tautology)
   2. $\vdash :q. \supset .p \lor q$ (addition)
   3. $\vdash .p \lor q. \supset .q \lor p$ (permutation)
   4. $\vdash .p \lor (q \lor r). \supset .q \lor (p \lor r)$ (association)
   5. $\vdash :.q \supset r. \supset :p \lor q. \supset .p \lor r$ (summation)
   Here the symbol $\dashv$ meant: 'It is true that ...'.
   $\supset$ meant 'implies' or 'if ... then ...'.
   $\lor$ meant 'or'.

43   P Bernays 'Axiomatische Untersuchungen des Aussagenkalkuls der *Principia Mathematica*' *Mathematische Zeitschrift* 1962 Vol 26 pp 305–20
44   J D P Nicod 'A Reduction in the Number of Primitive Propositions of Logic' *Proceedings of the Cambridge Philosophical Society* 1917 Vol 19 pp 32–41
45   For example, $p|q$ may be written as $p|q.|.p|q$; $\sim p$ as $p|p$; etc.
46   B Russell in Marsh ed *op cit* (note 24) p 75
47   *Ibid* p 79
48   I have discussed this matter elsewhere in relation to logical problems in the historiography of science. See my paper 'Sir Archibald Geikie (1835–1924), Geologist, Romantic Aesthete, and Historian of Geology: The Problem of Whig Historiography of Science' *Annals of Science* 1980 Vol 37 pp 441–62
49   For Russell's own lucid discussion of the matter, see his *Introduction to Mathematical Philosophy* (note 33) Ch 13 'The Axiom of Infinity and Logical Types'
50   Cf J O Urmson *Philosophical Analysis: Its Development Between the Two World Wars* Clarendon Press Oxford 1956 p 7. This book gives a more extended and very helpful account of many of the topics covered in this chapter, particularly the work of Russell and Wittgenstein.
51   B Russell 'The philosophy of logical atomism' in Marsh ed *op cit* (note 24) pp 177–281
52   *Ibid* p 179
53   *Ibid* p 182
54   *Ibid* p 199
55   *Ibid* p 209
56   For most interesting accounts of his life and times, see: W W Bartley III *Wittgenstein* Quartet Books London 1974; and A Janik & S Toulmin *Wittgenstein's Vienna* Weidenfeld & Nicolson London 1973
57   Wittgenstein kept philosophical notes in his knapsack during the First World War, and from these he prepared the text of the *Tractatus*. Some of the ideas of the *Tractatus* were developed before the war, and the well-known 'picture-theory' of language is thought to have been devised in 1914. The

manuscript was completed by 1918, and dispatched to Russell from a prisoner-of-war camp in Italy. After the war (1919), Wittgenstein and Russell met in Holland and discussed the ideas in Wittgenstein's manuscript and the development of his thought since he had last met with Russell in 1913. The *Tractatus* was published in German 1921, and in English in 1922 with a very helpful introduction by Russell. A good modern edition is found in the translation by D F Pears & B F McGuinness (Routledge & Kegan Paul London & Henley 1961). There are several useful commentaries on the *Tractatus*. For clarity, I recommend particularly: H O Mounce *Wittgenstein's Tractatus Logico-Philosophicus* Open University Press Milton Keynes 1976.

58 L Wittgenstein *Notebooks 1914–1916* Blackwell Oxford 1961 p 93

59 In further references to the *Tractatus*, we shall simply give the paragraph numbers, thereby referring the reader to the Pears & McGuiness translation.

60 Some commentators maintain that Wittgenstein was chiefly concerned with ethical [*sic*] matters in the *Tractatus*. And certainly the later sections of the book do give some warrant to this interpretation.

61 See above, p 220

62 This is, of course, not necessarily so in modern art: The Belgian surrealist painter René Magritte (1898–1967) was, like Wittgenstein, enormously interested in the relationship between the physical world, our ideas of the world, and our representation of the world in symbolic form. He worked out his views on such matters in many curious and interesting paintings and was in fact 'doing' philosophy that had much in common with that of Wittgenstein through the medium of paint and canvas, rather than philosophical disquisition. (See S Gablik *Magritte* Thames & Hudson London 1970 Ch 8.) I also find that the plays of Luigi Pirandello (1867–1936) are often concerned with analogous philosophical problems; also, more recently, the plays of Tom Stoppard.

64 Wittgenstein *op cit* (note 57) p xii. With characteristic wit, Russell then went on to say (p xxi) that 'Mr.Wittgenstein manages to say a good deal about what cannot be said, thus suggesting to the sceptical reader that possibly there may be some loophole through a hierarchy of languages, or by some other exit.' Russell here was obviously thinking of his theory of types. But he did not here commit himself to suggesting how the loophole might be found or where it might lie.

65 Bartley *op cit* (note 56) p 9

66 L Wittgenstein *Philosophical Investigations* trans G E M Anscombe Blackwell Oxford 1963

67 L Wittgenstein *The Blue and Brown Books* ed R Rhees Blackwell Oxford 1958 p 32

68 N R Hanson *Patterns of Discovery: An Inquiry into the Conceptual Foundations of Science* Cambridge University Press Cambridge 1958

69 On the history of the Vienna Circle, see: J Joergensen *The Development of Logical Empiricism* University of Chicago Press Chicago & London 1951; V Kraft *The Vienna Circle: The Origin of Neo-Positivism A Chapter in the History of Recent Philosophy* Philosophical Library New York 1953; O Hanfling *Logical Positivism* Blackwell Oxford 1981

70 This is the thesis that every descriptive term in the language of science is connected with terms that designate the observable properties of things.

71 A E Blumberg & H Feigl 'Logical Positivism: A New Movement in European Philosophy' *Journal of Philosophy* 1931 Vol 28 pp 281–96 (at p 281)

72 Joergensen *op cit* (note 69) p 6. Joergensen gives a number of other lesser-known names that I have omitted.

73 F Waismann 'Logische Analyse der Wehrscheinlichkeitsbegriffs' *Erkenntnis* 1930 Vol 1 pp 228–48

74 See above, p 122

75 See above, p 112

76 Upholders of the ontological argument would disagree on this point.

77 M Schlick *Gesammelte Aufsätze* Olms Hildesheim 1969 pp 219–30

78 A J Ayer *Language, Truth and Logic* Pelican Harmondsworth 1971 p 7 & *passim* (1st ed Gollancz London 1936)

79 Various forms of the principle, additional to those have been discussed here, are dealt with in some detail, but very clearly, by Hanfling *op cit* (note 69).

80 R Carnap 'Intellectual Autobiography' in P A Schilpp ed *The Philosophy of Rudolf Carnap* Open Court La Salle and Cambridge University Press London 1963 pp 1–84 (at p 3)

81 *Ibid* p 13

82 B Russell *Our Knowledge of the External World* Open Court Lasalle 1914

83 R Carnap *Der Logische Aufbau der Welt* Weltkreis Verlag Berlin 1928; *The Logical Structure of the World* [together with *Pseudoproblems in Philosophy*] trans R A George University of California Press Berkeley & Los Angeles 1967 (Note: The German word '*Aufbau*' literally means 'building up' or 'construction'. Carnap's book is often referred to in English texts simply as 'the *Aufbau*'.)

84 Carnap *op cit* (note 80) p 16.

85 *Ibid*

86 Carnap *op cit* (note 83 1967) p 178 para 108. Earlier empiricists such as Locke and Mach had found it necessary to allow the mind this 'constructive' capacity.

87 *Ibid* p 179 para 109

88 *Ibid* p 130 para 80 and p 180 para 111

89 *Ibid* p 181 para 112. See also pp 131–3 para 81

90   It was much worse when given in the formalism of symbolic logic:

$$\text{qual} =_{Df} \alpha\{(\gamma) : \gamma\text{esimilcirc} . \text{Nc}'(\alpha \cap \gamma)/\text{Nc}'\alpha > \tfrac{1}{2} . \supset . \alpha \subset' \gamma :. (x):x \sim \varepsilon \supset . (E\delta). \delta\varepsilon \text{ similcirc} . \alpha \subset \delta . x \sim \varepsilon \delta\}$$

91   *Ibid* pp 239–40
92   *Ibid* pp 120–1
93   This sketch (Figure 32) is my own interpretation of Carnap's thesis. It is not exactly clear where the 'bottom' of this 'curve' lies. See also, note 94.
94   Carnap *op cit* (note 83 1967) p 121. Carnap sought to make his procedure somewhat clearer with the help of a musical analogy (pp 114–6). He maintained that a chord cannot, as such, be analysed, for it does not have constituents. It just 'is'. Nevertheless, one can think of the chord *c-e-g* as belonging to three chord classes: those containing the note *c*, those containing the note *e*, and those containing the note *g*. These three classes are Carnapian 'similarity circles'. And they overlap in the set of chords, *c-e-g*, that one may hear from time to time. Carnap held that the abstraction of the *c*-containing, the *e*-containing and the *g*-containing chords from all the audible sensations was an act of *quasi*-analysis; not 'true' analysis, since, as I say, the chords as such were supposedly not strictly analysable.
95   *Ibid* p 287
96   One of the few is Nelson Goodman. See his volume *The Structure of Appearance* Harvard University Press Cambridge Mass 1951; and also his 'The Significance of *Der Logische Aufbau der Welt*' in Schilpp ed *op cit* (note 80) pp 545–58
97   R Carnap 'Die Physikalische Sprache als Universalsprache der Wissenschaft' *Erkenntnis* 1931 Vol 2 pp 432–65; *The Unity of Science* translated with an introduction by M Black Kegan Paul Trench & Trubner London 1934
98   Carnap *op cit* (note 80) pp 17–18
99   Carnap *op cit* (note 97)
100 *Ibid* (1934) pp 46 & 47
101 Carnap suggested: '[H]ere now pointer at 5, simultaneously spark and explosion, then smell of ozone there.' (*Ibid* p 44)
102 O Neurath 'Protokollsätze' *Erkenntnis* 1932–33 Vol 3 pp 204–28 at p 207; English translation in O Hanfling ed *Essential Readings in Logical Positivism* Blackwell Oxford 1981 pp 160–8 (at p 163)
103 Carnap *op cit* (note 97 1934) p 93
104 R Carnap 'Logical foundations of the unity of science' in O Neurath ed *International Encyclopedia of Unified Science* Chicago University Press Chicago Vol 1 No 1 1938 pp 42–62
105 *Ibid* p 46
106 *Ibid* p 52
107 *Ibid* p 53
108 R Carnap 'Testability and Meaning' *Philosophy of Science* 1936 Vol 3 pp 419–71 & Vol 4 1937 pp 1–40; partly reprinted in H Feigl & M Brodbeck eds *Readings in the Philosophy of Science* Appleton-Century-Crofts New York 1953 pp 47–92
109 The example is from Carnap *op cit* (note 104) p 50
110 Carnap *op cit* (note 108 1953) p 53
111 For a brief introduction to the topic, see, for example, C G Hempel *Philosophy of Natural Science* Prentice-Hall Englewood Cliffs 1966 Ch 4
112 R Carnap *Foundations of Logic and Mathematics* (*International Encyclopedia of Unified Science* Vol I No 3) University of Chicago Press Chicago 1939; partly reprinted in Feigl & Brodbeck eds *op cit* (note 108) pp 309–18
113 *Ibid* (1939) p 59; (1953) pp 310–11
114 *Ibid* p 315
115 R Carnap, 'The Methodological Character of Theoretical Concepts' in H Feigl & M Scriven eds *Minnesota Studies in Philosophy of Science* Vol 1 University of Minnesota Press Minneapolis 1956 pp 38–76
116 R Carnap *Philosophical Foundations of Physics: An Introduction to the Philosophy of Science* ed M Carner Basic Books New York & London 1966 Ch 24
117 *Ibid* p 233
118 See above, p 197
119 A Tarski 'The Concept of Truth in Formalized Languages' in J H Woodger (trans) *Logic, Semantics, Mathematics: Papers from 1923 to 1938 by Alfred Tarski* Clarendon Press Oxford 1956 pp 152–278. This paper, first read to the Warsaw Scientific Society in 1931, was published in German as 'Der Wahrheitsbegriff in den formalisierten Sprachen' *Studia Philosophica* 1936 Vol 1 pp 261–405. For a simplified account of Tarski's procedure, see, for example, H Putnam *Meaning and the Moral Sciences* Routledge & Kegan Paul Boston London & Henley 1978 Lecture 1.
120 See above, p 160
121 Neurath *op cit* (note 102 1981) p 168
123 F Suppe 'The Search for Philosophical Understanding of Scientific Theories' in F Suppe ed *The Structure of Scientific Theories* University of Illinois Press Urbana Chicago & London 2nd ed 1977 pp 1–241

124   For example, O Hanfling *Logical Positivism* Blackwell Oxford 1981 p 6
125   Reichenbach himself used the term 'logistic empiricism' as a generic term. It was, he asserted, a
      philosophy that grew from 'American pragmatists and behaviorists, English logistic epistemologists,
      Austrian positivists, German representatives of the analysis of science, and Polish logisticians'. (H
      Reichenbach *Experience and Prediction: An Analysis of the Foundations of Science* University of
      Chicago Press Chicago 1961 p v [lst ed 1938])
126   For descriptions of the life and work of Reichenbach, see: M Strauss 'Hans Reichenbach and the
      Berlin School' in his *Modern Physics and its Philosophy* Reidel Dordrecht 1972 pp 173–285; C
      Schuster 'Hans Reichenbach' in C C Gillispie ed *Dictionary of Scientific Biography* Vol 11 Scribner
      New York 1975 pp 355–9; various authors 'Memories of Hans Reichenbach' in M Reichenbach & R
      S Cohen eds *Hans Reichenbach: Selected Writings 1909–1953* Reidel Dordrecht 1978 Vol 1 pp 1–
      87; and W C Salmon 'The Philosophy of Hans Reichenbach' in W C Salmon ed *Hans Reichenbach:
      Logical Empiricist* Reidel Dordrecht 1979 pp 1–84
127   Laplace wrote (in English translation): 'We ought…to regard the present state of the universe as the
      effect of its anterior state and as the cause of the one which is to follow. Given for one instant an
      intelligence which could comprehend all the forces by which nature is animated and the respective
      situation of the beings who compose it — an intelligence sufficiently vast to submit these data to
      analysis — it would embrace in the same formula the movements of the greatest bodies in the
      universe and those of the lightest atom; for it, nothing would be uncertain and the future, as the past,
      would be present to its eyes.' (P-S de Laplace *A Philosophical Essay on Probabilities* translated from
      the 6th French Edition by Frederick Wilson Truscott and Frederick Lincoln Emory with an
      introductory note by E T Bell Dover New York 1951 p 4)
128   H Reichenbach *Ziele und Wege der Leutigen Naturphilosophie* Meiner Leipzig 1931, English
      translation in H Reichenbach *Modern Philosophy of Science: Selected Essays* Routledge & Kegan
      Paul London and Humanities New York 1959 pp 79–108; 'Kausalitat und Wahrscheinlichkeit'
      Erkenntnis 1930 Vol 1 pp 158–88, partial English translation *ibid* pp 67–78
129   See note 125 (1961)
130   *Ibid* p 7
131   This presumed role of certain metascientists has not endeared the breed as a whole to some
      scientists. One is reminded irresistably of the 'teaching-grandmother-to-suck-eggs syndrome'.
132   Reichenbach *op cit* (note 125 1961) p 54
133   Reichenbach supposed that propositions might be considered to have different 'weights' according
      to the reliance that might be placed upon them. Wittgenstein had supposed that propositions could
      have but two truth-values: true and false. But for Reichenbach the notion was of weight as a
      continuous quantity, 'running from the utmost uncertainty through intermediate degrees of
      reliability to the highest certainty'. (*Ibid* p 23)
134   *Ibid* p 24
135   *Ibid* p 273
136   *Ibid* p 308
137   *Ibid* p 309
138   *Ibid* p 313
139   For a summary statement of the elements of the probability calculus, see, for example: W C Salmon
      The Foundations of Scientific Inference University of Pittsburgh Press Pittsburgh 1966 pp 56–65; or,
      for a more detailed exposition, H Reichenbach *The Theory of Probability* University of California
      Press Berkeley 1949. The calculus contains such formulae as:
      $P(a\ b) = P(a) + P(b) — P(a.b)$ (meaning 'the probability of either a or b occurring = the
      probability of a occurring + the probability of b occurring — the probability of a and b occurring
      together');
      $P(a\ b) = 1 — P(a) + P(a.b)$;
      $P(a\ b) = 1 — P(a) — P(b) + 2P(a.b)$.
140   Reichenbach *op cit* (note 125 1961) p 350
141   *Ibid* p 353
142   *Ibid* p 355
143   *Ibid* op 357
144   *Ibid* pp 365–7
145   A posit to which one is able to assign some degree of probability is said to have 'appraised weight'.
146   This point will be discussed further in our concluding chapter, p 370.
147   Reichenbach *op cit* (note 125 1961) pp 6–7
148   It is well known that scientists' publications, whether in monographs or journal articles, rarely bear
      much relationship to the processes of discovery. Various forms of rhetoric are commonly deployed.
      Darwin's *Origin of Species* is a classic example of scientific 'apologetics'. But Darwin did not, of
      course, seek to make his views appear plausible by metascientific analysis. This is, naturally, the kind
      of activity customarily engaged in by metascientists when they seek to produce rational recon-
      structions of scientific theories or the thought processes of workaday scientists.
149   P K Feyerabend *Against Method: Outlines of An Anarchistic Theory of Knowledge* New Left Books
      London and Humanities Press Atlantic Highlands 1975 p 165
150   What this bit of rock might be equivalent to is hard to say. But it might, for example, be some

neighbouring, well-established theory which could provide an *analogy*, to be used in the construction of the new theory.

151   For a bibliography of Salmon's works, see R McLaughlin ed *What? Where? When? Why? Essays on Induction, Space and Time, Explanation inspired by the work of Wesley C. Salmon* ... Reidel Dordrecht 1982 pp 289–94.

CHAPTER 7

THE NEW PHYSICS AND ITS IMPACT ON PHILOSOPHY OF SCIENCE

In this chapter, I want to give at least an indication of the nature of Einstein's theory of special relativity, as a preliminary to some discussion of Einstein's own very interesting ideas on philosophy of science. This will lead us to a consideration of the work of P.W.Bridgman and his doctrine of operationalism, which was strongly influenced by Einstein's work and also linked up with the ideas of the logical positivists that were considered in the previous chapter. Next, we shall say a few words about the work of A.S.Eddington, likewise deeply influenced by Einstein. And then we shall say something about theories of models and analogies in science — their use in the formulation and understanding of scientific theories.

Einstein

Albert Einstein (1879–1955)[1] was born into a German Jewish family, and his early life was spent in Munich, where he attended a Catholic school. He was, however, never attracted to formal religion, and his life was that of a left-wing, agnostic, pacifist — though on occasions he did refer half humorously to 'the secrets of the Old One'. Einstein did not do particularly well at school. Moreover, he was unsuccessful in his first attempt to gain entrance to the prestigious Zurich Polytechnic. However, having spent a further year at a Swiss high school he was admitted to the Polytechnic, where he studied experimental physics. After graduating, it was not initially possible to gain suitable academic appointment, and so for several years he was employed as an examiner of patents in a government office in Berne. Here, in congenial surroundings, though in intellectual isolation so far as research work at the boundaries of theoretical physics was concerned, he sought to probe some of the most basic principles of classical (Newtonian) physics, as Mach had done before him. For at Zurich Einstein had come to the conclusion that some of these basic principles were fundamentally flawed, so that it was necessary to attempt a thoroughgoing reconstruction of theoretical physics.

The first major outcome of Einstein's labours was the celebrated paper of 1905, entitled 'On the Electrodynamics of Moving Bodies'.[2] This, together with some other fundamental papers on the theory of the Brownian motion,[3] the photoelectric effect,[4] and the quantum theory, soon secured Einstein a wide reputation, and he was appointed to teaching posts in Zurich and Prague and then at Berlin, where he became Director of the Institute of Physics at the Kaiser Wilhelm Institute. In 1915, Einstein published his first account of the 'general' theory of relativity[5] (as opposed to the 'special' theory of 1905), considering accelerated systems rather than systems moving with uniform relative velocity such as were treated in the theory of special relativity. The theory of general relativity, which offered a new account of gravitation, was given approximate empirical support in 1919 as a result of observations made by Eddington and others at the time of a solar eclipse.[6]

As a result of the subsequent political crises in Germany, Einstein resigned his position at Berlin in 1933 and moved to America, where he took American citizenship in 1940. His last years were spent in Princeton, where the Einstein archives are now held. As is well known, Einstein was active in the world peace movement in his later years. It is generally considered that from the point of view of scientific achievement Einstein's later years were not so successful as his earlier period. He came into fundamental disagreement with other physicists over the way in which the quantum theory ought to be interpreted. He also sought to establish a so-called 'unified field theory', which would embrace the phenomena of gravitation and electricity, and (hopefully) would incorporate his ideas on quanta. But this daunting task was never brought to a successful conclusion, and the question of whether Einstein was or was not correct in his interpretation of the quantum theory is still under discussion, though opinion tends to favour those who took the position opposed to that of Einstein.

We need not, however, concern ourselves with the work of Einstein's later years here. Let us return, then, to consider some of the ideas that underlay his early investigations in special relativity. To this end, we shall give first a brief account of one of the major 'problems' of late nineteenth-century physics — the 'null result' of the famous Michelson–Morley experiment.[7] As will become apparent later, there is some advantage for our exposition in this, even though Einstein was not, it seems, directly stimulated to undertake his theoretical investigations because of the Michelson–Morley result.

In the nineteenth century, it was commonly believed that one could envisage an 'aether' pervading the universe, which, through non-material (ie, lacking Newtonian inertial properties), supposedly acted as the medium through which electromagnetic radiations (such as light) were propagated. So, if light was a wave motion of some kind, it could be the aether that was doing the waving. This idea of an all-pervasive aether was very ancient, and it had also been given the weight of Newton's authority. In the nineteenth century, besides serving as a medium for the propagation of radiation, it was also thought of as serving as a kind of 'absolute', Newtonian frame of reference, to which the motions of other bodies might be referred. This was, no doubt, an improvement on God's sensorium playing a somewhat similar role in the structure of theoretical physics, but in some ways the aether was no less mysterious. It remained very much a hypothetical entity, yet an object of much speculation. It is not surprising, therefore, that strenuous efforts were made to try to detect the motion of the Earth relative to the aether. For this purpose, the celebrated Michelson–Morley experiment was devised, and first performed in 1881.[8] In simplified form we may represent the apparatus as in Figure 34.

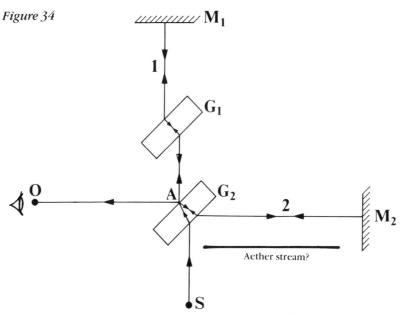

Figure 34

M_1 and M_2 are two perpendicular mirrors. G_1 and G_2 are two parallel-sided blocks of glass. A light ray is emitted from S and is divided into two rays at A. These pass to M_1 and M_2 respectively and are recombined at A, to pass to the observer at O. Each ray travels through G_1 and/or G_2 three times, so that the two path-lengths could in principle be exactly the same. But in journey₁ there is an 'external' reflection at A, whereas during journey₂ there is an 'internal' reflection. Hence, when the rays arrive at O they should be exactly out of phase, and if the two journeys are of exactly the same length no light should be visible at O. In practice, of course, it is impossible to make the path lengths exactly equal, so one sees a series of 'interference fringes' at O, produced by the 'interference' of the light arriving in the two wave-trains. If while the experiment is being carried out the path-lengths ('1' and '2') are gradually made very slightly different, this would cause an observable shifting of the interference fringes observed at O. Or if, for some reason, the light were to begin to travel at different velocities in the directions '1' and '2', then again this would manifest itself by an alteration in the fringes seen at O.

Of course, if the apparatus were standing still in the medium that is transmitting the light vibrations, then naturally we should expect no alt-erations in the journeys '1' and '2', and no alterations in the speeds with which the light rays make their journeys. But suppose the Earth is carrying the apparatus along, relative to the aether. Then the situation would be different. It would be as if the apparatus were flowing in a stream of aether. And then some observable effects might be detected.

To explain this, consider the model of a swimmer swimming in a moving stream of water. Suppose he/she swims the same distance — either across the stream and back, or downstream and back again. Then according to the kinematics of Galileo and Newton, the time for the two journeys is not the same. It takes longer for journey$_2$ than for journey$_1$[9] (see Figure 35).

Figure 35

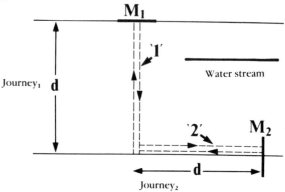

The reason for this is as follows. Journey$_1$ is the same, time-wise, as it would be if the water were motionless. Things are different for journey$_2$. In this case, the swimmer is speeded up on the way towards M_2 and slowed down on the way back again. But he/she swims for a longer time at the slower speed. Therefore, overall, journey$_2$ takes longer than journey$_1$.

Reconsidering the Michelson–Morley experiment, then, if the Earth is moving through the aether, the fringe pattern observed at O would be expected to be different (shifted) from the pattern that would exist if the Earth were stationary. And (more easy to detect) if the apparatus were swung round while the observations were being made, the fringe patterns should gradually alter, as a result of the gradual alteration in the effective path lengths, '1' and '2'. This was what was expected — what theory predicted; but when the experiment was actually carried out a *null result* was obtained. The fringes were unaffected as the apparatus was swung round on its Earth-anchored base.

The null result gave rise to much speculation among physicists, and various possible explanations were canvassed. The most interesting was the so-called Fitzgerald–Lorentz contraction hypothesis,[10] which supposed that the arm of the apparatus that was parallel to the presumed direction of the Earth's movement through the aether *contracted* by an amount just sufficient to counterbalance the longer time required for the light's back-and-forth journey in this arm. The required amount for the contraction was calculated to be given by the equation $l = l'(1 - v^2/c^2)^{1/2}$, where l' is the 'rest length', l is the length when the apparatus is moving at a velocity v through the aether, and c is the velocity of light.

As stated above, the problem generated by the null result of the Michelson–Morley experiment set off a great deal of discussion among physicists. But contrary to what was formerly supposed, it was not *this* which led Einstein to the dramatic new theory that he published in 1905. Rather, he seems to have arrived at this views chiefly as a result of Mach's positivistic critique of the fundamental principles of Newtonian physics, and by employing various 'thought experiments' of the kind that would have greatly appealed to Mach himself.[11] Let us, therefore, attempt to say something about the genesis and nature of Einstein's theory, and show how it related to a kind of Machian theory of knowledge.

During the nineteenth century, the velocity of light was measured in several different ways, and was found to have a constant value of about 3×10^{10} cm/sec. Now if Galilean/Newtonian kinematics were appropriate for dealing with the motions of bodies travelling at speeds comparable to the speed of light, if c is the velocity of light emitted from a stationary body and v is the 'absolute' velocity of the body, then the velocity of the light emitted by the moving body ought to be '$c + v$'. But such a state-of-affairs was unacceptable to Einstein,[12] and is incompatible with experimental evidence. For in the case of double stars (rotating round one another), if the velocity of light could vary with the velocity of the emitter then the star might be visible in two or more places simultaneously. But this is *not* the case. As Einstein said later: if the velocity of light could vary it would get all mixed up with itself. So unless light moved at an infinite speed (which experiment showed it did not) the world would have looked rather different from what it actually did.

Be this as it may, in 1905 Einstein simply introduced the principle of the constancy of the velocity of light (*in vacuo*) as an axiom or postulate of his theory. It appears that even as a teenager he had used a thought experiment to inquire what it would actually be like to travel at the speed of light. At such a speed, an observer should be travelling 'with' the light, so to speak, so that no light waves would pass him or her. Consequently no light would be visible. Einstein rejected this situation as unintelligible.

The other major postulate of Einstein's new theory was that all uniformly moving frames of reference are equivalent to one another. There are no absolute motions and no privileged frames of reference. The laws of physics are exactly the same, both for stationary observers and for those moving with uniform velocity with respect to the stationary observers.

There are many published accounts of Einstein's theory of special relativity, showing how the equation for the Fitzgerald–Lorentz contraction may be deduced from the postulates of the theory. A fairly simple derivation was subsequently given by Einstein himself in his popular book *Relativity*,[13]. Here I shall give what is, I think, the simplest possible version of the matter, but I do so emphasising that it is an ahistorical treatment, not as Einstein himself actually presented the matter.[14]

Imagine two frames of reference, S and S', moving with a relative velocity v with respect to one another. And imagine further that in each frame of reference, light impulses are being emitted at unit intervals of time, reflected off a mirror, and received by a light collector which records the reception of each pulse of light, as depicted in Figure 36.

Figure 36

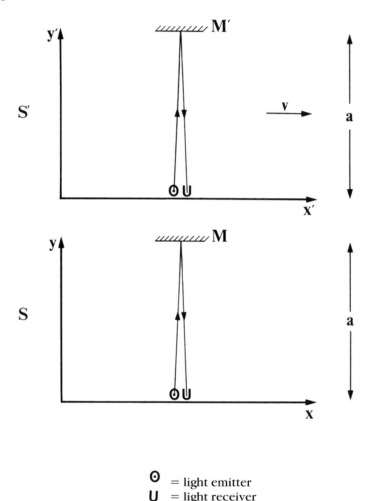

$$\odot = \text{light emitter}$$
$$\mathsf{U} = \text{light receiver}$$

Within their own frames of reference, two observers will make exactly the same observations in relation to such comparable states of affairs. But an observer in the first frame will see things differently in the second frame when he or she looks at what is going on there. What is seen is as in Figure 37.

Figure 37

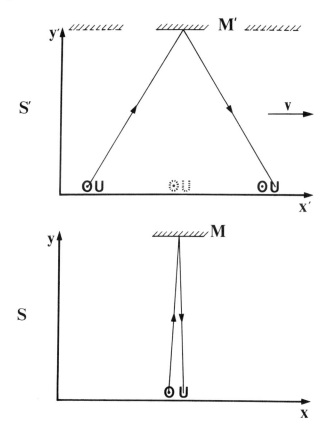

Let t' and t be the time intervals for the back-and-forth light journey in the frame of reference S', as determined in the frames of reference S' and S respectively. Within the frame S', we have the situation as in Figure 38.

Figure 38

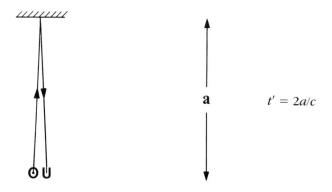

$$t' = 2a/c$$

But for frame S', as perceived from frame S, the situation is:

Figure 39

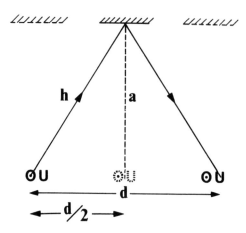

$t = 2h/c$
and $d = vt$
But by Pythagoras's theorem:
$h^2 = a^2 + (d/2)^2$.
So $(ct/2)^2 = (ct'/2)^2 + (vt/2)^2$;
or $c^2t^2 = c^2t'^2 + v^2t^2$;
or $t^2(c^2 - v^2) = c^2t'^2$;
or $t^2 = (c^2t'^2)/(c^2 - v^2)$;
or $t^2 = t'^2/(1 - v^2/c^2)$;
or $t = t'(1 - v^2/c^2)^{-1/2}$

In other words, the time intervals in one frame of reference, as measured from the viewpoint of another frame of reference, are different according to the formula:
$t = t'(1 - v^2/c^2)^{-1/2}$.

Of course, the effect is negligible unless v is large compared with the speed of light. This is not the case in our ordinary everyday experiences, but it may be relevant in astronomy, or for rapidly moving particles at the micro-level.

The equation just deduced tells us that time in the frame of reference S' seems to travel more slowly when determined from the frame of reference S than does time in frame S. Naturally, this has an effect on how lengths will appear in S' from the perspective of S. To calculate this effect, we may carry out the following thought experiment. The length of a rod, lying in the direction of the relative motion of the two frames of reference, is measured by observers in the two frames, using light signals for the purpose. Within the frame of reference S', we have the situation depicted in Figure 40.

So the time of travel for a light ray from one end of the rod and back again, as measured by an observer in this frame S', is: $t' = 2l'/c$. But when the measurement is made from the frame of reference S, the situation is as in Figure 41.

Figure 40

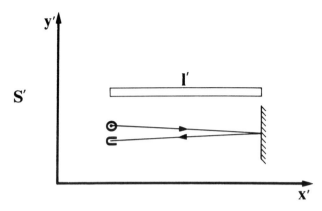

Figure 41

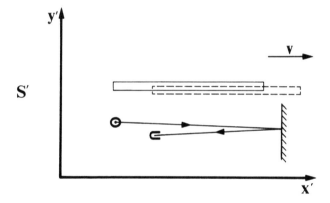

Here, the time for the back-and-forth light journey, as measured from S, is:

t = (time for forward path) + (time for return path)

$\quad = l/(c - v) + l/(c + v)$

$\quad = 2lc/(c^2 - v^2)$.

(An observer in reference frame S will measure the length as l, not l'.)

If we now substitute these two values for t' and t into the equation $t = t'(1 - v^2/c^2)^{-1/2}$, we have:

$\quad 2lc/(c^2 - v^2) = (2l'/c)(1 - v^2/c^2)^{-1/2}$.

From this, by simple algebra, one can readily obtain the equation

$\quad l = l'(1 - v^2/c^2)^{1/2}$.

This, it will be noted, is the equation proposed, *ad hoc*, by Lorentz and Fitzgerald to explain the null result of the Michelson–Morley experiment. We can see how the new arguments of relativistic theory enabled what was otherwise a considerable puzzle, or physical anomaly, to be accounted for most satisfactorily.

The strange new ideas about time are of particular importance. As simple-minded Newtonians, we may well be inclined to believe in a doctrine of absolute time. And consequently (as Newtonians) we believe that there can always be simultaneity of events, even in parts of the cosmos that are very remote from one another. But Einstein's theory showed, contrary to 'common (Newtonian) sense' that it is meaningless to talk about the simultaneity of events in two systems travelling with a uniform velocity with respect to one another. For Einstein asked himself what was really meant by the 'time' of an event. And he concluded that it referred to the simultaneous occurrence of two events in the same reference system. For example, to say that the train arrives at 7 o'clock means that the arrival of the train and the movement of the watch-hand over the figure '7' coincide.[15] This is the practical significance of the talk about time, in this case.

Perhaps the Newtonian may still assert that two events can occur simultaneously in some kind of 'absolute' sense. But consider the situation in Figure 42.

Figure 42

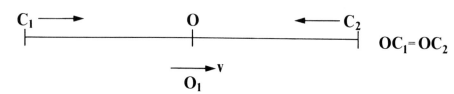

Suppose C_1 and C_2 are two synchronised clocks, equidistant from the observer O. Under such circumstances, the observer O will construe two events at C_1 and C_2 as simultaneous if light messages arrive at O simultaneously. But a moving observer, O_1, though he will receive signals from C_1 and C_2 simultaneously as he passes the point O, will construe this information differently (if he knows he is in relative motion with respect to the other frame). He will think that the signal from C_1 preceded the signal from C_2. So he will not think that the two clocks are synchronised. And consequently events that are simultaneous in one frame of reference are not so when determined from another frame. The clock of a person in one frame seems to run slow to a person in a different frame: *and vice versa*. All this is revealed by the formula: $t = t'(1 - v^2/c^2)^{-1/2}$.

So what Einstein asked for was the renunciation of the concept of absolute simultaneity of events. But he held to the principle that all laws of physics had the same form, no matter in which frame of reference they were determined or deployed.[16] There was to be no privileged frame of reference. Also, as we have seen, there was the doctrine — essential to the whole argument — of the constancy of the velocity of light, no matter what the velocity of the emitting source might be. Of course, if light travelled at infinite velocity, then there would be no principle of relativity to contend with. All light 'messages' would be transmitted instantaneously, so there would be the possibility of having a practical knowledge of simultaneous events. But *this* thought experiment refers to some other kind of cosmos — not the one we actually have to deal with.

Also present in the 1905 paper was a presentiment of the doctrine that the mass of an object was not constant, but dependent on its velocity, according to the equation $m = m'(1 - v^2/c^2)^{-1/2}$. And there was an adumbration of the equivalence of mass and energy according to the famous equation $E = mc^2$.

Having given so much attention to the Michelson–Morley experiment, the reader might suppose that it was the direct stimulus to Einstein's early work in relativity. But as indicated above, this does not appear to have been the case. Rather, Einstein developed his ideas by consideration of the work of Mach. Mach's positivism and phenomenalism, it will be noted, accorded well with Einstein's consideration of 'events' – such as the arrival of the train corresponding with the passing of the watch-hand over the figure '7'. Moreover, Mach himself was at first glad to accept the new relativity theory. But he changed his mind on this not long before he died.[17] Added to this, Einstein himself began to move away from his early Machism. And in fact the 1905 paper itself contained certain elements that were anything but Machian. The critical analysis of the concepts of absolute time, simultaneity, and the customary assumption of the equivalence of clocks in all frames of reference were entirely pleasing to Mach. So also was the kind of mental analysis of a physical situation that Einstein used to perform with the help of thought experiments (considering various observers synchronising and carrying their clocks, the behaviour of moving trains, lightning flashes, and so on). But the two basic postulates of the theory — the principle of the constancy of the velocity of light and the relativistic principle itself — seemed utterly to transcend the direct empirical experience of relationships between phenomena. And so Mach gradually came to draw back from Einstein's ideas. And, as we have said, Einstein gradually began to repudiate his early Machism. He came to think of physical principles as 'free inventions of the human intellect'[18] rather than convenient short-hand statements of the relationships between phenomena, as a true devotee of Mach would suppose.

But although the creative human intellect might be needed to break through the accustomed modes of thinking, this did not mean that Einstein supposed that any old principles would do. Rather, he believed that when a theory succeeded in giving a simple mathematical correlation and representation of experience then it was providing an adequate 'copy' of reality. To be sure, Einstein did not wish to assert that science could ever achieve a full and final description of the world. There was, nevertheless, a strong 'realist' component to his philosophy of science.

So Einstein believed[19] that a scientific theory was made up of a set of fundamental axioms or principles, which could be freely chosen by the creative act of the scientist. From these axioms, mathematical deductions might be made to theorems; and these had to be tested experimentally. Unlike Newton, Einstein did not believe that the axioms could be derived directly or logically from the data of experience — from phenomena.[20] A creative act of mathematical construction was required. The link with phenomena came at the end of the chain of deductions, where the theorems of the mathematical system were compared with the empirical evidence. The whole process was guided by a seemingly *a priori* assumption that there was a kind of 'pre-established harmony' between thought and reality — almost as the Aristotelians had supposed long before.

It can be seen, then, that Einstein's account of scientific methodology had a good deal in common with orthodox hypothetico-deductivism. There was 'no logical path to... [the laws of physics]; only intuition, resting on sympathetic understanding of experience'.[21] The processes involved could be compared with those of the painter or poet.[22] In particular, Einstein seemed to set his face against any kind of 'construction system' such as that of Carnap's *Aufbau*; for he wrote:

> [E]very attempt at a logical deduction of the basic concepts and postulates of mechanics from elementary experiences is doomed to failure.[23]

A valuable overview of Einstein's metascientific position has been given by V.F.Lenzen, and it will be helpful, therefore, to quote what he has to say at some length:

> The postulates of a mathematical system implicitly define the conceptual relations of a set of objects which are thereby realized in thought. Theorems are derivable from the postulates by logical deductions which may be viewed as certain. Cognition of reality, however, originates in sensory experience, is tested by sensory experience, and shares the uncertainty of such experience. Cognition of physical reality occurs through the media of concepts which express properties of objects in a spatio-temporal environment. Theoretical physics represents reality by models, such as particles and continuous fields. A model of reality serves to order sense-impressions; the theory based upon it is confirmed by approximate agreement of logical consequences with sense-impressions. The concepts of theoretical physics need to be correlated with sense-impressions, but in the last analysis concepts are created by the the spontaneity of thinking. Experience may suggest theories which operate with concepts which are intuitively related to experience, but for deeply lying objects the search for principles is guided by ideas of mathematical simplicity and generality. Methodical, intimate consideration of a problematic situation is the foundation of an intuition which gives insight into the order of reality. That the model of reality represents an order independent of experience is justified by the belief, founded on past successes, that nature embodies an ideal of mathematical simplicity. The pursuit of truth under the guidance of mathematical ideals is founded on the faith that a pre-established harmony between thought and reality will win for the human mind, after valiant effort, an intuition of the depths of Reality.[24]

Einstein himself put it as follows:

> Our experience hitherto justifies us in believing that nature is the realization of the simplest conceivable mathematical ideas. I am convinced that we can discover by means of purely mathematical constructions the concepts and the laws connecting them with each other, which furnish the key to the understanding of natural phenomena. Experience may suggest the appropriate mathematical concepts, but they most certainly cannot be deduced from it. Experience remains, of course, the sole criterion of the physical utility of a mathematical construction. But the creative principle resides in mathematics. In a certain sense, therefore, I hold it true that pure thought can grasp reality, as the ancients dreamed.[25]

We can see in the final sentence quoted here that in his view of the scientist's *modus operandi* Einstein envisaged a kind of Aristotelian theory of human rationality. There is also a passage in Einstein's Herbert Spencer lecture of 1933 that I find reminiscent of the Platonic 'arch of knowledge':

> Pure logical thinking can not yield us any knowledge of the empirical world; all knowledge of reality starts from experience and ends in it.[26]

I invite the reader to compare this remark with the closing section of the passage from Plato's *Republic* that was given above, p.10. Verily, the arch still standeth!

If we put together Einstein's various remarks on the nature of scientific theory, such as have been quoted or discussed above, it may readily be seen that his 'picture' of the 'structure' of a theory was much like that envisaged by Duhem, or later elaborated by Carnap and the logical positivists. It was, in fact, a version of what we have called the 'received view'.[27] There was, therefore, nothing so very remarkable in Einstein's metascience, and in a sense his account of it was rather misleading in that really it had applicability only to the work of those of great genius such as himself. It said little about the work of the humdrum, day-to-day scientist, or the structures of the social formations that shape science in such an important way — matters that will be dealt with in Chapter 9. What is, I think, particularly important for the history of metascience is the way in which Einstein used thought experiments, and also the manner in which he considered the concrete procedures and processes that take place when some fundamental property like length or time is measured. It was Einstein's actual procedure in these matters, helping him to grasp new ways of thinking about the world, that was to have profound influence on metascience, leading to the doctrine of operationalism — a topic to which we shall now turn our attention.

Bridgman

The metascientific term operationalism (or operationism)[28] is chiefly associated with the work of the Harvard physicist, Percy Williams Bridgman (1882–1961). Bridgman, a Nobel prize-winner, carried out important investigations concerning the properties of matter under high pressures, but today he is chiefly known for his contributions to philosophy of science and the aforementioned conception of operationalism. It is, of course, these philosophical contributions with which we shall be concerned.

As a practising scientist himself, Bridgman was particularly influenced by other philosopher–scientists such as Mach, Duhem, Poincaré, and above all Einstein. Bridgman claimed that he was merely making explicit what was implicit in the work of these men. But in fact he was doing a good deal more than this; he was developing a whole philosophical and methodological system. This system, while owing much to the influences of individual scientists such as those mentioned above, and to Bridgman's reflections on his own experimental investigations, may be seen as a particular sub-class of the tradition of American pragmatism. One may also think of it as a sub-class of positivism and an extreme form of empiricism.

The first and most influential statement of Bridgman's ideas appeared in his *Logic of Modern Physics* (1927).[29] Since the essence of his argument is to be found in the opening chapter of this book, we shall, in the first instance, look particularly at this.

According to Bridgman, the attitude of the physicist must be one of 'pure empiricism'.[30] Somewhat naively (it may seem today) he believed that: 'the *fact* has always been for the physicist the one ultimate thing from which there is no appeal, and in the face of which the only possible attitude is a humility almost religious'.[31] For Bridgman there were no *a priori* principles determining experience. Evidently, he was no Kantian. And nature could not be embraced by any single formula or conception (such as absolute idealism, one might suggest).

Bridgman further complained that Newton failed to define time, space, place and motion adequately, so that his system of mechanics was wholly abstract, like a pure geometry. This was true in a sense, though it will be recalled that Newton himself supposed that his 'first principles' were drawn directly from experience — by induction.[32] We have, however, felt sceptical about Newton's position on this point and may prefer to think of the structure of Newton's science in a manner like Duhem or Poincaré rather than Newton himself. Bridgman's solution to the problem was to define fundamental scientific concepts operationally. Let him speak in his own words on this:

> The new [ie, Bridgman's] attitude toward a concept is entirely different [from that of Newton, for example]. We may illustrate by considering the concept of length: what do we mean by the [concept of the] length of an object? We evidently know what we mean by length if we can tell what the length of any and every object is, and for the physicist nothing more is required. To find the length of an object, we have to perform certain physical operations. The concept of length is therefore fixed when the operations by which length is measured are fixed: that is, the concept of length involves as much as and nothing more than the set of operations by which length is determined. In general, we mean by any concept nothing more than a set of operations; the *concept is synonymous with the corresponding set of operations.* If the concept is physical, as of length, the operations are actual physical operations, namely, those by which length is measured; or if the concept is mental, as of mathematical continuity, the operations are mental operations, namely those by which we determine whether a given aggregate of magnitudes is continuous.[33]

The basic doctrine underlying operationalist theory of knowledge is contained in the italicised section of this passage: the meaning of a physical concept was identical with the corresponding set of operations required to make the measurements that were relevant to that concept.

To illustrate his position, Bridgman placed particular reliance on the work of Einstein as exemplar, giving great emphasis to it in the opening pages of his book, in order to make clear the nature of the operationalist thesis. As we have seen, before Einstein the concept of simultaneity of events had been generally accepted as intelligible in a kind of absolute sense. But Einstein showed that the simultaneity of two events was not a kind of 'absolute' property, but must involve the relation of the events to the observer. The observer's observations enter, so to speak, into the concept of simultaneity. In the simple case of the train arriving at the station, for example, the actual meaning of the train's arriving at 7 o'clock resided in the fact that this event coincided with a particular observed disposition of the hand on a watch.

Again, for ordinary everyday measuring, such as might be done by a surveyor, the concept of length is definable (according to operationalism) by the actual physical operations carried out with measuring rods, tapes, or whatever. But Einstein's work suggested different procedures for measuring the lengths of bodies moving with high velocities relative to the measuring instruments. Consequently, his lengths were conceptually quite different from those of the land surveyor. This accorded with the theoretical prediction of relativity theory that the length of a moving body was not the same as that of a stationary body; indeed one needs the 'Lorentz transformation' to convert one to another.

Bridgman's operationalism obviously accorded well with some of the chief doctrines of the logical positivists that we described in the previous chapter. It ruled out questions such as 'Does time have a beginning or an end?' as meaningless, for there was no way in which such questions could be treated empirically (or operationally). (Kant would certainly have agreed with this, and for reasons somewhat analogous to those of Bridgman.) But besides the elimination from the domain of science of certain questions of a speculative and metaphysical nature, beyond the reach of practical investigation, Bridgman's suggestion was intended to effect a considerable clarification of what was going on in science. It would do this by making clear the proper meaning of theoretical terms, with the aid of operational definitions. And it should be conducive to intellectual hygiene by, for example, preventing the reification of hypothetical entities. Nevertheless, it can be seen that pure empiricism could not be an adequate basis for a complete description of all scientific procedures. For example, to measure a velocity one has to measure a length (defined operationally) and a time interval (defined operationally), and then divide the length by the time. This last involves a mental operation which clearly had to be allowed for within Bridgman's system. This he was, in fact, prepared to do, though he also sought to give an empiricist interpretation of mathematics.

In 1937, a number of important criticisms of Bridgman's operationalism were made in the journal *Philosophy of Science* by R.B.Lindsay,[34] to which Bridgman replied the following year.[35] Lindsay objected that if the doctrine of operationalism were taken seriously it would be quite at odds with what anyone could see physicists were actually doing, namely working with concepts for which no operational definitions could be supplied — for example numbers[36] or the mysterious 'wave-functions' deployed in quantum theory or wave mechanics.

Bridgman had to acknowledge the truth of this objection, and so he introduced what he called 'paper-and-pencil' operations. He also admitted that 'verbal operations' played a significant role in science. These he subsequently (1952) described as follows:

> The paper-and-pencil world is a world in which free invention is possible, divorced from any immediate contact with the instrumental world of the laboratory... Very great

latitude is allowed to the verbal and paper-and-pencil operation. I think, however, that physicists are agreed in imposing one restriction on the freedom of such operations, namely that... [they] must be capable of eventually, although perhaps indirectly, making connection with instrumental operations. Only in this way can the physicist keep his feet on the ground or achieve a satisfactory degree of precision; instrumental contact affords the only 'reality' which he accepts as pertinent for him.[37]

This was a particularly thoroughgoing statement of the principles of empiricist philosophy. There could, Bridgman continued, be many others concepts in domains such as politics, philosophy or religion that were 'purely verbal', being unconnectable with the world of physical/empirical experience. They were, as Bridgman inelegantly put it, 'not capable of eventual instrumental emergence'. Such verbal concepts were, he acknowledged, of great importance in human life, but they were not the field of the physicist; they lay outside science.

It would seem that Bridgman had watered down his original position very considerably by the introduction of 'paper-and-pencil' operations. But in doing so he was, in effect, bringing himself into somewhat closer alignment with other positivist philosophers and the 'received view' of the structure of scientific theories. Once again our 'arch of knowledge' may help us to see what was going on. In the orthodox account of the structure of scientific theories, stemming from Duhem, and given in great detail by Carnap and other logical positivists, postulates were proposed and theorems deduced from these, both postulates and theorems being stated in terms of some 'theoretical language'. The theorems were tested against experience (which was articulated in an 'observation language') with the help of 'correspondence rules'. We thus had a nebulous ascending 'inductive leg' of the arch[38] and a strongly constructed deductive leg connected with experience only in its lower parts. Language of theory and experience were carefully distinguished. All this I have said several times before. But in Bridgman's system, as originally stated, *all* theoretical concepts were linked directly to experience, for their very meaning was defined in such a way that it was rigorously empirical. So Bridgman's model of a scientific theory might imaginatively be thought of as a zip-fastener, zipped right the way to the top — empirical and theoretical matters being hooked and linked at every segment of the zip. (As we shall see below, other commentators have also found the simile of a zip-fastener useful in metascientific discussions.) But when Bridgman introduced the notion of 'paper-and-pencil' operations, some parts at least of the zipper came apart. He spoke of all scientific concepts as 'capable of eventual instrumental emergence',[39] which marked him as an empiricist. But the radicalism of the earlier position was thereby lost. Some theoretical terms, at least, would find their meaning through what people *thought* about them, rather than in particular practical activities. In such cases there would, at best, only be an indirect connection with experience.

There has been much criticism of Bridgman over the flexibility with which he used the term operationalism. Thus we find: 'The concept is synonymous with the corresponding set of operations' (1927); 'Meaning is to be sought in operations' (1934); 'Operations are a 'necessary' but not a 'sufficient' condition for the determination of meaning' (1938); and 'The operational aspect is not by any means the only aspect of meanings' (1952). In these successive statements we can see the gradual watering down of the original thesis — or the unzipping of the zipper. In fact, it has sometimes been suggested that the idea of an operational definition is so fuzzy that it ought to be operationally defined!

I fear such a suggestion is calculated to lead one into the thickets of Russell-type paradoxes. So we shall not choose to follow this approach. It may be more useful to tackle the problem more directly and ask whether there is any sense in identifying concepts with operations in a one-to-one relationship. It would appear that to do so involves a considerable distortion of what is normally meant by concepts, scientific or otherwise. There is, I suggest, always a kind of 'halo' round any concept — all its many connotations — which transcends the particular operations that may be relevant or appropriate to that concept.

Another frequently rehearsed difficulty lies in the fact that some concepts seem to be able to be given several quite different operational definitions. For example, time can be measured by stop-watch, pulse, stellar observations, water-clock, atomic clock, and so on. It would seem that if we were to accept operationalism, a different concept would be involved for each method of measuring time. Yet this hardly meshes with our general understanding of the concept of time.

Bridgman sought to get round the difficulty mentioned in the previous paragraph by suggesting that two methods for measuring (say) time should give results that are closely similar. But this seems to suggest that there is some 'thing' — time — to which operationally defined concepts of time correspond, and which the operations for measuring time are performed. Yet this is at odds with the whole notion of operationalism.

Another difficulty is that operationalism seems to suggest that 'nature' and 'knowledge of nature' are one and the same thing. Thus, thoroughgoing operationalism seems to be equivalent to a hard-line phenomenalism. This might have been acceptable to positivists. But it is hard to agree with the suggestion that length is measuring, science is sciencing, or intelligence is intelligence testing. This is a flagrant misuse of language. It is like confusing a cake with the recipe for making a cake.

It is in the example of intelligence testing, however, that we see operationalism in the kind of context in which it is most frequently encountered today — that is, in connection with psychological and sociological investigations, where theories have for a long time been in a somewhat hazy state. For example, it is often said that the *meaning* of 'IQ' (intelligence quotient) is simply what is measured in an IQ test, neither more nor less than this. It really says nothing about some kind of essential intelligence that a person may or may not possess (since *innate* intelligence cannot satisfactorily be teased apart from what a person knows through his or her education and other life experiences and so remains a somewhat intuitive notion). This is an operationalist attitude to the problem. Operationalism also accords with the doctrines of behaviourist psychology, which eschews any direct search for mind as such, and rests content with observing behavioural phenomena and attempting to discover the laws governing these. But that does not necessarily recommend it as a sound philosophy of physics.

Operationalism is, I suggest, closely related to empiricism, positivism and pragmatism. We have already noted the empiricist and positivist aspects. The pragmatic component is revealed in remarks of Bridgman such as the following: 'what a man means by a term is to be found by observing what he does with it, not by what he says about it'.[40] Bridgman also maintained that the operational approach to science was in itself beneficial, or had pragmatic value. But he did not work out a detailed theory of truth, and there was a somewhat awkward balancing of correspondence theory and pragmatic doctrine. Truth, for Bridgman, had no absolute, static character. It was determined by needs and intentions. It was through science, language and thought that men sought to accommodate themselves to the conditions of their environments. Yet pragmatic success in this enterprise should not be taken as an indication of scientific knowledge bringing one closer to *reality*.

Nevertheless, in *The Nature of Physical Theory*, Bridgman was employing a simple kind of correspondence theory of truth. He stated, for example, that the success (albeit limited) of language in faithfully reproducing experience might be ascribed 'to its ability to set up and maintain certain correspondences with experience'.[41] Admittedly, he went on to say that it was meaningless to inquire how and why it was that these correspondences could be set up and maintained; it simply had to be accepted as a brute fact that it was so. This may sound simplistic, but Bridgman may well have been prudent to turn away from a problem that has given philosophers such a deal of trouble, and is still not satisfactorily resolved. In any case, Bridgman's correspondences were not thought to be 'perfect' where ordinary language was concerned. In mathematics, a subject which Bridgman (disregarding all the efforts of Frege, Peano, Russell, etc.) regarded as an *empirical* science,[42] a 'language' was specially created for the description of the external world. So in mathematics one might hope for exact correspondence. But, said Bridgman, mathematical descriptions transcend experience. So the equations of theoretical physics needed to be accompanied by a 'text' elucidating the significance of the equations, and how they were to be used.[43] This 'text' supposedly provided the correspondence rules of a physical theory. Thus Bridgman's account of physical theory, as propounded in 1936, was pretty much that of the positivists' 'received view'; and it would have been perfectly comprehensible to Duhem. But since, as I have just mentioned, Bridgman seemed inclined to give operational definitions to mathematical entities, and indeed to as many theoretical conceptions as possible (though drawing the line at such animals as the 'wave functions' of the quantum theorists), his 'received-view' zipper was pretty well zipped up to its top — though admittedly coming undone a bit in the talk of 'paper-and-pencil' operations.

Bridgman's contributions to metascience were not inconsiderable, though he can hardly be regarded as a philosopher of great subtlety. His term 'operationism' is now very much part of the language of metascience, even though as an 'ism' (or as a basis of a whole philosophy of science) its appeal has declined, along with the positivism of which it was a part. Bridgman's work is of special relevance to us because of the way in which it illustrated the impact of Einstein's science on meta-scientific thought. We shall now turn to consider the work of another philosopher/scientist, A.S.Eddington, whose ideas were also greatly influenced by Einstein. Eddington, however, derived conclusions very different from those of Bridgman.

Eddington

Arthur Stanley Eddington (1882–1944)[44] had one of the most brilliant mathematical minds in Britain in the early twentieth century. He became First Wrangler at Cambridge in his second year in 1904, and in 1906 he was appointed Chief Assistant at the Royal Greenwich Observatory. Later he became Plumian Professor of Astronomy and Director of the Observatory at Cambridge. He was a man of deep religious convictions, being born into a Quaker family; and the Quaker notion of an 'inner light' appears to have been a guide to his scientific and mathematical work.

It was Eddington who was one of the scientists whose experimental work (in 1919) seemed to lend support to Einstein's general theory of relativity (not to be confused with the special theory, discussed earlier in this chapter). As we shall see in Chapter 8, this work was to have a con-siderable influence on the thinking of Karl Popper, and thereby on a good deal of twentieth-century philosophy of science.

What made Eddington's work notorious in its day, and still highly intriguing, was his claim to be able to calculate certain physical constants, such as the mass ratio of the proton to the electron, directly from *a priori* principles without recourse to observational evidence. He even claimed to be able to calculate the number of protons and electrons in the universe, suggesting in his *Philosophy of Physical Science* that calculations showed that there were '15,747,724,136,275,002,577,605, 653,961,181,555,468,044,717,914,527,116,709,366,231,425,076,185, 631,031,296 protons and a like number of electrons'![45] Such calculations no doubt seem miraculous, yet in some instances (as in the case of the ratio of the mass of the proton to that of the electron) Eddington arrived at values that agreed remarkably well with those obtained by experiment.

It is beyond the scope of this book to attempt to give any account of the way in which Eddington performed such seeming intellectual miracles.[46] (I may say, however, that subsequent commentators have generally found his arguments unsatisfactory, though they have been hard put to say exactly where the empirical elements crept into Eddington's calculations.) Here we shall simply look at some features of Eddington's philosophy of science, to see how he thought it possible that factual knowledge could somehow be derived from formal knowledge. He labelled his position 'selective subjectivism', for reasons that I shall now try to make apparent.

The argument was carried forward in *The Philosophy of Physical Science* with the help of one of the engaging analogies of which Eddington was so fond. He imagined an ichthyologist armed with a net having two-inch holes, casting for fish in the ocean.[47] Using such a device, the ichthyologist would be likely to come to conclusions such as: (i) no sea-creature is less than two-inches long; (ii) all sea-creatures have gills.

Analogies with the procedures of scientific investigation might be spelled out as follows:

The net	The sensory and intellectual equipment used in obtaining scientific knowledge
The ichthyologist	The physicist
Casting the net	The act of observation
The catch	The body of knowledge making up physical theory

Eddington offered a further amusing example from the story of Procrustes (the figure in Greek mythology who liked to stretch or chop his guests until they exactly fitted his bed), writing a learned article for the Anthropological Society of Attica entitled 'On the Uniformity of Stature of Travellers'.[48] As an example from actual scientific practice, Eddington suggested that the New Zealand physicist Ernest Rutherford 'rendered concrete the nucleus [of the atom] which his scientific imagination had created'.[49]

The point about all this, of course, was that in Eddington's view one could find out about the nature of the world — as revealed by the physicist's inquiries — by examining his conceptual and methodological apparatus. This could be more fruitful than looking at the body of knowledge fished up by physical inquiry. In other words, according to Eddington a good deal of what passes for objective scientific knowledge of the laws of nature is really *epistemological* in character. What is fished up is determined by the net-mesh as much as what is actually in the sea that is being trawled.

The knowledge that could be ascertained by epistemological analysis was, said Eddington,[50] '*a priori* knowledge'. But he emphasised that such knowledge was not to be regarded as innate. Far from being born with it, we need to have experiences and observations, and intercourse with our fellowmen. Nevertheless, we are allowed to make certain *a priori* statements about the results that the physicist will achieve by considering the procedures that he uses — just as the onlooker can prejudge the ichthyologist's 'discovery' that all fish in the sea are more than two inches long.

Because of the supposed partial *a priori* character of the physicist's work, Eddington referred to his view as 'subjectivism' — for the physicist clearly fails to achieve pure 'objective' knowledge of the world as it is 'in itself'. Quite the contrary. And because the information collected by the physicist is, as it were, 'filtered' by the very process of inquiry, Eddington employed the term 'selective'. So we have his philosophical position of 'selective subjectivism'.

It is interesting to consider Eddington's ideas in relation to other philosophers of science that we have considered in this book. It should at once be evident that selective subjectivism had a good deal in common with Wittgenstein's position, as stated in his *Tractatus*. Indeed,

Eddington's analogy of the fish net, and Wittgenstein's of a mesh laid over a piece of marked paper, were remarkably similar. There is also similarity between Eddington and Wittgenstein in that they both, in effect, were espousing a correspondence theory of truth. Wittgenstein supposed (in the *Tractatus*, not the *Philosophical Investigations*) that there was a parallelism between the 'logical' structure of language and the structure of the world. So too was Eddington presuming that the subjective and objective components of knowledge were so intimately united that one could find out certain 'facts' of the world by epistemological inquiry.

Eddington was also strongly influenced by Bertrand Russell,[51] and it might be said that he was trying to do for physics what Whitehead and Russell had done for mathematics, namely reveal its 'logical foundation' in the realm of abstract thought. Thus Eddington could be said to be postulating a correspondence between *a priori* 'structural' or relational knowledge and the relational structure of the external world. In this, his position was also somewhat similar to that of Einstein, which, as I mentioned earlier,[52] was in this respect quasi-Aristotelian.

Then again, one can see a kind of Kantian element in Eddington's position, and Eddington did in fact state that his system had something in common with that of Kant.[53] But unlike Kant, Eddington did not try to offer any formal justification of the correspondence between the *a priori* forms of knowledge and the structure of the external world, as comprehended by man.[54] It remained an unsupported assumption in his system.

Perhaps less to be expected are the Machian, quasi-positivistic elements in Eddington's position. As we have seen,[55] Mach's philosophical system was thoroughly monistic, concerned only with sensations and their relations. Even the self, for Mach, was a mere bundle of sensations. There was no mental/physical dichotomy. In a way, Eddington's system was somewhat similar, in so far as he postulated an identity between the '*group-structure of a set of sensations* in a consciousness' and the structure of the external world.[56] Given this assumption, the world and thoughts about the world were supposedly in a condition of correspondence. Communication between scientists was possible, according to Eddington, by virtue of the fact that science imposes certain invariance principles (as Einstein had done in his relativity theory). And physical science — or at least the kind that Eddington was interested in and sought to practise — was concerned with those aspects of metrical operations that were invariant, or independent of the status of the observer. So public, or communicable, scientific knowledge could be held to be possible. This was supposed to deflect the accusation of solipsism.

It will be clear from all this that Eddington, like so many others before and after him, was striving to provide a solution to that perennial problem of how it is that theoretical descriptions of the world match up with phenomena. Why does the deductive structure of Euclidean geometry match so well with the external world? Why do Newton's laws enable us to deduce theorems which can be tested so successfully in the external world? The operationalist such as Bridgman would have his answer prepared and ready without any trouble: the axioms, indeed all scientific concepts, are in fact empirical in character. We have looked at a number of other solutions to the problem — from pragmatists, instrumentalists, logicists, conventionalists, phenomenalists, and proponents of a good many other 'isms'. Eddington's solution was at the extreme rationalist end of the empiricist/rationalist spectrum. He did not deny that there was an objective component to knowledge, but in some of his more obscure writings he seemed to be able to dispense with it to a remarkable degree! If the reader will for a moment, look back to Carnap's reconstruction of the logical structure of the physicist's account of the linear expansion of iron,[57] it will be seen that the second axiom in the system was totally empirical in character. Now Eddington, we might say, was seeking to carry out a deductive program without any such corresponding empirical input. As I have indicated, subsequent investigations have not lent much support to Eddington's program, and his work now appears as something of an intellectual *cul-de-sac*, rich in interest though it was. But his very unusual philosophical position certainly deserves a position of honour in our story.

Campbell

I should now like to say something about the writings of the physicist/ philosopher, Norman Robert Campbell (1880–1949), as a means of introducing a discussion of the interesting and important topic of the role of models (and other forms of analogical thinking) in science. Campbell is helpful for this expository purpose in that he provides a further opportunity to discuss the 'received view' of the structure of scientific theories. And in considering Campbell we shall find ourselves looking back once again at the work of Pierre Duhem.

In thinking about the 'received view' of scientific theories, it is fairly evident that some aids are needful for the process of theory *construction* – Einstein's remarks about scientific principles being 'free inventions of the human intellect' notwithstanding. And it has long been recognised that models and analogies can be extremely helpful for the purposes of hypothesis formulation and theory construction. We have already encountered some examples of this in the previous chapters. For example, the seventeenth-century mechanical philosophy was based on a model of the micro-world which assumed that it was made up of minute corpuscles with properties somewhat like those which scaled-down billiard balls might be expected to have. Perhaps the most famous series of models used in the whole history of science was that of the great British physicist, James Clerk Maxwell (1831–1879), in his development of the equations for the propagation of electromagnetic radiation.

Maxwell started off thinking of electricity as a kind of invisible fluid and worked out equations such as that of the inverse square law for electrostatics on this basis.[58] Not long after, he changed his model and thought of magnetic fields as being due to rotating lines (or tubes) or magnetic force. Small particles of electricity supposedly occupied the interstices between the 'magnetic' tubes, and when the magnetic field fluctuated the rates of rotation of the 'magnetic' tubes varied correspondingly. Consequently, the interstitial electrical particles were set in motion, and so there was a fluctuating electric field accompanying the fluctuating magnetic field. Starting from this rather extraordinary physical model, Maxwell succeeded in developing some very satisfactory equations for the propagation of electromagnetic waves.[59] Subsequently, he found it possible to arrive at a set of equations without making any use of the model at all.[60] It was simply assumed that electromagnetic vibrations could be transmitted, wave-like, through a hypothetical aether. The *equations* became 'the theory'. The various models which had been used, rather like crutches, in the journey towards the theory were simply thrown away. All that remained was the equations — which could then be set up (if one so desired) in the manner of the 'received view' of scientific theories, allowing predictions to be made and tested in the orthodox fashion.

Now Duhem, well aware of Maxwell's achievements in all this, agreed that models are of obvious use in the *construction* of a theory's axiom system; and they *might* be associated with the interpretation of the theoretical terms of the axiom system. But Duhem did not think that models were essential components of a scientific theory, once it was constructed.[61] The axiom system plus (what were subsequently called) correspondence rules were all that were required. An interpretation of the meaning of the theoretical terms in the axiom system with the help of some kind of model was not thought necessary.

It was this view that was questioned by Campbell in his work *Physics: The Elements* (1919).[62] According to Campbell:

> A theory is a connected set of propositions which are divided into two groups. One group consists of statements about some collection of ideas which are characteristic of the theory; the other group consists of statements of the relation between these ideas and some other ideas of a different nature. The first group will be termed collectively the 'hypothesis' of the theory; the second group the 'dictionary'.[63] The hypothesis is so called ... because the propositions composing it are incapable of proof or of disproof by themselves; they must be significant, but, taken apart from the dictionary, they appear arbitrary assumptions. They may be considered accordingly as providing a 'definition by postulate' of the ideas which are characteristic of the hypothesis. The ideas which are related by means of the dictionary to the ideas of the hypothesis are, on the other hand, such that something is known about them apart from the theory. It must be possible to determine, apart from all knowledge of the theory, whether certain propositions involving these ideas are true or false. The dictionary relates some of these propositions of which the truth or falsity is known to certain propositions involving the hypothetical ideas by stating that if the first set of propositions is true then the second set is true and vice versa; this relation may be expressed by the statement that the first set implies the second.[64]

All this will, I hope, be construed by the reader as Campbell's version of the 'received view' of the structure of scientific theories. Campbell then went on to clarify the matter by means of a useful and widely quoted example. Suppose, he suggested,[65] the 'hypothesis' is made up of the following mathematical propositions:

1. $u, v, w, ...$, etc — independent variables;
2. a – a constant for all values of these variables;
3. b – a constant for all values of these variables; and
4. $c = d$ – where c and d are dependent variables.

Then the 'dictionary' (or 'correspondence rules') might say:

1. $(c^2 + d^2)a = R$, where R is the resistance of a piece of metal;
2. $cd/b = T$, where T is the temperature of the metal.

Given these assumptions, one can make deductions from the 'hypothesis' — for example as follows:

$$[(c^2 + d^2)a]/[cd/b] \text{ (an arbitrarily chosen function)}$$
$$= 2c^2(ab/cd)$$
$$= 2c^2(ab)/c^2$$
$$= 2ab$$
$$= \text{constant}$$

So $R/T = \text{constant}$.

If this equation is tested experimentally, and if the tests are successful (as they would be in practice), then it might seem that we have corroboratory evidence favouring the truth of the 'hypothesis'. But, said Campbell quite rightly, this is an absurdity! An infinite number of 'hypotheses' could have been formulated, all of which could equally well have been used as a basis for the derivation of the expression $R/T = \text{constant}$. All might 'save the appearances' equally successfully. But such hypotheses, drawn out of a hat so to speak, with no particular reasons underlying their choice, would never be seriously regarded by physicists. To make a theory worthy of consideration, some kind of reason for the choice of the elements of the 'hypothesis' had to be given. That is, some kind of analogy ought to underpin the 'hypothesis'.

To make his meaning clearer, Campbell gave an example drawn from the kinetic theory of gases, which was acceptable in his view because some kind of analogy *could* be provided which made the chosen 'hypothesis' meaningful and not merely arbitrary. Campbell himself did not actually use the term 'model' in his discussion; he spoke of analogies with known laws. Later writers, however, have generally interpreted his views as being similar to — and indeed an important precursor of — the 'modellist' point of view: that is, the view that the meaning of theoretical terms can only be intelligible if understood in terms of some suitable model or analogy. This reading of Cambell seems not unreasonable. His position was that a theory could do nothing towards *explaining* phenomena unless it displayed an element of analogy with known laws — or, we would add for him, unless it could be interpreted and made intelligible with the assistance of some kind of model, or by analogy with some kind of phenomenon, state-of-affairs, theory, law, conception or whatever, that was felt to be intelligible.

The question of whether or not Campbell was correct in his analysis has led to considerable discussion in the literature, and the examination of the role of models in science constitutes an important feature of recent philosophy of science. To show some of the arguments that might be raised against Campbell's position, we may mention a criticism made by the positivist C.G.Hempel in 1965.[66] This critic pointed out that it is perfectly possible to set up a calculus in which the theoretical terms *do* bear analogy with previously established laws. But in itself this is not sufficient to make the theory meaningful. For the laws could be quite irrelevant to the theory at issue.

This is true enough, of course, but it does not follow from the fact that inappropriate analogies are drawn from time to time that they are not necessary for meaning to be attributed to the terms of a theory or to the theory as a whole. Otherwise, a physical theory might be no more than an instrument for 'saving the appearances'. This, Campbell found unacceptable. So it may be seen why 'Campbellians' are customarily set up as opponents of 'Duhemists' — regarding Duhem as an 'instrumentalist' — even though they had a somewhat analogous view of the structure or 'shape' of scientific theories.

The modellist/contextualist debate

A number of writers, looking at the way the game of physical science is actually played, have remained unimpressed by Campbell's arguments. For example, the Cambridge philosopher of science, R.B.Braithwaite (1900–), has written:

> [A]n understanding of a theoretical concept in a scientific theory is an understanding of the role which the theoretical term representing it plays in the calculus expressing the theory, and the empirical nature of the theoretical concept is based upon the empirical interpretation of the final theorems of the calculus.
> If such a *contextualist* account of the meaning of theoretical terms is adequate, *thinking of a model for a theory is quite unnecessary for a full understanding of the theory.*[67]

Thus for Braithwaite (who represents what is customarily known as the 'contextualist' position), the meaning that is to be attributed to a theoretical term arises not from its analogy with some laws or some physical system external to the theory in question, but from its *context* within the total framework of the theoretical system in which it is embedded.

Braithwaite's view of the structure of a theory is also revealed in the following passage, which has become well-known by reason of its 'zip-fastener' simile:

> What happens in an abstract science is that we use, as in all inference, a calculus which we interpret as a deductive system; but we do not interpret the calculus by attaching meaning to its formulae separately. We give direct meanings to those formulae of the calculus which we take to represent propositions about observable entities; we give indirect meanings to the other formulae as representing propositions in a deductive system in which the observable propositions are conclusions. Thus we do not interpret the calculus all in a piece, as it were; we interpret the final part of it first, and work backwards to the beginning. A zip-fastener is a better simile for the fitting of a deductive system to a calculus than is the measurement of a rod by the simultaneous superposition of its ends on points of a scale.[68]

It should be emphasised that the point at issue here is one of meaning — of semantics. Do the meanings of theories (or of theoretical terms) reside in the analogies or models that may be used to give intelligibility to the theoretical terms of the axioms themselves? Or is it possible for a theory to be set up in the manner characterised as the 'received view'; and only then, when the theory is successfully in use, do the theoretical terms acquire meaning, as they are used in the derivation of theorems, experimentally testable with the help of correspondence rules? (What is not at issue here is whether models are sometimes used in the processes of theory construction, as in the example of Maxwell, which we briefly described above. Both 'Campbellians' and 'Duhemians' — or modellists and contextualists — agree on the utility of models for this purpose.)

On first thinking about the matter, it might seem that Campbell's position ('modellism') is obviously the correct one, if science seeks intelligibility. For example, modern geological theory entertains the notion of 'plates' floating on an underlying fluid. We are helped to understand what this means by thinking of the plates on our dinner table or ice floes floating on the sea. Darwin's idea of 'natural selection' is made intelligible by analogy with the procedures of animal breeders, or the struggle for existence experienced in the world of business and commerce (and he was possibly helped towards his theory by utilising this analogy). But when attention is turned to abstract physical theories such as that of modern quantum theory and wave mechanics, we find that abstract theoretical terms such as 'wave function' are commonly deployed. And one can only get the hang of what they mean by seeing how they are used in the mathematical formulation of wave mechanics — that is, with the help of the *context* supplied by the theory in which they are deployed. Indeed, it may be misleading to use the term '*wave* function' for it suggests that some physical substance (the aether?) is actually waving — like a flag or a rope! Recognising this, modern physicists concerned with the micro-world sometimes deliberately choose theoretical terms which are so outlandish that no one will ever be seduced into thinking that they bear any analogy to the world of everyday experience. So terms like 'wave' are carefully avoided; instead we meet ones like 'charm', 'flavour', and various 'colour vectors'.[69] In so far as these terms possess (theoretical) meaning, they do so only through their 'context' within the formalism of the theories in which they are situated. One can, if one is so disposed, imagine the meaning 'filtering' upwards towards the abstract theoretical terms from the lower levels of the theory where the theorems are tested experimentally.

Thus, when one examines the actual state-of-affairs that exists in some branches of theoretical physics we find that it gives but little comfort to the modellist view. For example, we find that the distinguished theoretical physicist, P.A.M.Dirac (1902–1984), espoused what seems to accord with the contextualist view:

> [T]he main object of physical science is not the provision of pictures, but ... the formulation of laws governing phenomena and the application of these laws to the discovery of new phenomena. If a picture exists, so much the better; but whether a picture exists or not is a matter of only secondary importance. In the case of atomic phenomena no picture can be expected to exist in the usual sense of the word 'picture', by which is meant a model functioning essentially on classical lines. One may, however, extend the meaning of the word 'picture' to include any *way of looking at the fundamental laws which makes their self-consistency obvious*. With this extension, one may gradually acquire a picture of atomic phenomena by becoming familiar with the laws of the quantum theory.[70]

Evidently, Dirac would see the theoretical physics of the micro-world as an abstract formalism, successful as an 'instrumental' device for making empirical predictions. Questions about the 'reality' of the micro-world, as portrayed by the theoretical physicist, are perhaps better not asked! In so far as the theoretical terms of quantum theory have meaning, this meaning can only be apprehended by seeing how the terms are used within the physicists' formalisms.

This still does not imply, of course, that the role of models as aids towards the establishment of scientific theories is to be dismissed, and the 'contextualists' would not claim that it should be. The issue is the semantic one of the ascription of meaning to the theoretical terms of scientific theories. It must be emphasised, however, that *this* issue is one that arises within the constraints imposed by what is an essentially positivist view of science. It is hardly an issue if one refuses to allow that there are two radically distinct kinds of language, theoretical and observational; and recent writings in philosophy of science have tended to stress the interconnectedness of all meaning and language, both scientific and non-scientific.[71] So instead of having two different *kinds* of language, we have two different kinds of *use* of the same language. On this view, there can be no independent and certain observational language, usable by all competent observers: for observations, it is insisted, are *theory*-dependent. And different observers may hold different theories.

Perhaps the best-known discussion of the modellist/contextualist controversy is to be found presented in an engaging fashion — in the form of an imaginary dialogue between a Campbellian and a Duhemist — by the Cambridge philosopher of science, Mary Hesse (1924–).[72] In this work we meet an important addition to the literature on models. Hesse introduces the notions of 'positive', 'negative' and 'neutral' analogy, considering by way of illustration the well-worn example of the billiard-ball model for the kinetic theory of gases:

> When we take a collection of billiard balls at random motion as a model for a gas, we are not asserting that billiard balls are in all respects like gas particles, for billiard balls are red or white, and hard and shiny, and we are not intending to suggest that gas molecules have these properties. We are in fact saying that gas molecules are *analogous* to billiard balls, and the relation of analogy means that there are some properties of billiard balls which are not found in molecules. Let us call those properties we know belong to billiard balls and not to molecules the *negative analogy* of the model. Motion and impact, on the other hand, are just the properties of billiard balls that we do want to ascribe to molecules in our model, and these we can call the *positive analogy*. Now the important thing about this kind of model-thinking in science is that there will generally be some properties of the model about which we do not yet know whether they are positive or negative analogies; these are the interesting properties, because ... they allow us to make new predictions. Let us call this third set of properties the *neutral analogy*. If gases are really like collections of billiard balls, except in regard to the known negative analogy, then from our knowledge of the mechanics of billiard balls we may be able to make new predictions about the expected behaviour of gases. Of course the predictions may be wrong, but then we shall be led to conclude that we have the wrong model.[73]

This view is, Hesse has contended, implicit in the Campbellian position. It has the great advantage of allowing a *dynamic* aspect into the representation of scientific theories. A Duhemist instrumentalist (or an exponent of the 'received view' of scientific theories) offers a rational reconstruction of a scientific theory as a kind of static 'museum piece'. The Duhemist is unable to explain how scientific research progresses, and consequently is forced to postulate the formulation of a constant succession of *discrete* theories. But for a Campbellian, Hesse suggests, research may take place through the exploitation of the possibilities inherent in the neutral analogy of the model, in the manner indicated in the preceding quotation. It is in this aspect that the Campbellian view of theories had proved particularly attractive to some recent philosophers of science and has been developed by Mary Hesse in her various publications. Thus, even though the modellist's account of theories may seem to be incompatible with theories to be found in certain branches of modern physics, the modellist interpretation of the syntax and semantics of scientific theories does seem to offer a means whereby one may take due cognizance of the on-going, dynamic features of science. And so the positivist account of science, with prime emphasis on logical structures, the problem of meaning and of verification, began, in its literature on models and analogies, to take note of the dynamic nature of the scientific enterprise. But this kind of development certainly did not occur solely through the route I have just sketched. There were numerous other approaches, accounts of some of which will be given in the next two chapters.

Before we turn to these matters, however, it may be helpful, by way of recapitulation, to attempt to give some pictorial representation (as in Figure 43) of the descriptions of the structure of scientific theories that we have considered in this chapter.

Figure 43

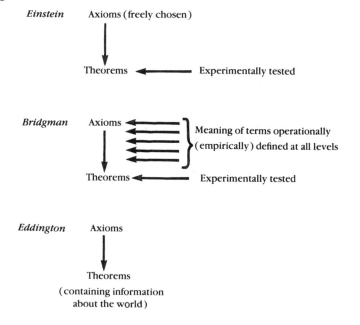

(Eddington's case is very unusual. It would seem that he assumed structural equivalences between the theoretical and the empirical. Therefore, in principle, empirical information could be 'unpacked' from the theory directly, so that experimental testing was, so to speak, an unnecessary luxury!)

Figure 43 (cont.)

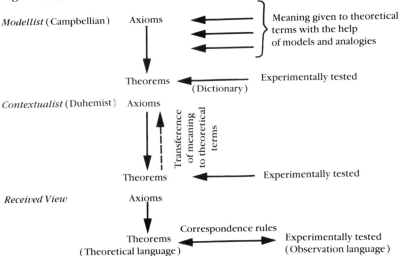

I should add that this does not by any means exhaust the list of descriptions of scientific theories that have been given by philosophers of science. For example, one such model given by Herbert Feigl[74] (and dubbed by Paul Feyerabend as the 'layer-cake' view) may be represented as in Figure 44.

Figure 44

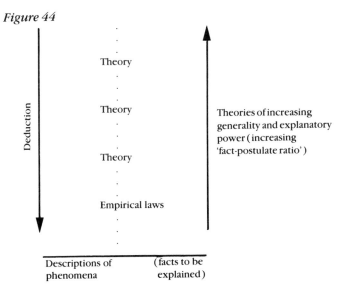

In all this we still find the deductive side of the arch of knowledge, standing firmly after many years of weathering; for the picture given above is roughly compatible with the account of theories given by Bacon, Descartes, Newton or Whewell. However, as indicated above, the whole positivist system for the representation of scientific theories would be thrown in doubt if the assumed distinction between the languages of observation and theory was rejected: or if Reichenbach's distinction between the context of discovery and of justification was found wanting. If, for example, it were the case that *all* observations are necessarily moulded, shaped (or 'infected', to use a more pejorative word that has gained wide currency in this context) by the theories held by the observer — if concepts necessarily influence percepts — then the 'received view' would be very much in question, and the positivists' account of theories might need to be substantially revised, or replaced altogether. We have already seen, in Chapter 6, that the work of the 'later' Wittgenstein tended in this direction, and with the work of Hanson, in particular, the positivists' distinction between separate 'theoretical' and 'observational' languages seemed increasingly implausible. So the positivists' picture of science has indeed come to be seriously questioned in recent years. And some of the developments in this line will be examined in Chapter 9. But first we shall look at another trenchant criticism of the metascientific 'orthodoxy' of the era of logical positivism — the work of Karl Popper.[75]

NOTES
1 The secondary literature on Einstein is immense. A useful introductory biography is J Bernstein *Einstein* Fontana/Collins Bungay (Suffolk) 1973. Or for a much larger volume, see R W Clark *Einstein: The Life and Times* Hodder & Stoughton London 1973. For the special theory of relativity, discussed in this chapter, see A I Miller *Albert Einstein's Special Theory of Relativity: Emergence (1905) and Early Interpretation (1905–1911)* Addison-Wesley Reading (Mass) 1980. For other discussions of the origins of Einstein's theory of special relativity, see, for example: M Wertheimer *Productive Thinking* Harper New York 1959 pp 213–33; G Gutting 'Einstein's Discovery of Special Relativity' *Philosophy of Science* 1972 Vol 39 pp 51–68; G Holton 'Einstein, Michelson, and the "Crucial" Experiment' *Isis* 1969 Vol 60 pp 133–97; A Grunbaum 'The Genesis of the Special Theory of Relativity' in H.Feigl & G Maxwell eds *Current Issues in the Philosophy of Science* Holt Rinehart & Winston New York 1961 pp 43–53 (with comments by M Polanyi pp 53–5). Readers may also wish to consult Einstein's 'Autobiographical notes' in P A Schilpp ed *Albert Einstein: Philosopher–Scientist* Open Court La Salle and Cambridge University Press London 3rd ed 1970 pp 3–94

2 A Einstein 'Zur Electrodynamik Bewegter Körper' *Annalen der Physik* 1905 Vol 17 pp 891–921. For an English translation, see H A Lorentz A Einstein H Minkowski & H Weyl *The Principle of Relativity: A Collection of Original Memoirs on the Special and General Theory of Relativity with notes by A Sommerfeld* (trans W Perrett & G B Jeffery) Dover New York 1923 pp 37–65. For a summarised English exposition of the paper, see C Lanczos *The Einstein Decade (1905–1915)* Elek London 1974 pp 131–9. The title of Einstein's paper might not indicate that it had anything to do with relativity theory, but in fact the connection with this topic was very intimate. Einstein was interested in Maxwell's equations for the propagation of electromagnetic waves — about which just a few words are said on, pp 284–85 — which contained the term 'c' for the velocity of light. He supposed that the Maxwell equations must have the same form in all frames of reference, and that the velocity of light must be constant for all observers in all frames of reference. Later he undertook an 'operational' (see p 276) analysis of the concept of simultaneity. The outcome of an attempted reconciliation of these several requirements was the special theory of relativity. By contrast, H A Lorentz had suggested a theory of 'deformable' electrons, such that they contracted in their direction of motion as they moved through a hypothetical aether. This model was supposed to give a physical (or 'classical') explanation of the Fitzgerald/Lorentz contraction hypothesis, discussed on p 266.

3 *Very* small particles, such as pollen grains, or even more minute colloidal matter, can be seen to be in constant random motions when observed under a microscope. The phenomenon was first noticed by the English botanist, Robert Brown (1773–1858). It is thought today that the observed motions of the minute particles are due to the random impacts of the surrounding molecules.

4 Some elements, such as selenium, emit electrons when light is shone onto them. This phenomenon of the photoelectric effect is utilised in photography in the common light-meter. Einstein's paper was chiefly concerned with the quantisation of light as much as the photoelectric effect.

5 A Einstein 'Zur Allgemeine Relativitätstheorie und Bewegungsgesetz' *Sitzungberichte der Preussischen Akademie der Wissenschaften zu Berlin* 1915 Part 2 pp 778–99 & 799–801 (See also Lanczos *op cit* [note 2].)

6 For Eddington's own semipopular exposition of his experimental investigations and their theoretical import, see his *Space, Time and Gravitation: An Outline of the General Relativity Theory* Cambridge University Press Cambridge 1921 (esp Chapters 6 & 7). For further discussion of Eddington's work, see below, pp 281–84.

7 For an accessible account of the scientific collaboration of A A Michelson (1852–1931) and E W Morley (1838–1932), see B Jaffe *Michelson and the Speed of Light* Heinemann London 1961.

8 See A A Michelson 'The Relative Motion of the Earth and the luminiferous Ether' *American Journal of Science* Series 3 1881 Vol 22 pp 120–9. A further paper, written collaboration with E W Morley, and published in the same journal in 1887, was thought to provide the decisive experimental evidence.

9 Journey$_1$ will not, of course, take place exactly as shown in the figure. The swimmer will be swept downstream by the current to arrive on the original riverbank some distance below the point of departure. But assuming that there is no component of the water's motion perpendicular to the riverbanks, the swimmer's time for journey$_1$ will be the same, no matter whether the water is moving or still. On the other hand, the motion of the water does make a significant difference for journey$_2$.

10 G F Fitzgerald (1851–1901) and H A Lorentz (1853–1928) proposed the contraction hypothesis independently: G F Fitzgerald 'The Ether and the Earth's Atmosphere' *Science* Vol 13 p 390; H A Lorentz *Versuch einer Theorie der elektrischen und optischen Erscheinungen in bewegten Körpern* Brill Leiden 1895 paragraphs 89–92. The relevant section of Lorentz's work is to be found in English translation in Lorentz *et al op cit* (note 2) pp 1–7. Fitzgerald's 'paper' consisted of a letter of one paragraph, and expressed the contraction formula in words only.

11 See above, p 182

12 In his 'Autobiographical notes' (*op cit* note 1 p 53) Einstein wrote: '[There was] a paradox upon which I had already hit at the age of sixteen: If I pursue a beam of light with the velocity *c* (velocity of light

in a vacuum), I should observe such a beam of light as a spatially oscillatory electromagnetic field at rest. However, there seems to be no such thing... From the very beginning [therefore] it appeared to me intuitively clear that, judged from the standpoint of such an observer, everything would have to happen according to the same laws as for an observer who, relative to the earth, was at rest. For how, otherwise, should the first observer know, i.e., be able to determine, that he is in a state of fast uniform motion?'

13 A Einstein *Relativity: The Special and the General Theory A Popular Exposition* trans R W Lawson 15th ed Methuen London 1954 pp 115–20

14 See J Schwartz & M McGuinness *Einstein for Beginnners* Writers and Readers Publishing Cooperative London 1971. This is a witty illustrated exposition of Einstein's theory, written from a Marxist perspective.

15 This example was given by Einstein in the celebrated paper of 1905: Einstein *op cit* (note 2 1923) p 39

16 Einstein *op cit* (note 13) p 42

17 This is evident from the preface of Mach's *The Principles of Physical Optics*, dated July 1913. For discussion of the gradual divergence of view between Mach and Einstein, see G J Holton 'Mach, Einstein, and the Search for Reality' in R S Cohen & R J Seeger eds *Ernst Mach: Physicist and Philosopher* Reidel, Dordrecht 1970 pp 165–99

18 A Einstein *Ideas and Opinions...* ed Carl Seelig Crown New York 1964 p 272 (from Einstein's Herbert Spencer lecture, delivered at Oxford in 1933)

19 *Ibid; ibid* pp 221–3 (from Einstein's inaugural address to the Prussian Academy of Sciences in 1914); *ibid* p 234 (from a lecture to the Prussian Academy in 1921)

20 *Ibid* p 226 (from an address to the Physical Society in Berlin in 1918)

21 *Ibid*

22 *Ibid* p 225

23 *Ibid* p 274 (from the Herbert Spencer lecture, 1933)

24 V F Lenzen 'Einstein's Theory of Knowledge' in Schilpp ed *op cit* (note 1) pp 357–84 (at p 384)

25 A Einstein *The World as I see It* Covici Friede New York 1934 pp 36–7

26 Einstein *op cit* (note 18) p 271

27 See above, p 248

28 For a detailed discussion of this 'ism', see A C Benjamin *Operationism* Thomas Springfield Ill 1955.

29 P W Bridgman *The Logic of Modern Physics* Macmillan New York 1927. See also his writings: *The Nature of Physical Theory* Princeton University Press Princeton 1936 (republished Science Editions New York 1964); *The Nature of Some of our Physical Concepts* Philosophical Library New York 1952; *The Way Things Are* Harvard University Press Cambridge Mass 1959; and *Reflections of a Physicist* Philosophical Library New York 1950 (2nd ed 1955)

30 *Ibid* (1927) p 3

31 *Ibid* pp 2–3

32 See above, p 84

33 Bridgman *op cit* (note 29 1927) p 5 (emphasis in original)

34 R B Lindsay 'A Critique of Operationalism in Physics' *Philosophy of Science* 1937 Vol 4 456–70

35 P W Bridgman 'Operational Analysis' *Philosophy of Science* 1938 Vol 5 pp 114–31

36 As defined by Russellian set theory, not J S Mill's empiricist account of mathematics.

37 Bridgman *op cit* (note 29 1952) p 9

38 In a 'construction' system such as that of Carnap's *Aufbau*, this side was strong too — except that it purported to be making deductions from phenomenal experience. If, however, one freely chose the fundamental 'principles' (as Einstein, for example, appeared to think was possible) then the inductive climb was achieved simply by 'jumping', not laboriously mounting an inductive staircase. Carnap's later presentations of the structure of science did not, as we have seen, seek an inductive construction from phenomenal experience. He did, however, give considerable attention in his later writings to 'inductive logic'.

39 See Bridgman *op cit* (note 29 1952) p 10

40 Bridgman *op cit* (note 29 1955) p 5

41 Bridgman *op cit* (note 29 1964) p 19

42 *Ibid* p 52

43 *Ibid* p 59

44 For discussions of Eddington's work, see, for example: A V Douglas *The Life of Arthur Stanley Eddington* Nelson London 1956; J W Yolton *The Philosophy of Science of A.S.Eddington* Nijhoff The Hague 1960; J Witt-Hansen *Exposition and Critique of the Conceptions of Eddington Concerning the Philosophy of Physical Science* Gads Copenhagen 1958. Also useful is E T Whittaker *From Euclid to Eddington: A Study of Conceptions of the External World* Cambridge University Press Cambridge 1949. A famous savage critique of Eddington and another philosopher–scientist, James Jeans, is to be found in a book by the logical positivist, Susan Stebbing: *Philosophy and the Physicists* Pelican Harmondsworth 1944. Eddington wrote works of immense mathematical complexity, and also highly successful popular accounts of the results of the new science of his day. Chief among the popular works were *Space, Time, and Gravitation* Cambridge University Press Cambridge 1920; *The Nature*

of the Physical World Cambridge University Press Cambridge 1928; and *New Pathways in Science* Cambridge University Press Cambridge 1935. His major text on the philosophy of science was *The Philosophy of Physical Science* Cambridge University Press Cambridge 1939.

45 Eddington *op cit* (note 44 1939) p 170

46 Witt-Hansen (*op cit* note 44) will give the reader some help. See also E T Whittaker 'Eddington's Theory of the Constants of Nature' *Mathematical Gazette* 1945 Vol 29 pp 137–44

47 Eddington *op cit* (note 44 1939) p 16

48 *Ibid* p 109

49 *Ibid* p 111

50 *Ibid* p 24

51 Eddington quoted Russell approvingly in *The Philosophy of Physical Science* (note 44 1939) p 152

52 See above p 274

53 Eddington *op cit* (note 44 1939) pp 188–9. (But Eddington said: 'We do not accept the Kantian label.')

54 See above pp 126–32. Kant, as we have seen, sought to give such a formal justification; but it is not thought today that it was carried through successfully. On the other hand, the Kantian view that there is always — and necessarily so — a subjective and an objective component to knowledge is widely accepted.

55 See above, pp 176–82

56 Eddington *op cit* (note 44 1939) p 148 (emphasis in original)

57 See above p 244

58 J C Maxwell 'On Faraday's lines of force' in W D Niven ed *The Scientific Papers of James Clerk Maxwell* 2 vols Cambridge University Press Cambridge 1890 Vol 1 pp 155–229 (The paper was first read in two parts in 1855 and 1856.)

59 J C Maxwell 'On Physical Lines of Force' *ibid* pp 451–513 (This paper was first published in 1861–62.)

60 J C Maxwell *Treatise on Electrictiy and Magnetism* Cambridge University Press Cambridge 1873

61 For our discussion of Duhem's view of the structure of scientific theories, see above, pp 194–201. Duhem, while conceding that models could play a role in theory construction, was not prepared to grant much in this direction, which he regarded as a manifestation of the rather crude kind of theoretical physics commonly practised in Britain (as opposed to France). He did, however, acknowledge the value of analogies between the mathematical formalisms in two distinct branches of theoretical physics, and the recognition of such analogies could, he thought, be very useful for theory construction. As to models (which he distinguished from 'analogies'), they were often (he claimed) made up after a theory was constructed, not before. See P Duhem *The Aim and Structure of Physical Theory* Princeton University Press Princeton 1954 pp 93–104

62 N R Campbell *Physics: The Elements* Cambridge University Press Cambridge 1919. When republished by Dover Publications, New York in 1957, it was given a new title: *Foundations of Science: The Philosophy of Theory and Experiment.*

63 This set of statements constitutes what other metascientists commonly have called the 'correspondence rules'.

64 Campbell *op cit* (note 62 1957) p 122

65 *Ibid* p 123

66 C G Hempel *Aspects of Scientific Explanation and other Essays in the Philosophy of Science* Free Press New York and Collier-Macmillan London 1965 p 444. Hempel also pointed out that the trouble with Campbell's 'imaginary' theoretical system, reproduced above, p 285, was that it had no empirically testable consequences other than the particular law that Campbell deduced.

67 R B Braithwaite 'Models in the Empirical Sciences' in E Nagel *et al* eds *Logic, Methodology and Philosophy of Science: Proceedings of the 1960 International Congress* Stanford University Press Stanford 1962 pp 224–31 (at pp 230–1) (emphasis added). See also R B Braithwaite 'The Nature of Theoretical Concepts and the Role of Models in an Advanced Science' *Revue Internationale de Philosophie* 1954 Vol 8 pp 34–40

68 R B Braithwaite *Scientific Explanation: A Study of the Function of Theory, Probability and Law in Science* Cambridge University Press Cambridge 1953 p 51

69 For a useful collection of papers surveying fairly recent work in this field, see, for example: *Scientific American: Particles and Fields* (with an introduction by William J Kaufman III) Freeman San Francisco 1980.

70 P A M Dirac *Principles of Quantum Mechanics* Clarendon press Oxford 4th ed 1958 p 10 (emphasis in original)

71 See, for example: M B Hesse *The Structure of Scientific Inference* University of California Press Berkeley & Los Angeles 1974.

72 M B Hesse *Models and Analogies in Science* Sheed & Ward London 1963; 2nd ed University of Notre Dame Press Notre Dame 1966

73 *Ibid* (1966) pp 8–9 (emphasis in original)

74 H Feigl 'The "Orthodox" View of Theories: Remarks in Defense as well as Critique' in M Radner & S Winokur eds *Minnesota Studies in the Philosophy of Science* Vol IV *Analyses of Theories and*

Methods of Physics and Psychology University of Minnesota Press Minneapolis 1970 pp 3–16 (at p 11)

75   In some respects Popper's work is itself positivistic in character. His criticisms, therefore, were generated within the positivist camp, or at least at its periphery, rather than without.

CHAPTER 8

THE REACTION TO INDUCTIVISM: POPPER AND FALSIFICATIONISM

In Chapter 6, I gave a general account of a broad sweep of early to mid-twentieth-century philosophy of science, and we looked at some of the manifestations of the union of interests in logic and empiricist epistemology which gave rise to logical empiricism and the contributions of the so-called Vienna Circle. A general feature of the work of this school was its program of giving a formal account of how scientific knowledge might be built up logically from observational experiences of phenomena. We have seen that this program ran into considerable difficulties, and what might have appeared, at first blush, a fairly straightforward enterprise turned out to be something of a philosophical nightmare.

It was, I suppose, almost inevitable that there should have been a reaction against the efforts of philosophers such as Carnap, trying to formulate a coherent blend of logic and empiricism. But the particular form that the reaction would take could hardly have been foretold, *a priori*. And the reaction, when it came in the work of Sir Karl Popper and some of his 'disciples', was in a sense something of a palace revolution. For many of the general features of the logical-empiricist approach to metascience were retained in Popper's work, particularly the strong emphasis on logic, as the means whereby *bona fide* philosophy of science should be pursued. On the other hand, through his criticism of induction, Popper (hearkening back to Hume) broke decisively with the general program of logical empiricism. In doing so, however, he was, in a sense, only drawing once again upon the age-old tradition of hypothetico-deductivism, elements of which, as we have seen through the course of this book, can be traced right back to antiquity.

In this chapter, then, we shall seek to give an account of some of the broad features of Popper's work; and this will be the only chapter that is devoted to the work of one man. Such an emphasis will no doubt seem idiosyncratic to one nurtured in any philosophical tradition other than that of the Anglo-Saxon — if I may call it that. But there can be little doubt that such an expansive account is warranted by virtue of Popper's very considerable influence within the English-language philosophy of the twentieth century. Whether such emphasis will seem warranted a hundred years hence, is of course impossible to say. But under the present circumstances Popper does seem to deserve a chapter of his own. Also, by examining his philosophy of science and the subsequent difficulties and controversies in which it has become engaged, we may conveniently trace the pathway from logical empiricism to present-day interests in the social aspects of scientific knowledge and the sociological dimension of epistemology. Popper also merits our attention for the respect accorded to his views by numerous practising scientists, most of whom have found the writings of logical empiricists unintelligible, patently absurd, or totally irrelevant to their interests.

Karl Popper was born in Vienna in 1902, son of a distinguished lawyer, and is still with us at the time of writing.[1] As a young man, he studied at the University of Vienna, later becoming a school teacher, and in 1937 a professional philosopher. In his youth, Popper was attracted to Marxist ideas, to the extent that for a time he chose employment as a manual worker, but he fairly soon renounced his far left-wing views in favour of something akin to Fabian socialism, and then to what one might loosely call small-l liberal views. In his various political writings he has placed great emphasis on the importance of democratic principles. He is strongly opposed to all forms of 'revolutionary' political theory and practice, and urges instead what he calls 'piecemeal social engineering' — that is, the suggestion and implementation of relatively small social changes, followed in each case by consideration of whether the changes have or have not been beneficial, and adjustment of the plans accordingly.

Popper left Austria at the time of the rise of Nazism, and during the Second World War he held a lectureship at Canterbury College, Christchurch, New Zealand, a country where his views have for long been particularly influential. In this wartime interlude in the southern hemisphere, Popper wrote his well-known *Open Society and its Enemies* (his 'war effort') — a polemic chiefly directed against the philosophical systems of Hegel and Marx, whose thinking he characterised as 'historicist' in that (according to Popper) they purported to be able to make statements about general trends or patterns in history, seeking thereby to influence people's beliefs and behaviour.[2] After the war, Popper made his home in England, where he occupied a chair in the logic and methodology of science at the London School of Economics. It was at the L.S.E. that his work achieved international recognition and renown — to the extent that he was knighted in 1965.

Popper's chief writings have been:
Logik der Forschung (1935)[3] — English translation *Logic of Scientific Discovery* (1959);
The Open Society and its Enemies (1945);[5]
The Poverty of Historicism (1957);[6]
Conjectures and Refutations (1963);[7]
Objective Knowledge (1972);[8]
Unended Quest: An Intellectual Autobiography (1976);[9]
(With Sir John Eccles) *The Self and its Brain* (1977);[10]
The Open Universe (1982);[11]
Quantum Theory and the Schism in Physics (1982);[11]
Realism and the Aim of Science (1983).[11]
His first book, written during his early years in Vienna, is still to appear.

Though living in Vienna in his early days, and greatly interested in metascientific issues, Popper was never a direct participant in the deliberations of the Vienna Circle. He was, however, in communication with some of the members of the Circle, and there was interaction between their ideas. As we have seen, the members of the Circle were concerned to establish a satisfactory criterion that would, they hoped, enable them to distinguish between meaningful and meaningless propositions. Thereby, they sought to eliminate metaphysics from philosophical discourse, dismissing it as meaningless. The young Popper was concerned with somewhat similar issues, though they were by no means identical. He was interested in establishing a clear line of demarcation between science and pseudo-science. It appears that this led to some confusion at the time, for the logical positivists understood Popper to be concerned with the problem of meaning — which was not the case. However, this misunderstanding was soon cleared up.

It was in 1919 that Popper attended a lecture by Einstein in Vienna and was dazzled by the introduction to the new physics that was presented by the great iconoclast of modern science. Also, it was in 1919 that Eddington's observational data, which seemed to support Einstein's predictions concerning the gravitational deflection of light, were announced. If light is gravitationally attracted by massive bodies such as the Sun, then the effect may be detected at times of total solar eclipse by examining the patterns of appropriate stellar configurations in the sky, as Figure 45 should make clear.

Figure 45

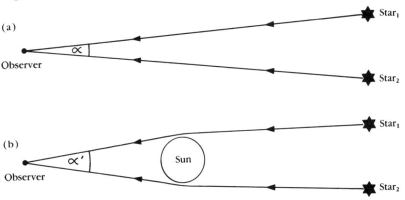

Suppose two stars normally subtend the angle α at the surface of the Earth. The angle will be modifed to α' on occasions when the light has to pass the Sun to reach the Earth — if it is the case that light is subject to slight deviations owing to the gravitational attraction of massive bodies such as the Sun. (If there were no such gravitational pull on light, when the Sun is in the position shown in Figure 45(b) the two stars would be invisible, being obscured by the Sun.) Under normal circumstances, it is hopeless to try to measure α', for the starlight is negligible compared with the light emitted by the Sun. But on the rare occasions of solar eclipse, α' may be measured. This is what Eddington did. And his observations seemed to support the 'risky' predictions made by Einstein, as deductions from the general theory of relativity — namely that light is subject to minute gravitational effects.

Popper compared the very 'risky' and precise predictions made by the new physics with the situation that obtained in three putative sciences: the Marxist theory of history, Freud's psychoanalysis, and Adler's individual psychology. These would-be sciences signally failed to make precisely formulated predictions that could be subjected to direct (or indirect) empirical test. Indeed, they seemed to be compatible with *any* states-of-affairs that might obtain, or any events that might occur! So in 1919–20 Popper came to the following conclusions:

1. It is easy to obtain confirmations, or verifications, for nearly every theory — if we look for confirmations.
2. Confirmations should count only if they are the result of *risky predictions...*
3. Every 'good' scientific theory is a prohibition: it forbids certain things to happen. The more a theory forbids, the better it is.
4. A theory which is not refutable by any conceivable event is non-scientific. Irrefutability is not a virtue of a theory...but a vice.
5. Every genuine *test* of a theory is an attempt to falsify it, or to refute it...
6. Confirming evidence should not count *except when it is the result of a genuine test of the theory*; and this means that it can be presented as a serious but unsuccessful attempt to falsify the theory...
7. Some genuinely testable theories, when found to be false, are still upheld by their admirers — for example by introducing *ad hoc* some auxiliary assumption, or by reinterpreting the theory *ad hoc* in such a way that it escapes refutation. Such a procedure is always possible, but it rescues the theory from refutation only at the price of destroying, or at least lowering, its scientific status...[12]

So in summary, said Popper:

[T]he criterion of the scientific status of a theory is its falsifiability, or refutability, or testability [not its confirmability].[13]

Thus, early in his life, Popper came to establish his famous criterion of demarcation which (so he supposed) could be used to distinguish science from pseudo-science. But, as indicated above, contrary to the logical positivists, Popper was concerned with the falsifiability of theories as a criterion of their scientificity. He was not seeking to deal with the actual meaning or meaningfulness of propositions.

In 1923, Popper tells us,[14] he became interested in the problem of induction, which like so many others before he approached through the writings of Hume. Popper accepted Hume's refutation of inductive inferences as being 'clear and conclusive'. But he was dissatisfied with the 'psychological' explanation that Hume offered to account for our claimed propensity to make and accept inductive inferences.[15] Popper contended that the philosophy of science of his day (for example latter-day Millian empiricism, or the products of the logical positivist/empiricist movement) was based upon the supposed legitimacy of inductive inferences. So he concluded that philosophy of science within the empiricist tradition had reached an impasse. The only way to put matters right was to turn the whole system upside down, so to speak, and place the main emphasis on scientists' efforts to *test* their theories — to show in what respects they were wrong — rather than to show in what manner they were verified, confirmed, supported or whatever, by the empirical evidence. So according to Popper's new way of thinking, metascience should place emphasis on the procedures of *falsification*, rather than *verification* – the showing of a theory to be false rather than correct. 'I thought', he wrote:

> that scientific theories were not the digest of observations, but that they were inventions — conjectures boldly put forward for trial, to be eliminated if they clashed with observations; with observations which were rarely accidental but as a rule undertaken with the definite intention of testing a theory by obtaining, if possible, a decisive refutation.[16]

Thus did Popper come to *his* version of the hypothetico-deductive methodology of science — the method, as he subsequently called it, of 'trial and error', or of 'conjecture and refutation'. Moreover, he claimed that by adopting *this* approach to scientific investigation the problem of induction could be circumvented. In fact, in 1972 he claimed directly that he had 'solved a major philosophical problem: the problem of induction'.[17] Unfortunately, however, it is not the case that the adoption of a particular methodology in scientific inquiry 'solves' the problem of induction. To adopt a particular method does not show inductive arguments can be rendered deductive, or otherwise certain. At best, to invoke a new methodology *might* show how science can get along without inductive inference — although as we shall shortly see, falsificationism doesn't even achieve that satisfactorily. Popper's own approach was, in effect, to turn away from the problem rather than solve it.

Putting the matter in what is, admittedly, a somewhat oversimplified way, the Popperian methodology can be construed as a straightforward instance of hypothetico-deductivism. Thinking back to the 'arch of knowledge' once more, it is illustrated yet again by Popper's description of science. One starts with a problem, and conjectures are made — in the light of known information and data that may be collected relevant to the problem — as to ways in which the problem may be solved. This involves formulating (or conjecturing) an hypothesis (or theory) that may account satisfactorily, or deal with, the problem or difficulty. The hypothesis is then rigorously tested. This is done by drawing the logical consequences from the hypothesis and testing these consequences experimentally. Thereby, the hypothesis may be shown to be false (using the logic of the *modus tollens*;[18] but it cannot be shown to be true by reason of the fallacy of affirming the consequent.[19] The data collected in the course of the experimental testing of the hypothesis may in turn lead to new problems, and so science may continue on its merry way, with an unlimited succession of 'conjectures and refutations'. In terms of our arch model, then, Popper's attention is directed almost exclusively towards the 'downward', 'deductive' limb. Indeed, he has claimed that the philosopher of science can say little that is of interest about the mysterious processes of hypothesis formation. Such creative acts, if they are to be examined at all, lie in the domain of the psychologist rather than the philosopher of science.[20]

All this is something of a caricature of the full Popperian position, in all its sophistication and complexity. Nevertheless, Popper's system can undoubtedly be accommodated to the traditional arch model. But as we have indicated, his emphasis is asymmetrical, yielding another of our ungainly and distorted structures, in this time with a strong deductive leg and a very insubstantial inductive side. We shall (later on) want to know whether Popper does in fact have some play with inductive reasoning. But for the moment we shall examine some other issues. In particular, we should consider whether the program of falsification allows a clear-cut, logical certainty to the Popperian scientist from which the inductivist is debarred.

We have already noticed, in our discussion of the work of Duhem,[21] that the logic of the *modus tollens*, though valid in itself, does not allow *isolated* hypotheses to be tested. One always has to test hypotheses in clusters, so to speak. The situation is always such that (even at best) one is testing a system that consists of some hypothesis under investigation, *together with* a greater or lesser number of auxiliary hypotheses — that is to say, hypotheses drawn from other branches of science, or related branches of the science into which the investigation is being made. Thus, the situation (as we have previously noted) should be represented thus:

H[ypothesis under test] + A[uxiliary assumptions or hypotheses]

Deduction

I[mplications — which are to be tested empirically]

So, in the event of the tests proving unsatisfactory, they do not point unambiguously to the falsity of H. It is the combination of $H + A$ that is shown to be wrong. And the trouble may lie in A, just as much as H, so far as the logic of *modus tollens* can show.[22]

For our present purposes, it will be helpful to illustrate this important point further by means of an historical example. In the Michelson-Morley experiment[23] no shifts of the interference fringes were observed when the apparatus was rotated through 90 degrees. The hypothesis being tested was that the Earth was moving through 'the aether', which served as an absolute frame of reference; and an attempt was being made to determine the Earth's absolute rate of motion. The auxiliary hypothesis was that the Earth was actually in motion round the Sun, as the Copernican theory stated. And there were, besides, various assumptions drawn from optical theory needed in order to carry out the test. The experimental results provided no evidence of any motion of the Earth relative to an aether, contrary to what was expected. But from this alone, it was not possible, by application of the logic of *modus tollens*, to tell whether the difficulty lay in the optical theory, the assumption that the Earth was in motion with respect to the aether, or whether the Earth was not in fact in motion round the Sun, as Copernican theory required.

The first possibility considered was that the Earth might, by chance, *not* be moving with respect to the aether. For it could have been in such a position in its orbit that the motion of the Earth round the Sun and the motion of the Solar System as a whole exactly cancelled out. However, the experiment has been repeated at other times of the year, and always with a null result. The possibility that the Earth was not moving round the Sun at all was not, so far as I am aware, ever given consideration. The optical theory seemed satisfactory. There was, at the time, no alternative on offer to the assumption of motion with respect to the aether (the physicists of the late nineteenth century mostly still living in a Newtonian world of absolute space and time, so far as their theories were concerned).

Faced with such a situation, the physicists were in a considerable quandary, and the logic of the *modus tollens* was of scant assistance. They could not be *sure* whether their experiments were conducted adequately or inadequately, whether the optical theory was mistaken, or whether the Earth was not in fact moving with respect to the aether. None of these alternatives seemed very attractive, and as we have seen in Chapter 7, what happened was that the so-called Lorentz–Fitzgerald contraction hypothesis was proposed, whereby the physical apparatus itself was supposed to contract when in motion with respect to the aether! Further, it was suggested that it contracted by just such an amount as would account for the null result of the experiment. The thought of shattering the whole edifice of Newtonian physics, without being able to offer anything else to put in its place, was just too awful to contemplate, thought it must be admitted that the 'uncaused' contraction of physical objects was probably as unpalatable as the thought of there possibly being no absolute frame of physical reference.

Of course, no one was particularly happy with the Lorentz–Fitzgerald hypothesis, and from the perspective of Popper's metascience it might seem a thoroughly bad thing. It was, in Popper's terminology, a typical '*ad hoc*' hypothesis, dragged in to shore up a theory that was getting into difficulties.[24] As we have seen, Einstein subsequently restructured the most basic assumptions of physical science when he introduced his relativity principles. And then the *ad hoc* Lorentz–Fitzgerald contraction became unnecessary; or rather it was subsumed by the new relativistic mechanics. But at the time it was proposed, the contraction hypothesis was the best that could be offered, even though the authors could suggest no theoretical reason as to *why* such a contraction should occur.

Now, according to Popper's description of science, the introduction of such an *ad hoc* hypothesis was most unsatisfactory. It was reprehensible, for though the hypothesis might deal with the difficulty in hand it did not offer any wider 'empirical content' for the theory. The theory of absolute motion was not 'allowed' to be falsified. So, in Popper's language, with the introduction of the Lorentz–Fitzgerald hypothesis, the doctrine of absolute motion was being *conventionalised*. A 'conventionalist strategem' was being deployed, which would be a 'bad thing' according to the Popperian description of science. Scientists *ought* to expose their theories to the most rigorous test, and accept falsification if that is what the experimental results dictated.

However, as we have seen, the message dictated by the data may often be ambiguous. We do not know exactly where the arrow of the *modus tollens* should be directed. Hence, it is not surprising that scientists do not always accept the verdict of the data, and resort to conventionalist stratagems, or suspend judgment, awaiting new theoretical and experimental developments. And given the obvious difficulty in applying falsificationist directives in practice, it is scarcely surprising that for many years (say, until the 1960s) scientists tended to react adversely to Popper's urgings that they should engage in attempts at falsification rather than verification — though more recently some well-known scientists such as Sir Peter Medawar and Sir John Eccles have publicly tied their colours to the Popperian masthead.

One may suppose that the problem of auxiliary hypotheses — which is at the centre of Duhem's thesis — is such that a falsificationist approach to philosophy of science offers little advantage over an inductivist or verificationist position. The thesis indicates that one cannot hope for clear-cut falsification of any given hypothesis. Moreover, the empirical evidence *against* an hypothesis, or a system of hypotheses, will have to be built up by successive observations, and this process is *itself* subject to the problem of induction. For a single observation *against* an hypothesis would not in itself suffice and one would need to build up a strong empirical case, for a falsification, just as much as for a verification.[25]

Are we to suppose, then, that Popper was unaware of such problems, even in his earliest work? He may have been, in the 1920s, when he was first developing his philosophical position in some detail. But he did make considerable efforts to deal with some of the difficulties sketched above, even in the early *Logic of Scientific Discovery*; and in looking at some parts of this work we may usefully be carried somewhat more deeply into the Popperian system.

Popper, as we have seen, has been and is concerned that science should make highly falsifiable statements — that is, precise predictions that can be tested experimentally. In this way, *bona fide* science is supposedly distinguished from pseudo-science. The making of precise, highly falsifiable statements is, in fact, for Popper more important than the acquisition of the 'the truth'. It may be true to say that it will rain some time next year. But this is not a particularly useful piece of information. To say, however, that it will rain this afternoon conveys much more. Such a statement has, in Popper's language,[26] a greater 'empirical content'. On the other hand, it is much less probable than the imprecise statement that it will rain some time next year. And according to Popper, 'empirical content' in science is more important than truth. So science, on this view, should be concerned with statements of high content and low probability, which, however, by repeated testing may gradually be revised and brought closer to the truth.

Here an analogy with gunnery target practice may be helpful. By continual repetition of shots, and adjustment of the aim after each firing, the aim of the gun may be altered so that eventually it exactly hits the bull's eye. The analogy is imperfect, however. Science, for Popper, certainly involves repeated conjectures and refutations — which may be compared with the directions of aim and the corresponding tests of those aims by examination of the marks on the target. But according to Popper, although such a method of conjecture and refutation will bring the scientist *nearer* to the truth, it will not bring him or her there exactly. Or if it does, the scientist has no means of knowing that he or she has arrived at the truth. There is, so to speak, no bull's eye in science — no spatial (or conceptual) region can be known to correspond with 'the truth' *per se.*

So Popper has maintained[27] that the higher the 'empirical content' of a statement the greater is its degree of falsifiability. The more it rules out (that is, the more precise it is) the more it says in a useful scientific way; the more a statement rules out, the more is it saying about the world of experience. Thus, one can supposedly compare the 'empirical content' of two theories by comparing their falsifiabilities, the 'empirical content' being *defined* by Popper as the class of statements which constitute its potential falsifiers.[28] (Popper's 'empirical content', by the way, should be distinguished carefully from Carnap's 'logical content'. For Carnap, the 'logical content' of a statement was equal to the class of non-tautological statements that could be derived from it.)

But how are the 'falsifiabilities' of two theories to be compared? One cannot merely compare the amounts of empirical data that they entail, since they will inevitably be infinite (in a universe infinite in space and time) in both cases. Popper's procedure, therefore, has been to seek to compare the sizes of the classes of basic empirical statements[29] which might clash with rival theories, rather than in some manner seeking actually to count the number of potentially falsifying statements that may be possessed by a particular theory or hypothesis. Thus, if the class of potential falsifiers of theory T_1 includes the class of potential falsifiers of T_2 and is thereby larger than T_2, then we may say that T_1 is more falsifiable than T_2 and has a higher 'empirical content'. To quote Popper's own illustration,[30] the theory that planetary orbits are circular has more potential falsifiers than does the theory that they are elliptical — for the class of circles is a subclass of the class of ellipses. Consequently, if we suppose that the orbits of planets are circles we rule out more observational possibilities than if we propose the hypothesis that they are ellipses. There are more potentially falsifying observations for circles than there are for ellipses. That is, the class of potentially falsifying statements for T_{circle} is greater than the class of potentially falsifying statements for $T_{ellipse}$. So the 'empirical content' of T_{circle} is greater than the 'empirical content' of $T_{ellipse}$. The example is no doubt rather contrived; but it should display Popper's position well enough in a general way.

Where such subclass relationships of circle and ellipse cannot be used, Popper has suggested that the 'dimensionality' of two theories may be compared. If we compare the number of observational statements required in principle to eliminate (or test) two hypotheses, the one that requires less may be said to have lower dimension, and thereby it is regarded as more falsifiable or testable, or (as before) of greater 'empirical content'.[31] But, as mentioned before, the higher the 'empirical content' of a theory, the higher its falsifiability.

It is important to note that it is within his doctrine of 'empirical content' that Popper sought to provide his answer, such as it is, to the challenge posed by the Duhem thesis. As we have noted, the arrow of the *modus tollens* cannot be directed unambiguously to a single, isolated hypothesis. And, according to Quine any hypothesis can be 'saved' by the introduction of suitable *ad hoc* hypotheses (such as the Lorentz–Fitzgerald contraction hypothesis). Popper urged, therefore, that the new auxiliary assumptions brought into a theory *ought* to be such that they increase the theory's observational falsifiability. If they do not, they are of the reprehensibly *ad hoc* variety, and should be eschewed. Of course, in saying this, Popper moved further from the realm of strict logical analysis to one of moral exhortation. In keeping with this, we find that his system contains a considerable number of methodological rules (which we examine collectively below), which rather detract from its would-be logical purity.

It should also be recognised that Popper's system did not envisage scientists as being engaged solely in the formulation of hypotheses and their testing to destruction. For Popper, a theory is falsified by a combination of discordant facts *and* an alternative theory which explains those facts satisfactorily. So his falsificationist system is closely tied to the experimental *comparison* and *critical discussion* of competing theories. For such reasons, Popper does not regard the views of Duhem and Quine as insurmountable objections to his system.

Another important term within the Popperian vocabulary is '*corroboration*'. As we have seen, throughout his work Popper has placed emphasis on falsification rather than verification. He urges scientists to examine the weaknesses of their theories rather than their strengths. But it seems that experimentation does, in some manner, lead scientists towards the acceptance of theories.[32] So Popper introduced the term 'corroboration', meaning 'failure to falsify'. A theory that has been well tested is said to be well corroborated, in that all attempts to prove it wrong have so far been unsuccessful. Also, a theory of high corroborability is one that is highly testable and of high content, but of low probability.[33]

Popper acknowledged in *Logic of Scientific Discovery* that he could not quantify the notion of 'corroborability' precisely. Nevertheless, some formulae that were supposed to give a measure of the conception were proposed.[34] These were:

1. 'Explanatory power' of theory x with respect to evidence y

$$= \frac{\text{Probability of y with respect to x}}{\text{Probability of } y \text{ with respect to } x}$$

2. 'Degree of confirmation [or corroboration]' of theory x with respect to evidence y
   = ('Explanatory power' of x with respect to y) ×
   [1 +(Probability of x) (Probability of x with respect to y)]

Working with these equations, we may now proceed as follows. If the theory x is a universal generalisation, according to Popper its probability would be zero.[35] Also, if the evidence y is logically deducible from the theory x, then 'Probability of y with respect to x' will be 1. So, considering the test implications of a claimed universally valid theory, the equations reduce to:

3. 'Degree of confirmation [or corroboration]' of theory x with respect to evidence y
= 'Explanatory power' of theory x with respect to evidence y
$$= \frac{1 - \text{Probability of y}}{1 + \text{Probability of } y}$$

Qualitatively, the significance of this formula within the Popperian system seems to be as follows. A highly probable theory is one of low corroborability. The repeated testing of a theory does not raise its corroborability. For high corroborability one requires low probability. That is, the theory must make 'risky' predictions that can readily be falsified — something like Einstein's relativity theory, as opposed to Freud's psychoanalysis. However, I suggest that Popper's equations here were really 'invented', so to speak, to give the appearance of quantitative expression to his qualitative appraisal of the relative merits of different theories. In any case, they don't really do the work that is required of them. If a theory becomes well supported by the accumulation of empirical evidence, *presumably* the probability of this evidence rises. But by Equation 3 above, this pushes down the corroboration of the theory. And the equation tells us that about the corroboration of the theory, rather than its corroborability. So we don't know, from some experimental results and the use of Equation 3, whether the theory's corroborability is improving or deteriorating as a result of the experimentation or what its standing is.

It is further worth noting that Popper has stated[36] that the efforts to refute the theory must be made severely and genuinely. Sincere attempts must be made to refute theories, rather than verify them, if the notion of corroborability is to be sound. This introduces an additional element of moral exhortation into the discussion, which can scarcely be captured by formal equations such as those given above. It is hardly surprising, therefore, that the theory of corroborability has been deemed unsatisfactory. It does not reconcile the *kind* of theory with which one is dealing (for example, 'risky', 'bold', 'timorous', 'cautious' or whatever — features that Popper seems to think highly relevant to the question) and the actual status of that theory at any given epoch, according to whether or not it has been well supported by experimental investigation.

It will be recalled[37] that the logical positivists sought to provide a construction of physical theory grounded on observation of 'things'; or, in Carnap's *Aufbau*, an attempt was made to build up physical theory starting at the very lowest level with sensations and the relations between sensations. Popper's writings don't display quite so much interest in the 'thing level' as do the writings of Carnap and his colleagues, but Popper (despite his repudiation of inductivism) is an empiricist, just as were the original members of the Vienna Circle, against whose work he reacted so strongly.

In *The Logic of Scientific Discovery*, Popper dubbed the view that all statements must ultimately be justified by perceptual experience with the label 'psychologism'.[38] He maintained that it was unsatisfactory, for it foundered on the rocks of induction and universals. Universals, he rightly maintained, could not be logically 'constituted' in the way that Carnap had attempted in his *Aufbau* from 'sense data' alone. Nevertheless, Popper's falsificationism required the use of empirical elements in order that the test-implications of theories might be checked out. So he introduced the notion of '*basic statements*'. Their role was somewhat similar to the 'protocol sentences' of Neurath and Carnap.

'Basic statements', we read in *The Logic of Scientific Discovery*, 'have the form of singular existential statements'.[39] (For example: 'There is an apple on the plate'; or 'The pointer points to the reading 0.75 amp on the ammeter'.) Such statements must be testable, inter-subjectively, by observation. That is, different people must be able to check on the statements by observation and agree with one another as to what is observed. A single claimed sighting of Nessy in Loch Ness, which others could not confirm, would *not* count as a basic statement. So the notion of 'observable event' becomes, as it were, a primitive concept in Popper's system: 'Basic statements are ... statements asserting that an observable event is occurring in a certain individual region of space and time.'[40] But there are, of course, an infinite number of events occurring all the time. It becomes a matter of *decision*, therefore, as to which particular event is selected for the purpose of making a corroboration or a falsification. This makes Popper's 'basic statements' substantially different from the 'protocol sentences' of Neurath and Carnap. For the Popperian, it is not a question of opening up one's senses, which may then serve as the first rungs for an ascent of the inductive ladder. Rather, Popper's basic statements are accepted or rejected in the light of the application of a particular theory or group of theories. So certain methodological rules have to be provided for this purpose. The process of acceptance or rejection of a basic statement may, says Popper, be compared to the process whereby members of a jury formulate their verdict on a case.

We see, then, that in the discussion of basic statements, in the discussion of auxiliary hypotheses and of corroboration, and elsewhere in Popper's writings, there is need for a set of methodological rules for the conduct of science. For despite Popper's liking for strict logical thinking and the empirical aspects of science, he recognises that there must be more to science than this. A normative component is also present. And this he has been active in attempting to supply over the years.

Popper's various methodological rules are to be found scattered through *The Logic of Scientific Discovery*, but they have been conveniently gathered together for us by Johansson, who gives them as follows:

1. *Demarcation rule*:
A theory is scientific if and only if it is falsifiable, ie if it is inconsistent [or incompatible] with at least one singular existential statement that can be tested by intersubjective observation.
2. *Rules against conventionalist stratagems*:
(i)   [A]dopt a rule not to use undefined concepts as if they were defined [by the theory].
(ii) [O]nly those [auxiliary hypotheses] are acceptable whose introduction does not diminish the degree of falsifiablity or testability of the system in question, but, on the contrary, increases it.
(iii) Surreptitious alterations of usage [are forbidden].

(iv) Inter-subjectively testable... [theories] are either to be accepted, or to be rejected in the light of counter-experiments...

(v) [A]uxiliary hypotheses should be used as sparingly as possible.

3. *Rules demanding a high degree of falsifiability*:

(i) [R]egard natural laws as synthetic and strictly universal statements.

(ii) [T]hose theories should be given preference which can be most severely tested.

(iii) [T]he number of... axioms — of... most fundamental hypotheses — should be kept down.

(iv) [A]ny new system of hypotheses should yield, or explain, the old, corroborated, regularities.

(v) [A] new theory should proceed from some *simple, new, and powerful, unifying idea* about some connection or relation...between hitherto unconnected things... or facts... or new 'theoretical entities'.

(vi) [A] new theory should be *independently testable*.

4. *Acceptance rules for basic statements*:

(i) [A] theory is [taken to be] falsified only if we discover a *reproducible effect* which refutes the theory. In other words, we only accept the falsification if a low-level empirical hypothesis which describes such an effect is proposed and corroborated.

(ii) [One] should not accept *stray basic statements* – ie logically disconnected ones — but... [one] should accept basic statements in the course of testing theories; of raising searching questions about these theories, to be answered by the acceptance of basic statements.

5. *Acceptance rules for theories*:

(i) [A] theory is to be accorded a positive degree of corroboration if it is compatible with the accepted basic statements and if, in addition, a non-empty sub-class of these basic statements is... accepted as the results of sincere attempts to refute the theory.

(ii) [I]t is not so much the number of corroborating instances which determines the degree of corroboration as the *severity of the various tests* to which the hypothesis in question can be, and has been, subjected.

(iii) [One must not] accord a positive degree of corroboration to a theory which has been falsified by an inter-subjectively testable experiment based upon a falsifying hypothesis.

(iv) [The] theory should pass some new, and severe, tests.

6. *Rule for falsifying probability statements*:

(i) [R]egard highly improbable events as ruled out... [or] prohibited.

7. *Rules for the social sciences*:

(i) [A]ll theoretical or generalizing sciences [should] make use of the same method, whether they are natural sciences or social sciences.

(ii) [T]he task of science is...to describe how things behave... [T]his is to be done by freely introducing new terms wherever necessary, or by re-defining old terms wherever convenient while cheerfully neglecting their original meaning [sic]. For...words [are] merely...*useful instruments of description*.

(iii) [A] *preconceived selective point of view* [should be introduced] into one's history... [Or] *write that history which interests* [you]....

(iv) [T]he task of social theory is to construct and analyse... sociological models carefully in descriptive or nominalist terms, that is to say, *in terms of individuals*...[41]

All this demonstrates that Popper's metascience involves more than logic. However, when I was a brash young student, I had the good fortune to attend a series of lectures given by Popper at University College, London, and in the discussion period after one of the lectures I asked Popper whether he considered the description of science he was giving was provisional, subject to correction and test, falsification, and so on. The answer I received was that 'There is nothing provisional, conjectural or falsifiable about *logic*'. No doubt, Popper was thinking that the force of the argument of *modus tollens* was such that falsificationism was the correct metadescription of science, and that inductivism was wholly unsatisfactory, for the kinds of reasons that Hume had given in the eighteenth century. However, I take it that he did not think of the problems raised by Duhem and Quine on that occasion. And as can be seen from the foregoing list, when Popper's methodological rules are brought out into the open, what may at first sight have seemed to be a carefully wrought, logical system turns out to be rather fuzzy, and open to a number of different possible interpretations.

On the other hand, it must be acknowledged that Popper has never made the slightest attempt to conceal the 'conventional' character of his 'methodological rules',[42] and he regards them as subject to change — to be criticised and improved if possible. Even so, the very presence of these 'methodological rules' within the Popperian system is an indication that it offers much more than a logical analysis of science. Popper is telling scientists how he thinks they *ought* to conduct their inquiries. To be fair to Popper, however, he would maintain that he can see from the history of science that the kind of rules that he advocates have, in fact, been effective for the prosecution of science. So he would want to argue that the rules that have been *successful* in sustaining the scientific enterprise are the ones that ought to be pursued. (This, I suppose, gives Popper's position a pragmatist flavour.) There is, it would seem, an element of paradox here, for one who claims to be deeply concerned about 'Hume's problem'; but presumably it would be answered that methodological rules can be followed without contravening any logical principles. Anyway, it is clear that the Popperian meta-level has a distinct normative aspect; and obviously there can be differences of opinion on matters of value — on questions of 'ought'. So Popper's enterprise cannot be represented as one that is wholly dispassionate and neutral, even if it is claimed that it is grounded in logic, about which (one may suppose) there is nothing that is 'provisional, conjectural or falsifiable'.

To illustrate some of the difficulties associated with Popper's methodological rules, consider for example the ones associated with 'basic statements'. We note that there is nothing said about the question of observability. So we have little guidance as to what happens (or should happen) when a theory is refuted (by a basic statement), for we are forced to rely on some intuitive notion of observability. Again, Rule 3(v) enjoins us to favour simple, new, powerful and unifying ideas; but Rule 3(ii) asks us to give preference to those theories that may be most severely tested. Now it is rather difficult to test Einstein's theory of general relativity, and it is certainly anything but simple. On the other hand, it is (relatively speaking) new and powerful and brings about a remarkable consilience or unification of ideas. So Popper's rules would not have given a clear indication as to what to do when considering whether or not to subscribe to the theory of general relativity. All this might not, perhaps, be of great import were we not hoping for science to be a strictly rational and logical enterprise, based upon empirical sources of information. But Popper undoubtedly holds that science does have this character. Why else would he be so concerned about the problem of induction?

In considering Popper's work, it is most important to keep at the front of one's mind that he regards himself as a realist, both in his ontology and in his epistemology. He believes that there *is* a real, external world (he is no phenomenalist), and that by the formulation and rigorous testing of hypotheses, science enables human beings gradually to acquire an increasingly accurate knowledge of this real world. For this reason, Popper has been a vigorous critic of instrumentalism throughout his career. But as we have seen above, he does not claim that the never-ending formulation and testing of hypotheses will generate truth or certainty. One may shoot closer to the centre of the target with the arrow of the *modus tollens*, but one can never know when one has hit it. So Popper does not seek to mount his realism directly on a doctrine of truth — perhaps some correspondence theory of truth or a coherence theory. Rather, he has devised a new conception,[43] named 'verisimilitude', which is intended as a measure of the 'truthlikeness' of hypotheses or theories. If we have an empirical statement, then the class of true statements that it entails is called its 'truth content', and the set of false statements it entails is called its 'falsity content'. The 'verisimilitude' is defined as 'truth content' *minus* 'falsity content'. The verisimilitudes of two theories which offer competing explanations of the same phenomena may thus supposedly be compared if one of them explains or accounts for that which the other explains and *also* some additional phenomena; and the first theory has stood up to tests in this extra domain where the second theory is unsuccessful, or which it does not cover.

Unfortunately, however, it has been shown by Tichy[44] that the verisimilitudes of two theories cannot be satisfactorily compared if one of them is false, and this has led Popper[45] to attempt a modified definition of verisimilitude. We shall not seek to trace the details of these contemporary controversies for the matter is still under active debate. I believe, however, that at the time of writing no generally acceptable doctrine of verisimilitude has been found, so that there is a distinct lacuna in the Popperian system. Yet this is scarcely surprising. If one had a satisfactory doctrine of verisimilitude, one would in fact have a kind of 'hook' (as I have previously called it) whereby theories about the world might be connected with the world. Or at least one would have a kind of vantage point from which the success or otherwise of the attempted hooking-together of ideas and states-of-affairs in the world might be appraised. Besides, if one is *really* worried to death about the problem of induction, one must also, I think, be displeased with any theory that purports to give measures of the truthlikeness of scientific statements. For such a theory can always be subjected to the kinds of criticisms that were directed against induction by Hume, since 'truth content' and 'falsity content' cannot be determined *a priori*, only by empirical investigation.

There are numerous other interesting aspects to Popper's work, but we only have space to touch on them briefly here. Popper has for many years mounted a strong campaign against the doctrines of essentialism and instrumentalism.[46] It is interesting to note that his nominalism has led him to discount the twentieth-century philosophers' customary interest in the analysis of language, which was a major concern of some of the logical positivists. To look for exact definitions and formulations of meanings of terms is, in Popper's view, philosophically fruitless. Rather, one should be striving for theories that approach the truth (that is, theories that are of increasing verisimilitude). The distinction between Popper's views and those of analytical philosophers and logical positivists is easily seen from Figure 46, a table that was published in Popper's autobiography.[47]

Figure 46

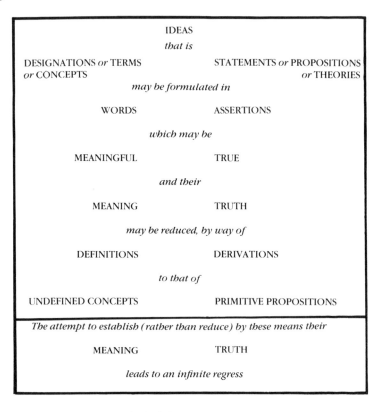

	IDEAS	
	that is	
DESIGNATIONS *or* TERMS *or* CONCEPTS		STATEMENTS *or* PROPOSITIONS *or* THEORIES
	may be formulated in	
WORDS		ASSERTIONS
	which may be	
MEANINGFUL		TRUE
	and their	
MEANING		TRUTH
	may be reduced, by way of	
DEFINITIONS		DERIVATIONS
	to that of	
UNDEFINED CONCEPTS		PRIMITIVE PROPOSITIONS

The attempt to establish (rather than reduce) by these means their	
MEANING	TRUTH
leads to an infinite regress	

According to Popper, the left-hand side of the table is quite trivial; but the right-hand side is philosophically all-important. It will be seen that the left-hand side of the table is a rough outline of the sort of philosophical program favoured by the logical positivists. So Popper, although a nominalist, repudiating the search for philosophical essences (for THE TRUTH cannot be found, he thinks), does not suppose that enlightenment is to be found by analysis of language. Rather, it is the processes of experimental inquiry, conjecture and (attempted) refutation, and critical discussion that provide an endless road of progressive inquiry.

Among other things, Popper has been particularly influenced by the Darwinian theory of evolution, as is specially evident in *Objective Knowledge*, subtitled *An Evolutionary Approach* (1972). But even in his earliest formulations of his doctrine, Popper believed that each person is equipped with certain inborn expectations, and gradually builds up a more truthful picture of the world by trial and error — by conjecture and refutation. There is, then, the notion of a kind of 'survival of the fittest' among ideas, and Popper's doctrine is made to mesh quite well with that of Darwinian evolutionism:

> All *organisms* are constantly, day and night, *engaged in problem-solving*; and so are all those evolutionary *sequences of organisms* – the *phyla* which begin with the most primitive forms and of which the now living organisms are the latest members.[48]

We have already encountered various uses of the analogy between the evolution of ideas and the evolution of living organisms in nineteenth-century philosophers of science such as Peirce and Mach. The analogy has continued to be used in the present century by various philosophers, perhaps in the most unguarded (or flagrant) way by Stephen Toulmin.[49] Popper is only one of many who find the biological analogy useful in epistemology. But it is, after all, only an analogy, with many 'negative' aspects as well as 'positive'. Perhaps the most important 'negative' feature of the analogy is that ideas are put forward consciously, and with particular ends or purposes in view — to solve problems, as Popper would say. But the variants upon which natural selection acts among living organism are, according to the Darwinian theory, produced randomly, with no particular end or purpose in view.

Be this as it may, Popper sees the process of problem-solving — which he thinks has characterised the evolutionary history of the human race — as being involved with language (critical discussion) and also with ethical systems, law, religion, science, philosophy and various social institutions. These together form vitally important aspects of the environment in which man evolves, over and above the physical environment provided by the Earth. And such systems form part of what Popper calls 'World 3'. Let us look briefly what he means by this.

For Popper, there are three so-called 'worlds'. 'World 1' consists of material things such as sticks and stones, bodies, brains. 'World 2' is made up of all the various thoughts in people's minds. It is the world of mental phenomena, but not the world of brain cells which (according to the traditional dualist view of the mind–body problem) merely serve as a kind of 'substrate' for thoughts, which are non-material entities. 'World 3', according to Popper, consists of 'objective' structures that are the products of minds or living beings. It includes such things as information in libraries, works of art, language, music, scientific theories, and so on. It constitutes the cultural heritage that is preserved in 'World 1' objects such as books, gramophone records or brain cells. The contents of this third world are, on Popper's view, man-made, yet autonomous. They constitute what he calls 'objective knowledge'.[50] It is through this 'objective knowledge' that it is possible for ideas to have histories. Ideas, therefore, can evolve without needing some 'historicist' (in Popper's sense of the word[51]) explanation to account for their evolution. It is, says Popper,[52] the process of *criticism* that provides the motor for the evolutionary development of 'World 3' objects.

The whole idea of a 'World 3' — a domain of 'objective knowledge' —

has proved unattractive to many modern critics,[53] and it is clear that the doctrine has a decidedly Platonist aspect. However, whether or not we like his metaphysics at this point, we should take note of the way in which Popper has used the notion of 'objective knowledge' to mount an attack on orthodox theories of knowledge. For example, even an author who has just completed a book does not know (in a 'World 2' sense) every item of information that is contained in that book. Nevertheless, the knowledge is there (encoded in 'World 1' form), objectified in that book, and other persons may gain access to it if they so wish. As a result of such arguments, Popper regards traditional theories of knowledge such as those of Descartes, Locke or Kant as misconceived, for they quite overlook what he regards as a vitally important aspect of the way in which human knowledge is gained and deployed. Popper, then, seeks to rectify such 'mistaken' views, taking due regard of the 'World 3' features of human culture. His ideas on epistemology and the mind-body problem are further developed in his *The Self and its Brain*[54] which adopts a very strong dualist approach to the relationship between mind and body.

The influence of Karl Popper on twentieth-century philosophy of science has been very considerable. Certainly, in his emphasis on the need for critical examination of ideas, he has had a beneficial influence amongst the scientific community; and the plea for scientists to seek to falsify their ideas, rather than verify them, has not gone unnoticed. Also, the suggested line of demarcation between science and pseudo-science has had a salutary effect, even though, as we shall see in a moment, it no longer seems as satisfactory as once was the case. In his emphasis on critical discussion, and by drawing attention to the fact (in stating his various methodological rules) that there is a definite normative dimension to science, Popper has been — perhaps unwittingly — preparing the ground for the present-day interest in the sociology of science, with epistemologies based on sociological principles, about which we shall say something in our next chapter.

On the other hand, I think it has to be said that falsificationism, and Popper's account of the 'logic' of scientific method,[55] have to be accounted as failures. The incubus of the Duhem/Quine thesis has still not been exorcised successfully, so that there is really no more certainty in the process of falsification than in the processes of verification or induction. Indeed, as recent commentators such as David Stove have shown,[56] the whole logic of falsificationism is something of a shambles. Also, as Anthony O'Hear has pointed out,[57] in his efforts to counter the objections of his critics, and even in his original position, Popper has allowed inductivism into his system by the back door, so to speak. Following a kind of Kantian transcendental argument, O'Hear maintains that one cannot make sense of the external world at all unless one assumes *some* kind of stability therein. Certainly, one could not make a falsifying observation without some such assumption. Thus some (perhaps weak) inductive assumption of uniformity of nature is necessarily presupposed in scientific work, as in all human activity. O'Hear claims, in fact, that our customary distinction between the subjective and the objective — between us as persons and the external world as objects — would be impossible without some inductive presupposition. It is interesting, therefore, that in one of his later writings[58] Popper has acknowledged that his realist assumption that the results of scientific inquiry lead one to greater verisimilitude is infected by 'a "whiff" of inductivism'. In saying this, Popper has in fact conceded a great deal, for the whole elaborate apparatus of falsificationism, empirical content, corroboration, verisimilitude, and so on, was created in order to circumvent 'Hume's problem' entirely. Now, it seems, the effort may not really have been worth it! Inductivism is creeping in again, even within Popper's own writings.

It must also be acknowledged that the much-vaunted distinction between science and pseudo-science is now beginning to look somewhat tatty. One can accept that a theory that is formally unfalsifiable cannot be scientific, but it is often the case that a falsifiable theory is not falsified because its proponents choose not to press home the falsifying evidence — Popper's methodological rules against conventionalist stratagems notwithstanding. They prefer to invoke *ad hoc* hypotheses, or to conventionalise the theory. But this has to do with aspects of human behaviour, rather than some logical feature of the structure of the theory — its inherent unfalsifiablity. So whether or not theories are falsified on any given occasion is as much a social question as anything else. So, as I say, the distinction between science and pseudo-science should be made by examination of social practices just as much as by examination of the logical structures of theories.

In fact, Popper himself acknowledged this long ago when he wrote:

> *Only with reference to the methods applied* to a theoretical system is it at all possible to ask whether we are dealing with a conventionalist or an empirical theory.[59]

With hindsight, this can perhaps be seen to have been the thin end of the wedge of the sociology of knowledge cutting into Popper's theory. Thus, if we want to know whether a theory is or is not scientific, we should look and see how it is handled by people, rather than consider its logical structure.

The movement towards a widespread recognition of the social dimension of scientific knowledge will be treated in our next chapter. The chief historical roots of the sociological approach to scientific knowledge certainly do not lie in the work of Karl Popper. Nevertheless, we can proceed from Popper to our final discussions by considering the work of some of his followers and the controversies that have broken out in recent years between Popper and his critics. To these tasks we may now address ourselves.

NOTES

1 For details of Popper's life beyond the very brief account given in this chapter, readers are referred to his most interesting autobiography. (See below, note 9.)

2 Popper's use of the word historicism is generally regarded as somewhat idiosyncratic. Generally speaking, the word refers to the position which holds that the best way to understand something is to be achieved by examining its history. Historical determinism or historical inevitability are not usually thought to be essential components of the historicist position. But Popper seems to think otherwise — or rather he uses the word in a sense rather different from that which is customary.

3 K R Popper *Logik der Forschung zur Erkenntnis-theorie der modernen Naturwissenschaft* Springer Vienna 1935

4 K R Popper *The Logic of Scientific Discovery* Hutchinson London 1959 (We may mention here that the first German edition of this work had rather little impact on the philosophical world. It was only with the publication of the English edition, following the success of *The Open Society*, that Popper's views on the philosophy of science began to be really influential.)

5 K R Popper *The Open Society and Its Enemies* 2 vols Routledge London 1945

6 K R Popper *The Poverty of Historicism* Routledge & Kegan Paul London 1957

7 K R Popper *Conjectures and Refutations: The Growth of Scientific Knowledge* Routledge & Kegan Paul London 1963

8 K R Popper *Objective Knowledge: An Evolutionary Approach* Clarendon Press Oxford 1972

9 K R Popper *Unended Quest: An Intellectual Autobiography* Fontana/Collins Glasgow 1976

10 K R Popper & J C Eccles *The Self and Its Brain* Springer Berlin & London 1977

11 K R Popper (ed W W Bartley III) *Postscript to The Logic of Scientific Discovery* Rowman & Littlefield Totawa and Hutchinson London: I *Realism and the Aim of Science* (1983); II *The Open Universe: An Argument for Indeterminism* (1982); III *Quantum Theory and the Schism in Physics* (1982) (These recent texts are not discussed in the present book.)

12 Here Popper explained that he subsequently referred to such a 'rescuing operation' — involving the *ad hoc* introduction of hypotheses to help keep an endangered theory afloat — as involving a 'conventionalist twist' or 'conventionalist stratagem'. The point is, of course, that when *ad hoc* hypotheses are introduced in this way there is a refusal to allow a theory to submit to falsification. Nothing is allowed to count against it. Thus, in language reminiscent of that of Poincaré, the theory may be said to be 'conventionalised'.

13 Popper *op cit* (note 7) pp 36–7

14 *Ibid* p 42

15 See above, p 114

16 Popper *op cit* (note 7) p 46

17 Popper *op cit* (note 8) p 1

18 See above, p 45

19 *Ibid*

20 Popper *op cit* (note 4) p 31

21 See above, p 201

22 For detailed discussions of this important issue, see S G Harding ed *Can Theories be Refuted? Essays on the Duhem-Quine Thesis* Reidel Dordrecht & Boston 1967

23 See above, p 266

24 For discussion of the concept of *ad hoc* hypotheses, see J Leplin 'The Concept of an *Ad Hoc* Hypothesis' *Studies in History and Philosophy of Science* 1975 Vol 5 pp 309–45 (It may be noted in passing that the simple contraction hypothesis was tested independently in 1932 by the so-called Kennedy–Thorndike experiment.)

25 This argument against Popper was adduced by Reichenbach long ago. See his review 'Induction and Probability: Remarks on Karl Popper's *The Logic of Scientific Discovery*' in M Reichenbach & R S Cohen eds *Hans Reichenbach: Selected Writings 1909–1953* Reidel Dordrecht 1978 Vol 2 pp 372–87 (at p 374) (first published in German in Vol 5 of *Erkenntnis* 1935)

26 Popper *op cit* (note 4) p 120 & *passim*

27 *Ibid* p 113

28 *Ibid* p 120

29 For Popper's doctrine of 'basic statements', see p 308

30 Popper *op cit* (note 4) p 122

31 *Ibid* pp 126–30

32 In these present strange philosophical days even this apparently innocuous statement is by no means universally acceptable. See, for example, the work of P K Feyerabend discussed below, pp 334–42

33 Popper *op cit* (note 4) pp 269–70 & Appendix IX

34 *Ibid* p 400

35 This is so, in Popper's view, because of the lack of an acceptable principle of induction. Personally, I should be happy to suppose that some scientific statements of universal scope have a probability greater than zero.

36 Popper *op cit* (note 4) p 414

37 See above, p 239

38 *Ibid* p 94

39 *Ibid* p 102
40 *Ibid* p 103
41 I Johansson *A Critique of Karl Popper's Methodology* Scandinavian University Books Stockholm 1975 pp 16 17 18–19 19 20–1 21 22
42 Popper *op cit* (note 4) p 53
43 Popper *op cit* (note 7) p 234 & *passim*
44 P Tichy 'On Popper's Definition of Verisimilitude' *British Journal for the Philosophy of Science* 1974 Vol 25 pp 155–60
45 K R Popper 'A Note on Verisimilitude' *British Journal for the Philosophy of Science* 1976 Vol 27 pp 147–59
46 For his discussions of these, see particularly K R Popper 'Three Views of Human Knowledge' in *Conjectures and Refutations* (note 7) pp 97–119; and 'A Long Digression on Essentialism' in *Unended Quest* (note 9) pp 18–31
47 Popper *op cit* (note 9) p 21
48 Popper *op cit* (note 8) p 242
49 S E Toulmin *Human Understanding* Clarendon Press Oxford 1972
50 Popper *op cit* (note 8) pp 108–9 & *passim*
51 See note 2, above
52 Popper *op cit* (note 8) p 121 & *passim*
53 See, for example, P K Feyerabend's review of Popper's *Objective Knowledge* in *Inquiry* 1974 Vol 17 pp 475–507
54 Popper & Eccles *op cit* (note 10)
55 It is a curious fact that the *Logik der Forschung* was presented to English-language philosophers as *The Logic of Scientific Discovery*. Yet Popper says almost nothing about the processes of discovery. As we mentioned above, his 'arch of knowledge' is a grievously asymmetric structure with an insubstantial ascending, inductive leg.
56 D C Stove 'Popper on Scientific Statements' *Philosophy* 1977 Vol 55 pp 81–8
57 A O'Hear *Karl Popper* Routledge & Kegan Paul London 1980 pp 62–7
58 K R Popper 'Replies to my Critics' in P Schilpp ed *The Philosophy of Karl R Popper* 2 vols Open Court La Salle 1974 Vol 2 pp 959–1197 (at p 1193)
59 Popper *op cit* (note 4) p 82

CHAPTER 9

SCIENCE AS A DYNAMIC SOCIAL SYSTEM: KUHN, LAKATOS AND FEYERABEND — SOCIOLOGY OF KNOWLEDGE THEORISTS

In this last expository chapter, we shall seek to show how certain important ideas in the work of Popper were taken up and developed by some of his distinguished students or co-workers, and also how interesting philosophical positions have, in recent years, been put forward in more or less direct opposition to his work. Consideration of these matters leads us gently and conveniently towards an examination of some of the recent developments in metascience that lay particular emphasis on the social aspects of scientific inquiry, and hence on to the social dimension of epistemology. In doing this, however, to bring our account more or less up to the present, we shall necessarily have to omit major areas of the metascientific literature. The pathway that I shall choose is dictated chiefly by considerations of historical continuity with topics that have been under discussion in previous chapters, the notice that the particular matters selected have excited in recent years, and the inherent interest and accessibility of the topics that are selected for exposition. Such a course inevitably entails the omission of numerous topics of historical and philosophical interest and importance. But we cannot hope to accomplish everything within the compass of these few pages.

One of the most distinctive features of Popper's work, apart from its obvious enthusiasm for falsificationism rather than verificationism, has been the emphasis upon science possessing a particular *scientific method* – a way of proceeding and thinking characteristic of scientists and epitomising the rationalism of the scientific movement. In a sense, Popper's thinking on this marks him as a latter-day positivist and we may compare *his* rationalism with the enlightened thinking that was supposed by Comte to characterise each science when it achieved its positivist phase.

As we have seen, however, when one lists Popper's methodological directives, they form a strangely assorted collection, distinctly arbitrary in appearance. And a good deal of recent philosophy of science has concerned itself with the question of whether in fact science can be said to have a definite set of methodological rules — or whether there are any such rules at all. Indeed, some commentators have gone so far as to question whether science is in fact a rational enterprise. Thus, in the closing stages of our account we find the venerable 'arch of knowledge' beginning to collapse in a heap. It had, for so very long, been supposed that there *was* a right and proper method for the successful conduct of inquiry — analysis and synthesis, Mill's canons, critical discussion and rigorous efforts towards falsification, or whatever. To be sure, there had been rather little success in giving an adequate account of this 'method'; or there had been remarkably little agreement as to what it might be. But at least there was a sort of agreement that there were certain common features that characterised the scientific mode of inquiry. And it was hardly supposed that science was an irrational enterprise. Yet, as we shall see, this is the pretty pass to which some recent commentators on science would wish to bring us.

Just as the particular form of the reaction to logical empiricism — Popper's falsificationism — could not very well have been predicted in advance, so also the form that the reaction against Popper was to take could hardly have been foreseen. And in fact the response to Popper was not really directly dialectical in character. Rather, certain ideas were developed in America which soon came to be seen to be antithetical to the ideas of Popper and his co-workers. Matters came to a head at an International Colloquium on the Philosophy of Science held at Bedford College, London, in 1965, and a good deal of the subsequent work in English-language philosophy of science has grown from the published versions of the papers presented at this meeting.[1]

At the 1965 meeting, the debate was chiefly between 'Popperians' on the one hand and the American scholar Thomas S.Kuhn (1922–) and his supporters on the other. It appears from the published version of the meeting that there were, at that time in London, more Popperians on the ground than Kuhnians, though as time has passed the situation has tended to be reversed. At the time of writing, I may add, the momentum of Kuhn's efforts in philosophy seems to be declining, though he is continuing to make important contributions in the historiography of science. So, in order to examine some of the more recent developments in philosophy of science, with the fairly rapid disintegration of the belief that there is a solid 'arch of knowledge', we may usefully give our attention to the work of Kuhn. I should emphasise that this is not intended to imply that Kuhn himself believes that science is an irrational enterprise (though some critics have attributed this position to him). It is, however, fair to say, I think, that his work is an important component of the trend *towards* that way of thinking on the part of some commentators. Also, Kuhn's work has tended to draw attention to the importance of social considerations in the conduct of scientific inquiry, though I do not suppose that Kuhn would regard himself as a full-blooded exponent or adherent of the 'sociology of knowledge' school.

Kuhn

Kuhn is chiefly an historian of science, not a philosopher, even though greater use has been made of his work by philosophers, sociologists, economists, etc, than by historians. He originally trained as a physicist, and published a few papers in theoretical physics in the early 1950s;[2] but he soon turned to history of science, then to philosophy, and in recent years back to history of science once again. In history of science, Kuhn has written particularly on the Copernican Revolution[3] and the history of quantum mechanics[4] but the work by which he is best known is his *Structure of Scientific Revolutions*, which has been read very widely by practitioners in all sorts of disciplines since its first publication in 1962.[5] The second edition was published in 1970 and included an important 'Postcript', which offered some significant modifications of the author's earlier views. A somewhat different revision was presented at a symposium held in Urbana in 1969, and may be found in a volume edited by F.Suppe in 1974.[6] More recently, however, Kuhn seems to have lost interest to some degree in the controversies that his book has generated, and he has been concentrating his attention on theoretical issues in historiography and in direct historical research. What is particularly interesting (and perhaps encouraging) to practitioners of history and philosophy of science is that Kuhn's work, which has had such a considerable influence in philosophy, sociology and elsewhere, was based on, and grew from, researches in history of science, and is as much an essay in history as in (say) philosophy or sociology.

In discussing Kuhn's work, let us first give a brief sketch of the main points that are to be found in *The Structure of Scientific Revolutions*. The author gives us a thumb-nail sketch of what he considers to be the leading features of the histories of the several sciences in general.[7] In each science, he suggests, there is a so-called 'pre-paradigm' period. (I shall say shortly what is meant by the word 'paradigm', according to Kuhn's vocabulary.) In this pre-paradigm period, facts are gathered, almost randomly, without reference to any accepted plan or theoretical structure. Francis Bacon's 'natural histories'[8] provide good examples; the work of the Roman naturalist Pliny — a vast assemblage of assorted information, gathered together with negligible theoretical basis — is even better.[9] At the pre-paradigm state of a science, says Kuhn,[10] there may be a number of competing schools of thought, but no one of them achieves general acceptance. Gradually, however, one theoretical system begins to receive general acceptance, and so the discipline's first *'paradigm'* is established. This term, which can also be used to refer to a Platonic 'form' and can therefore also be interpreted as an *'exemplar'*, was defined by Kuhn as referring to a model 'from which spring particular coherent traditions of scientific research'[11]; or as a 'universally recognized scientific... [achievement] that for a time provide[s] model problems and solutions to a community of practitioners'.[12]

Kuhn's language is not, I fear, as felicitous as one might wish, so some elucidation of these rather odd-sounding definitions is probably called for. As examples of paradigms (or paradigmata, if you are particularly pedantic), Kuhn suggested such things as Ptolemaic astronomy, the 'new' chemistry of Lavoisier and his school, Aristotelian dynamics, or Newtonian corpuscular optics.[13] So a paradigm is much more than a theory, or even a group of interrelated theories, in a particular branch of science. It is almost a world-view — a way of seeing the world through the spectacles provided by a particular branch of science. But it isn't simply a world-view. Young scientists in training, preparing themselves for membership of a scientific community, study the appropriate paradigm, thereby assimilating themselves to the relevant research tradition. They work on problems in text-books, which lead them to think in accordance with the demands of the paradigm of the day. The whole educational process, in fact, moulds their thinking in such a way that it meshes with the generally received views that are dictated by the paradigm. The paradigm is made up of theories, yes, but also particular techniques suited to solving research problems within a given area of research. For example, the Ptolemaic paradigm stipulated that one should try to account for the observed motions of heavenly bodies by means of circles. And the student of Ptolemaic astronomy would have to learn the specific techniques for the geometrical representation of the motion of heavenly bodies and the ways appropriate for gathering the data in order to support the Ptolemaic hypothesis; also the use of astronomical methods for divination, calendrical work, navigation, and so on. Or in Lavoisier's chemistry,[14] one tried to find which substances were the reactants and which were the products in a reaction. Then, weighing reactants and products, and applying the principle of conservation of mass, one could notice if any substance got missed in the accounting, so to speak, and hunt for it accordingly. This was what you had to *do* if you were an aspiring student of Lavoisierian chemistry. And Lavoisier's doctrine of elements and the oxygen theory of combustion provided the basic theory. So a paradigm, according to Kuhn, is a kind of amalgam of theory and method, which together constitute something amounting almost to a world-view.[15] But it is a rather curious entity, nonetheless, of somewhat uncertain ontological status, and one often hears a lot of loose talk about paradigms as if they were real things under which one might 'shelter', like a person under an umbrella. This, of course, is nonsense if taken literally, but may be useful nonetheless in informal discourse or writing.

Suppose, then, that following a pre-paradigm period, one particular paradigm becomes well established. Then, according to Kuhn, there follows a period of '*normal science*', in which research is conducted in accordance with (or following the example of) the model provided by the previously successful research work carried out in this area.[16] In this way, the promise of the paradigm is fulfilled, so to speak; its potential is rendered actual. However, in the process, nature is forced into the 'inflexible box' of the paradigm.[17] The paradigm is assumed to be correct and appropriate to its task (for the scientist has been educated to think in this way), and the parts of nature that do not fit satisfactorily are simply ignored, for the scientist's very way of seeing the world is somehow 'shaped' by the paradigm according to which (or 'under which', to use our previous analogy) he works.

Because of the characteristics outlined above, the general function of the scientist in a period of 'normal science' becomes that which Kuhn calls *'puzzle solving'*.[18] Under such conditions of research, the results obtained are more or less anticipated; the anticipated is merely achieved in new ways. Problems become 'puzzles' because (like cross-word puzzles) they have assured solutions, guaranteed as it were by the potency of the paradigm, which provides appropriate 'rules' for the successful conduct of the inquiries. Such a state-of-affairs is conducive to the rapid progress of research, with a gratifying production of a steady flow of published papers and reports. Whether, however, it is producing knowledge that corresponds with the way things are in the world is, of course, another question altogether. For in Kuhn's 'normal science' it is the investigator that is under trial, as much as nature being interrogated or the paradigm being subjected to test.

Of course, during a period of 'normal science', unexpected or anomalous results may occur; but these are customarily suppressed for a while, or dealt with by the use of *ad hoc* hypotheses. Nevertheless, a time may come when a paradigm is so encumbered by these *ad hoc* hypotheses that it begins to break down, and the science then enters a period of *'crisis'*.[19] In such a period, there is a questioning of the basic experimental techniques (the paradigm's methodology), and one finds earnest discussions of the most fundamental assumptions of the theoretical structure of the paradigm. Often metaphysical questions are raised that would never be brought into the open in periods of normal science; and thought-experiments abound. Also, in a period of crisis one may find disagreement as to what the paradigm actually *is*.

Thus the situation in a period of crisis is not unlike that which obtains in the original pre-paradigm period. But eventually, from among the several possible alternative paradigms that may be proposed, one seems to achieve general acceptance and the science settles down again into a further period of normal science and puzzle solving. The text-books get re-written and curriculums are revised in the light of the newly adopted paradigm.

All this constitutes what Kuhn calls a 'scientific revolution' — the subject of his book. Obvious examples are provided in the history of science by the transition from Aristotelian to Galilean dynamics, from Ptolemaic to Copernican astronomy, from phlogistic to Lavoisierian chemistry, and so on. And each science, according to Kuhn's thesis, passes through a succession of scientific revolutions, the general structure of which is anatomised in his book. Thus, contrary to historians such as Duhem, Kuhn sees the history of science as punctuated by major discontinuities, being anything but a process of steady development by accumulation, as the inductivist (perhaps to some degree in caricature) would have us believe. Indeed, it appears that Kuhn would compare the process whereby a scientist abandons one paradigm and adopts another in its stead as analogous to what the psychologists call a *Gestalt* switch.[20]

But if this is so, it has some important implications for Kuhn's thesis. If one cannot hold two competing paradigms in one's head simultaneously — that is, if one cannot be committed to, and work according to, more than one paradigm at a time — then two scientists working 'under the aegis' of two different paradigms will find their thoughts mutually incomprehensible (in detail), for they will *mean* different things by the same words. That is, according to Kuhn, different paradigms are 'incommensurable'.[21]

This thesis of incommensurability has led to a deal of discussion which has by no means died away. It has been linked with the position that science is not concerned with discovering 'the truth'. Or in so far as it is, what counts as 'the truth' is determined by what the scientific community *allows*, rather than some relationship between theories about the world and the world itself. Indeed, some critics have gone so far as to accuse Kuhn of suggesting that science is an *irrational* enterprise, governed by social commitments, rather than struggles with nature in attempts to discover the truth. Actually, there doesn't seem to be much textual evidence to support charges against Kuhn that he presents science as an irrational enterprise; indeed, he has taken pains to disallow any suggestion that his thesis was or is one of irrationalism.[22] But he certainly wants to deny that there is any definite 'scientific method' or set of 'methodological rules' according to which scientific inquiry ought to be undertaken. In this, needless to say, there lies a strong bone of contention with the Popperians.

Kuhn's thesis is, I suggest, both sociological and epistemological. The scientist is trained within, and works according to, the edicts of a paradigm, which is itself the product of a scientific *community*. And in a period of 'normal science' scientific work can only be generated successfully according to some accepted paradigm. (Work outside it will simply remain unpublished.) But the thesis is also epistemological. Kuhn wants to claim that the very way the scientist *sees* the world is shaped by his commitment to some paradigm or other. His very knowledge is governed and conditioned by this commitment. There will be no set of Popperian 'basic statements' to which workers with different theoretical presuppositions can turn. If they try to do so, they will simply interpret the 'facts' differently; and in that sense their thoughts may be said to be incommensurable. We need not say, therefore, that Kuhn's thesis is *either* sociological *or* epistemological. It is, in fact, both, and his ideas may be seen as an important contribution to 'sociology of knowledge' (though, as I say, Kuhn is not really a fully paid-up sociologist of knowledge, according to the most recent developments in that area of inquiry).

As we have indicated, the publication of the first edition of Kuhn's book in 1962 caused a great deal of interest, and considerable critical discussion. His ideas have been applied in numerous fields, such as political science, economics and education, and even in theology and art. But perhaps surprisingly, not many have chosen to rewrite the history of science according to the Kuhnian formula, though some attempts in this direction have certainly been made.[23] Rather, there have been some vociferous criticisms of Kuhn's theses from the history and philosophy of science community.

The Popperians, of course, took particular exception to Kuhn's work, as may be seen from the papers given at the 1965 Bedford College

Colloquium. Popper himself found the Kuhnian notion of 'normal science' particularly distasteful, being contrary to all that he held dear in the matter of critical discussion, and the falsificationist testing of hypotheses to destruction:

> The 'normal' scientist, as described by Kuhn, has been badly taught. He has been taught in a dogmatic spirit: he is a victim of indoctrination. He has learned a technique which can be applied without asking for the reason why... As a consequence, he has become what may be called an *applied scientist*, in contradistinction to what I should call a *pure scientist*. He is... [merely] content to solve 'puzzles'. ...
> I admit that this kind of attitude exists... [But] I can only say that I see a very great danger in it and...the possibility of its becoming normal...a danger to science and, indeed, to our civilization.[24]

Much criticism has also been levelled against the imprecision of Kuhn's main theoretical feature — the paradigm. The incommensurability thesis has disturbed numerous critics; and as mentioned above, Kuhn has been charged with peddling irrationalism. In addition, there has been scepticism as to whether the history of science actually looks the way it would if Kuhn's description of the history of science were correct. On the other hand, the claim that there is a strong social component to scientific knowledge has been widely accepted.

It was in response to criticisms of this kind that Kuhn added his 'Postscript' to the second edition of *The Structure of Scientific Revolutions*, and also published his paper, 'Second Thoughts on Paradigms', in 1974.[25] These publications involved substantial revisions of his earlier position, and in particular an attempt was made to clarify the meaning of the term 'paradigm'. He gives up the idea of paradigms being anything like overarching world-views, and the history of science is no longer presumed to be riven from time to time by huge discontinuities. Even the number of people adhering to a particular paradigm at any one time shrinks to about a hundred or less, for Kuhn's thesis now refers to micro-communities, and the grand revolutions of 1962 are now but micro-events, which may be occurring more or less regularly.

With such watered-down claims, the problems of incommensurability begin to evaporate, and the accusation of irrationalism is specifically repudiated. If paradigm shifts are only relatively small-scale affairs, adherents of opposed views can 'recognize each other as members of different language communities and then become translators'.[26] Thereby, they can engage in rational discussion. So Kuhn has really climbed down on the radical incommensurability thesis — which is gratifying, since it always seemed somewhat implausible that the problem between, say, the phlogistonists and the antiphlogistonists was that they simply didn't *understand* one another, though it would be equally unreasonable to suppose that there is no language problem, or mutual incomprehension, between members of different social groups, in science any more than any other kind of community.

Kuhn's revisions also involved a splitting of the meaning of the old term, paradigm, with the coining of two new ones: '*disciplinary matrix*' and '*exemplar*'. The first of these refers to shared group commitments, or 'the entire constellation of beliefs, values, techniques, and so on, shared by the members of a given community'.[27] The processes of socialisation are involved in the acquisition of such commitments, and the 'disciplinary matrix' is that social and cognitive structure in which a would-be scientist must embed himself if he or she is to become an accredited practitioner of science and produce what is sometimes called 'certified knowledge' — that is, papers published in scientific journals, which carry the seal of approval of the scientific community.

'Exemplars', on the other hand, are 'concrete puzzle-solutions which, employed as models or examples, can replace explicit rules as a basis for the solution of the remaining puzzles of normal science.'[28] This notion has survived without so very much change from the first edition of *The Structure of Scientific Revolutions*. An exemplar is provided by a successful piece of scientific research (using some particular theory or theories, experimental techniques, etc.), which is then used by others as a model for their own work. For example,[29] a particular sequence of rocks was suggested as a general stratigraphical arrangement by the eighteenth-century mineralogist A.G.Werner (1749–1817) as a result of his investigations in Saxony. This work (and Werner's empirical technique) was used as a model or exemplar by his students such as Robert Jameson (1774–1854), who attempted to describe the rocks of Scotland according to what he had learnt from his teacher's theories and techniques in Germany. The example that Kuhn himself gives is that of the Newtonian equation $F = ma$, used as an exemplar for solving different mechanical problem situations, such as an object in free fall, the motion of a pendulum, or the behaviour of coupled harmonic oscillators.[30]

The way in which research exemplars are used in science seems to be rather similar to the way in which a student uses the worked examples in a text-book for solving other problems at the end of the chapter. This, implies Kuhn,[31] is essentially the way in which the research scientist operates under 'normal' conditions in the process of 'puzzle solving'. The student (or fully-fledged scientist) learns how to solve problems by applying the exemplar to the new situation, treating the matter first symbolically and then experimentally. In so doing, he or she comes to perceive a similarity or resemblance relation between the experimental set-up and the paradigmatic exemplar. It is in this way, thinks Kuhn,[32] that a 'hook' is forged between theory and reality. It is not achieved by the use of selected correspondence rules, as the upholders of the 'received view' of the structure of scientific theories formerly supposed.[33]

I may say here that I find the concepts of disciplinary matrix and exemplar useful so far as they go, but in themselves not sufficient to deal with the chief problems in philosophy of science. In talking about disciplinary matrices, Kuhn is underscoring the fact that there is a strong social dimension to scientific knowledge, and he is giving some of the several components of that social dimension. But in itself this does not take us very far into the sociology of knowledge. As for the notion of exemplar, I accept that a great deal of research is carried out according to examples provided by other successful pieces of work. And the process is, no doubt, somewhat analogous to that which occurs when a student works through some of the problems (puzzles) in his or her text-book. But, I do not see how this account will serve for the creative work that goes on when radically new theories are propounded — where there are no existing models of research to work from. Kuhn may account in some degree for the functioning of 'normal science', but his revolutionary episodes remain somewhat mysterious. Further, I do not see why there should not be non-mathematical exemplars (as in my example from the history of geology above), in which the process of application would be something other than the application of the symbolic representation of the exemplar to new experimental problem situations. (Actually, Kuhn gives an illustration of his ideas by means of an imaginary process in which a child learns the meanings of bird words in a visit to a park. So Kuhn does not perhaps stipulate that the articulation of each paradigm always involves mathematical/physical symbolism.) More importantly, I do not see that the use of paradigmatic exemplars resolves the problem of the relationship between theory and reality. For this, one needs something else altogether, such as a coherence or a correspondence theory of truth, difficult though it may be to give such theories satis-factorily. If Kuhn's description of science is to be taken seriously, it would seem that at best he can only offer us an account of *why* we believe there is correspondence between theory and reality. He doesn't provide any *warrant* for such beliefs as we may come to hold as we perform our work under the aegis of some paradigm. *However*, for one who sees knowledge as essentially social, there might seem to be no need for any such warrant: all knowledge would be 'relative' to the social formation within which it is generated. Kuhn tells us how 'corres-pondence rules' come to be constructed by the student or researcher and he could well say that he is under no obligation to guarantee that they are well founded.

Yet Kuhn is anxious to deny the accusations of relativism that have been levelled against him. He doesn't want to suggest that one paradigm is just as good as another, or that no progress is made in science as one paradigm gives way to another. More 'advanced' theories and methods may solve problems more successfully than others. They may be superior in terms of 'accuracy, simplicity, fruitfulness, and the like'[34] or 'internal and external consistency'.[35] But Kuhn's system has nothing equivalent to Popper's doctrine of verisimilitude. So, for Kuhn, a theory's success in 'puzzle solving' cannot be taken as a mark of its truth-likeness. He doesn't want to claim that progress through scientific revolutions brings scientists nearer 'the truth'. In a sense, then, his system has quite strong instrumentalist features.

Thus, while Kuhn's modified account of the structure of scientific revolutions certainly seems to bring it closer to what historians of science will recognise as the history of science, and it rightly emphasises (and tells us something about) the social dimension of science, it does not, I think, provide a complete solution to the age-old problem of the relationship between theory and reality. The thesis of Kuhn$_2$ is somewhat less interesting than that of Kuhn$_1$ (being less provocative!); and neither of them seems to have the full and final story. But Kuhn has certainly left his mark in metascience. No student of the subject can ignore his insights, and the vocabulary that he has bequeathed to us is now widely used, even by those who do not subscribe to Kuhn's theses.

As we have seen, an important feature of the philosophy-of-science landscape in the 1960s was the debate between Kuhn and Popper and their respective disciples, Popper wringing his hands at the advent of Kuhnianism, with the complaint that it seemed to undermine all that he held dear in the world of metascience — mob rule and the imperatives of the bandwagon superseding the cool-and-calm critical discussion that Popper so admired and held to be the foundation of Western democratic society. Given this deep split within the structure of English-language metascience, it is not surprising that someone should have tried to find a middle way between the two opposing positions. This task was attempted by Imre Lakatos, whose work we shall now consider.

Lakatos

Lakatos (1922–1974) was a Hungarian by birth, and in his early years was a member of the anti-Nazi resistance movement. For a time after the war he was a high-ranking official in the Hungarian Ministry of Education, but was gaoled for three years for alleged 'revisionism'. He managed to move to the West after the 1956 Hungarian uprising. Going to Cambridge, he wrote a Ph.D. thesis entitled 'Essays in the Logic of Mathematical Discovery', which was subsequently published in the *British Journal for the Philosophy of Science*.[36] The work took the unusual form of an imaginary discussion between a teacher and a group of students on the question of the proof of the Descartes/Euler conjecture that for all polyhedra the number of corners *minus* the number of edges *plus* the number of faces is equal to two. The actual history of the mathematical discovery was recounted in the bulky footnotes — Lakatos's rather idiosyncratic way of doing history and philosophy of science.

Later, Lakatos moved to the London School of Economics, where he came under Popper's influence, and eventually occupied a chair. His most widely read paper, entitled 'Falsification and the Methodology of Scientific Research Programmes',[37] has excited considerable interest and comment and has served as the basis of a number of historical investigations. Lakatos's untimely death in 1974 was widely regretted, and there seems little doubt that had he lived he would have made many more significant contributions to the literature of metascience. Even so, his writings were quite numerous; but we only need focus attention on 'Falsification and the Methodology of Scientific Research Programmes'.[38]

Before discussing Lakatos's doctrines, let us remind ourselves of a point made in our chapter on Popper. The Duhem thesis tells us that the falsification of a hypothesis can never be made with certainty for hypotheses always come in clusters, so to speak; so the arrow of *modus tollens* can never be directed infallibly towards a single hypothesis. Further, a theory (or hypothesis) in difficulties can always be 'saved' by bringing in additional auxiliary hypotheses. According to Popper, this is permissible if a new auxiliary assumption (or assumptions) *increases* the number of observational consequences. That is, it must increase the 'empirical content' of the theory. If it does not, the auxiliary hypothesis is regarded as reprehensibly *ad hoc* and is disallowed according to Popper's preferred methodological rules.

It was chiefly this point that Lakatos took up in order to develop a description of the 'dynamics' of theories. That is to say, he sought to analyse not only the structures of theories and the way in which they may be falsified, but also the processes whereby one theory (or hypothesis) gives way to another in a gradually evolving 'research program.' In his examination of the problem, Lakatos introduced a number of new terms, which are, unfortunately, not particularly felicitous, but which we must seek to explain nonetheless. A *'research program'* consists of a sequence of researches, carried out by one or more investigators, in which certain methodological rules are adhered to. In fact, Lakatos referred to the rules as if they were the program itself, thereby conflating, as it were, the car-driver's manual with the car, the driving of the car, or the journey that it makes. He wrote:

> [A] programme consists of methodological rules: some tell us what paths of research to avoid (*negative heuristic*), and others what paths to pursue (*positive heuristic*).[39]

What does this mean? In normal parlance, the word 'heuristic' means 'serving to find out'. It is that which facilitates discovery. The word (which was coined by Whewell for philosophy of science) is most commonly used in reference to systems of education in which pupils (supposedly) find things out for themselves. So according to Lakatos, each research program is characterised by two sets of rules, one set telling the researcher what paths of research to avoid, the other telling him or her which to follow. Presumably, then, the application of such rules will facilitate discovery.

Lakatos further noted that the hypotheses or theories of a research program are not all of equal status. Some are treated as sacrosanct, so to speak. Or in Poincaré's language they are conventionalised; for they are deliberately placed beyond the reach of the arrow of the *modus tollens*. Others, however, are accepted as being susceptible to modification and change, and *will* be so modified and elaborated as the research program evolves. The inviolate cluster of hypotheses at the heart of the research program Lakatos called the *'hard core'*.[40] The peripheral hypotheses, subject to change and modification, he called the *'protective belt'*.[41]

The 'negative heuristic' of the program stipulates that the assumptions of the 'hard core' are not to be questioned or altered. ('Hands off the "hard core"', one might say.) If the 'hard core' is altered, then one has given up one research program and changed to another; or in Kuhn's language one has changed paradigms. The 'positive heuristic' of the program, on the other hand, is made up of various methodological guidelines saying how the program is to be developed. Or in Lakatos's inelegant prose:

> [T]he positive heuristic consists of a partially articulated set of suggestions or hints on how to change, develop the 'refutable variants' of the research-programme, how to modify, sophisticate, the 'refutable' protective belt.[42]

Presumably, one learns the 'art' of working according to a research program in one's period as a research student, when one is 'inducted' into the 'game' of scientific research. The work on the research program — at the level of the 'protective belt', not the 'hard core' — may be compared to the processes involved in 'puzzle solving' in Kuhn's periods of 'normal science'.

To illustrate all this in a simple way, one can consider the Ptolemaic theory in the history of astronomy. We can interpret the geocentric hypothesis, and the hypothesis of the necessity of circular motion on the part of heavenly bodies, as elements of the 'hard core' of this research program. The details of the various epicycles and deferents[43] constitute the 'protective belt', and it is the task of the Ptolemaic astronomer to devise the various geometrical figures in such a way that the appearances are satisfactorily saved.

Now according to Lakatos, research programs may on different occasions be developing in such a way that they are either *progressive* or *degenerating*.[44] If a research program is going well, leading to the discovery of new phenomena which are successfully accounted for in terms of the program's various hypotheses, then we have a 'progressive *problem shift*',[45] for the fresh assumptions made in the 'protective belt' increase the empirical content of the theory (or the program). But in less favourable circumstances, when the program is running out of steam, the new hypotheses that have to be added are *ad hoc*. They allow the aberrant observations to be accounted for, and they save the presuppositions of the 'hard core'; but they do not allow the prediction of new testable phenomena. That is (in Popper's language), they do nothing to improve the empirical content of the theory. In such circumstances, the research program is said to be undergoing what Lakatos called a 'degenerating problem shift',[46] and it may well be superseded if some alternative, more attractive and 'progressive' program is available or can be constructed. However, a superseded theory may always be resuscitated. Lakatos quoted the case of the so-called 'Prout's hypothesis'[47] which emerged like a phoenix from the ashes at the beginning of the twentieth century, with the suggestion that different elements were made up of different numbers of basic particles of charged matter, namely electrons and protons (and later, neutrons and many other strange beasts).

It will be seen that Lakatos's model is quite a subtle synthesis of the ideas of Popper and Kuhn. The Kuhnian notion of the paradigm has been transformed into the idea of the 'hard core', and the rules of the game ('positive heuristic') that obtain in the 'protective belt'. And clearly there is a strong 'dynamic' element in Lakatos's model of science of which Kuhn might approve. On the other hand, the Popperian notions of conjecturing, testing, probing, and falsifying are also present. Lakatos has captured the important Popperian concern with the 'ad hocness' of theories and the notion of empirical content. Yet the propensity in science towards conventionalism is given due weight. Also, Lakatos's system is certainly not inductivist.

The Lakatos model of science has the special advantage over Popper's that it gives reasons why particular hypotheses may be accepted in the early stages of a research program, even though it may be known at that time that they actually conflict with detailed experimental evidence. For example, Newton (according to Lakatos[48]) first worked out his gravitational theory by considering a point planet moving around a stationary point Sun in an elliptical orbit. Then he considered the Sun and planet revolving around a common centre of gravity. Then he allowed for the gravitational effects of the other planets in the solar system. Then he considered the planets not as point masses but as spheres. Then he regarded them as spinning objects with certain wobbles (due to their spheroidal shapes). And only at about this stage did he begin to look seriously at the relationship between theory and observations. But when the observational tests did not yield exactly what was expected, Newton refined his model, rather than throwing it away as naive falsificationism would require. So gradually theory and evidence were brought into better and better fit with one another.

This whole process constituted a 'progressive problem shift', since the increasingly precise predictions were supported by further observational evidence. The theoretical modifications made were not *ad hoc*, since they led to further predictions, which were themselves found to be successfully supported by further tests and observations. Empirical content increased rather than decreased.

All this seems very fine, and at first blush, Lakatos's model of science seems rather attractive, even though what constitutes the 'positive heuristic' of a research program seems hopelessly vague. He appears to allow conventionalism limited play (at the level of a program's 'hard core') and there is a certain realist tenor to Lakatos's argument, even though he does not discuss the realist/instrumentalist issue specifically. However, Lakatos's model has been subjected to damaging criticism by Alan Musgrave (a former student of Popper), and we may look briefly at some of the arguments that he has advanced.[49]

The first point to note is, perhaps, that an examination of the history of science shows startling weaknesses in Lakatos's doctrines of 'hard core' and 'negative heuristic'. For example, if we accept the history of Newtonian celestial mechanics as an exemplar of scientific research programs, exhibiting a 'progressive problem shift',[50] we find that there was little agreement amongst Newtonians as to what *was* the 'core' of the Newtonian program. Thus in the case of the observations of the motion of Uranus, which eventually led to the discovery of Neptune, the astronomers Airy and Bessel were willing to contemplate modifications to Newton's inverse-square law of gravity to account for the discrepant observations, whereas Adams and Leverrier preferred to suppose that some hitherto unseen massive body was responsible. Subsequently, some difficulties were found with the observed motion of Mercury — observations which were eventually accounted for in terms of Einstein's relativity theory. But the edition of the *Encyclopaedia Britannica* of 1910 thought a possible explanation might be that gravitational forces of attraction might be inversely proportional not to the squares of distances, but to the power 2.0000001612![51] And, as William Berkson has pointed out,[52] though it may be true that any individual scientist holds to a theoretical 'core' that is hard and sacrosanct to him, different scientists may have substantially different views as to what is and what is not 'hard'. The historical evidence seems to be, then, that scientists do *not* always, as a group, choose a certain portion of their theoretical beliefs and render it unfalsifiable by fiat. Or if there is a 'hard core' it may emerge only some time *after* the period when it might have been methodologically useful and the subject of some 'negative heuristic'.

What, then, of the notion of 'positive heuristic', which I have referred to above as 'hopelessly vague'? Musgrave construes Lakatos's doctrine of 'positive heuristic' as a means whereby scientists can 'anticipate' empirical refutations of a research program during the course of its evolutionary development. The 'hints' of the 'positive heuristic' 'save the scientist from becoming confused by the ocean of anomalies'[53] and will (somehow) direct him how to conduct his theoretical and empirical investigations in a fruitful way. In this manner, we are to presume that Newton was guided by a 'positive heuristic' as he gradually refined his theory and brought it into closer correspondence with observations.

So Musgrave suggests that the 'positive heuristic' is not some strategy laid down in advance for 'producing and digesting empirical refutations' but rather (for the Newton example) 'a strategy for solving, by a method of successive approximation, the difficult mathematical problem of calculating what Newton's theory asserts about planetary motions'.[54] But this is still very imprecise, and one might wonder indeed whether Newton *himself* ever had any consciously or unconsciously held 'hints' in his mind as to what he was to do next in his investigations. Certainly, if there were such he must have invented them himself and did not receive them from some already existing research program. Thus Lakatos's notion of 'positive heuristic' tells us little about what a man of genius has to do in order to carry through his investigations successfully — though admittedly it would be an extraordinary thing if a metascientist could give an account of such matters satisfactorily.

After Newton, however, when Newtonian science had become 'normal' in Kuhn's sense, then certain 'hints' that might be said to constitute a 'positive heuristic' could be and were identifiable. For example, Musgrave suggests, an important thought that the Newtonian astronomer might constantly bear in mind could be: 'Blame anomalies upon the disturbing effects, either mechanical or gravitational, of masses not hitherto taken into account'.[55] So in general Musgrave believes that the 'positive heuristic' is not something definitely laid down in advance. Rather, it may develop in the course of a research program's evolution. Musgrave thinks that in cases where an elaborate 'positive heuristic' appears to have been laid down in advance (as in the work of Bohr and the theory of the hydrogen atom) one is really dealing with an extended theoretical analogy, which is gradually explored in the course of the investigation.

Now a major difference between Popper and Lakatos is that the former is concerned with the proposal, comparison and testing (to destruction) of theories, whereas the latter thinks that the chief point of comparison lies between two competing research programs. The problem for Lakatos is how one program gets preferred at the expense of another. Basically, programs that are 'progressive' are to be favoured at the expense of those that are 'degenerating'. And what is 'progressive' or 'degenerating' is to be determined in a quasi-Popperian way by considering whether the new hypotheses brought in from time to time are or are not *ad hoc*; or by considering whether the new hypotheses add to or detract from the program's falsifiability and 'empirical content'.

Curiously, however, in a paper published in 1971, Lakatos seemed to draw back from his earlier position, saying that whether a research program was progressing or degenerating should not be taken as a directive to a would-be investigator, stipulating that he must necessarily prefer the 'progressive' alternative.[56] The only methodological advice that Lakatos then felt able to give was that scientists should be honest about the relative merits of the competing programs, and should keep a public record of the known anomalies and inconsistences of each. Again, we get the kind of moral exhortation that is a common feature of the Popperian school. But there is, nevertheless, a retraction of some of Lakatos's earlier methodological zeal. The relative merits of different programs may be appraised; but we are still told little, ...i general terms, of what the 'positive heuristic' of a program actually is, and how it should be applied.

The difficulty lies partly in the fact that a choice between two competing research programs is often no more clear-cut than is the choice for an orthodox Popperian between two competing theories, given the problems of the Duhem thesis and the difficulties involved in assigning empirical contents, degrees of corroboration, verisimilitude, and so forth. And given that the fortunes of different programs may wax and wane very considerably, there seem to be no general rules that can be offered which will allow one program to be preferred, firmly and securely, over another. This does not mean, of course, that it is impossible to give *some* sensible advice. Obviously one can. But if it is merely the best advice that can be given, and not necessarily sound, then we could (so far as logical worries need be concerned) unpack the whole Popperian/Lakatosian metascientific program, forget about Hume's problem and go back to inductivism.

Musgrave's way out of this difficulty has been to accept Kuhn's recognition of the importance of the scientific *community*, as well as that of the individual scientist. Then preference for one program or another might be expressed at the level of the community. This would not preclude individual scientists from working on what may, at a particular moment, seem to be an unsatisfactory and degenerating program. So Musgrave is unwilling to give up the idea that rational choice may be made between competing programs in a broad and general way, even if it should not be pressed upon individual investigators. He further offers some attractive illustrations of what the 'positive heuristic' might actually have looked like in the late eighteenth century for chemists working under the aegis of Lavoisier's new chemical paradigm.[57] It looks like the *kind* of advice that might be given by a research supervisor to a research student embarking on a new research project or entering into one that is already in full swing. In particular, it suggests particular *problems* that are in need of investigation, and which might have a fair hope of success, under the research program then under way. However, it is clear from Musgrave's example that each program would have its own quite idiosyncratic 'positive heuristic', and I do not see how any general account of 'positive heuristics' can be given that would cover all cases.

There is in Lakatos's work an acknowledgement of Kuhn's emphasis on the social dimension of science. Further, as we have seen, there is in Lakatos's later work a kind of admission that no particular research program is unambiguously to be preferred as compared with another. Indeed, there may be advantage in having more than one program running simultaneously, and having even the most 'degenerate' of programs kept ticking over. But if this be granted, we find ourselves moving towards the views of Paul Feyerabend and his doctrine of 'methodological anarchism', considered below. We have previously noted that Kuhn has been accused of seeing science as an essentially irrational enterprise, and of giving no weight to the gradual approach of scientific knowledge to 'the truth'. As I have said, Kuhn himself has sought to repudiate the charges of irrationalism and relativism, and has watered down his incommensurability thesis. Some recent philosophers of science, however, have taken up the idea of science as a relativistic, nay irrational, enterprise with considerable enthusiasm, and have sought to support their interpretations by appeal to the history of science. In such writings, it would seem, the 'arch of knowledge' has altogether collapsed. So at the end of our story we find ourselves describing a condition of intellectual anarchy rather than robust strength. Let us be not deterred, however, but have a look at some of these recent strange developments in metascience.

Feyerabend

I am thinking here particularly of the work of Paul K.Feyerabend (1924–), whose ideas have certainly enlivened recent philosophy of science, outraging some critics and delighting others. Feyerabend's early writings were relatively conventional compared with his two recent books, *Against Method*[58] and *Science in a Free Society*.[59] In 1962, for example, he published a paper entitled 'Explanation, Reduction and Empiricism'[60] and in 1963 one entitled 'How to be a Good Empiricist: A Plea for Tolerance in Matters Epistemological'.[61] These were concerned with standard issues in philosophy of science. Yet on looking back on them now, one may discern many hints of the iconoclasm that has since made Feyerabend's work so notorious, and in particular there are adumbrations of his 'incommensurability' doctrine.[62] In the 1960s, however, it would have been difficult to foretell the rather extraordinary direction that his work was later to take.

Nevertheless, when we look at some of the details of Feyerabend's career,[63] the remarkable path that his work has followed in recent years may not seem so surprising. After the Second World War he was in Weimar with a State Fellowship at the Institute for the Methodological

Renewal of German Theatre. Leftist plays such as those of Brecht were performed, and after the performances the audience used to discuss and evaluate what they had seen. After a year in Weimar, Feyerabend became a history student in Vienna. But he also studied physics and astronomy, and attended philosophy lectures, becoming a founder member of a philosophy club, the 'Kraft Circle', under the leadership of Victor Kraft, a former member of the Vienna Circle. Meetings were held, among other places at the village of Alpbach in the Tirol. 'Here', says Feyerabend:

> I met outstanding scholars, artists, politicians and I owe my academic career to the friendly help of some of them. I also began suspecting that what counts in a public debate are not arguments but certain ways of presenting one's case. To test the suspicion I intervened in the debates defending absurd views with great assurance. I was consumed by fear — after all, I was just a student surrounded by bigshots — but having once attended an acting school I proved the case to my own satisfaction.[64]

Thus Feyerabend's dramaturgical instincts were brought into play, leading him to comprehend a useful social truth and lay a foundation stone for some of his later intellectual eccentricities.

In Vienna, Feyerabend also met the maverick physicist, Felix Ehrenhaft,[65] and was impressed with his willingness to take up unorthodox positions, quite at odds with those of paradigmatic physics. Apparently, Feyerabend's respect for the rationality of science suffered at Ehrenhaft's hands, though he did not at that time come to regard it as an enterprise that depended for its progress on acts of irrationality. Feyerabend also at that time came in contact with the distinguished physicist Philipp Frank (formerly of the Vienna Circle), and with various Marxist intellectuals.

Besides these contacts, Feyerabend met and was influenced by the British philosopher, Elizabeth Anscombe, who had come to Vienna to learn German for her translation of the works of Wittgenstein. She and Feyerabend discussed Wittgenstein's ideas in detail, and Feyerabend became persuaded, from consideration of Wittgenstein's later philosophy, that certain generally accepted principles may change from one generation to the next; and they may be substantially different for different languages and cultures. Feyerabend tells us that he conjectured that:

> principles ... might change during [scientific] revolutions and ... deductive relations between pre-revolutionary and post-revolutionary theories might be broken off as a result.[66]

Thus he was feeling his way towards the doctrine of 'incommensurability'.

Popper and Feyerabend met at Alpbach in 1948 and the latter was greatly impressed. He recalls that in his discussion group falsificationist philosophy of science was quite taken for granted, and he was at a loss to know why it occasioned such a stir.

In the 1950s, Feyerabend worked with Popper in England and obtained a lectureship in philosophy of science at Bristol, where he extended his previous studies of quantum theory. On his thoughts at this period, he has subsequently written with his accustomed irreverence:

I found that important physical principles rested on methodological assumptions that are violated whenever physics advances; physics gets authority from ideas it propagates but never obeys in actual research, methodologists play the role of publicity agents whom physicists hire to praise their results but whom they would not admit access to the enterprise itself.[67]

From 1958, Feyerabend held a philosophy chair at the University of California. Again with his tongue out at the usual social norms, he has written:

My function was to carry out the education policies of the State of California which means that I had to teach people what a small group of white intellectuals had decided was knowledge. I hardly ever thought about this function and I would not have taken it very seriously had I been informed. I told the students what I had learned, I arranged the material in a way that seemed plausible and interesting to me — and that was all I did. Of course, I had also some 'ideas of my own' — but these ideas moved in a fairly narrow domain (though some of my friends said even then that I was going batty).[68]

In the 1960s, Feyerabend inevitably became caught up in the student protest movement at Berkeley, and he became interested in the so-called alternative society and the ideas and ideals of non-European cultures and races. 'These cultures', he writes:

have important achievements in what is today called sociology, psychology, medicine, [and] they express ideals of life and possibilities of human existence. Yet *they were never examined with the respect they deserved* except by a small number of outsiders, they were ridiculed and replaced as a matter of course first by the religion of brotherly love and then by the religion of science or else they were defused by a variety of 'interpretations'.[69]

All this, as we might expect, Feyerabend thoroughly deplored and to offer some account of it, he set out to examine the rise of 'intellectualism' in Ancient Greece, and its subsequent alleged baneful influence on Western culture. At the same time he began to re-examine the Dadaist school of painting and the theatre of the absurd.

Later, in England, Feyerabend was closely associated with the work of Lakatos, and entered into a long-running intellectual debate with him. But Lakatos died prematurely, so their intended joint book on 'rationalism' never appeared. What we do have, of course, is *Against Method*; but this only offers the anti-rationalist part of the argument.

I mention these biographical details in relation to Feyerabend somewhat expansively, since without them his views might seem even more outlandish and unintelligible than would otherwise be the case. Also, by seeing thus briefly some of the major elements of his social/ intellectual life, we may gain some preview of the sociology of knowledge hypothesis, which is discussed in the closing part of this chapter. Feyerabend — himself a subscriber to sociology of knowledge doctrines — provides through his own life and work a rather pleasing witness to their plausibility.

Feyerabend begins *Against Method* by telling us that he is engaged in an *'anarchistic'* enterprise. He wants to argue — contrary to Popper and the Popperians (or even Lakatos, no doubt) — that there is no privileged method of scientific inquiry which, if followed, will lead to the successful acquisition of knowledge. There are innumerable *different* methods, and each is worth trying. So, looking at the matter from the context of the present book, Feyerabend would say, I presume, that the whole notion of a (rational) 'arch of knowledge' is an illusion, for he favours 'methodological anarchy'. He acknowledges that anarchism may not be a very attractive political philosophy, but it is, he thinks, 'excellent medicine for *epistemology*, and for *philosophy of science*'.[70] In keeping with such a stance, Feyerabend describes himself as a Dadaist[71] — preferring the term to ·'anarchist', since he does not (he would have us believe) have the seriousness of purpose of a true anarchist.[72]

Noting, then, that according to his own prescriptions Feyerabend is not to be taken seriously, what is the 'argument' that is to be found in *Against Method*? First, of course, Feyerabend holds that science has no special method of its own that makes it a privileged form of activity, worthy of esteem because it can produce true knowledge. Indeed, in his view science cannot be regarded as a strictly rational enterprise; for when an important theoretical advance is made the new ideas are often 'irrational', judged by the canons of thought of the previous theoretical position. Thus progress in science may be dependent upon people thinking counter-intuitively, that is, at odds with previously practised norms of thought. (Using Kuhnian language, Feyerabend's view would be that the position of Paradigm$_2$ is irrational from the perspective of Paradigm$_1$.) Thus, for example, when Galileo urged the Copernican theory upon the Catholic clergy, it was *he* who was being irrational, not they (looking at the matter from the standpoint of traditional Aristotelian physics and cosmology).

In fact, much of Feyerabend's argument depends upon the particular historical case-study of the Galileo episode, so let us examine in a little more detail what he has to say on this. Aristotelian physics and cosmology were grounded in 'common sense'. The Earth does not appear to move: therefore it is at rest. To suppose otherwise is absurd — irrational. But that the Earth *does* move was what Galileo wished to persuade his readers. Consequently — according to Feyerabend[73] — Galileo had to resort to 'propaganda' and 'psychological tricks' in his writings. For example, Feyerabend quotes the following passage from Galileo's *Two Chief World Systems*:

> *Salviati:*.... |I|magine yourself in a boat with your eyes fixed on a point of the sail yard. Do you think that because the boat is moving along briskly, you will have to move your eyes in order to keep your vision always on that point of the sail yard and to follow its motion?
> *Simplicio*: I am sure that I should not need to make any change at all; not just as to my vision, but if I had aimed a musket I should never have to move it a hairsbreadth to keep it aimed, no matter how the boat moved.
> *Salviati*: And this comes about because the motion which the ship confers upon the sail yard, it confers also upon you and upon your eyes, so that you need not move them a bit in order to gaze at the top of the sail yard, which consequently appears motionless to you.[74]

By various illustrative arguments of this kind (which I personally would not want to describe as 'psychological tricks' or propaganda'), Galileo was able to persuade[75] his readers of the possibility of the motion of the Earth, and that a stone falls from the top to the foot of a tower is evidence *in favour* of the Earth's motion, rather than evidence to the contrary.

Another interesting and important part of Feyerabend's argument is as follows. According to Aristotelian doctrine, the laws of nature are different in different parts of the cosmos. Accordingly, it was believed that different physical laws obtained in the region above the Moon (the superlunary realm), as compared with the region below.[76] Now an important part of Galileo's argument for the Copernican heliocentric hypothesis was concerned with his telescopic observations made on the phases of Venus, the mountains of the Moon, and the satellites of Jupiter (the so-called Medician stars). Galileo had no difficulty in convincing people that his instrument (consisting of a tube, a convex and a concave lens) made distant terrestrial objects clearly visible, and the economic and military possibilities of the instrument were readily appreciated. However, given an Aristotelian world-view, the evidence that one might gather by the application of the telescope to the heavens was not necessarily accepted as relevant to the question of the Earth's motion, and hence to the Copernican hypothesis, given that the laws of the superlunary and the sublunary realms were believed to be essentially different. Moreover, some of the evidence gathered from the heavens with the help of the telescope appeared incongruous and inconsistant, even to Galileo. (For example, the Moon appeared magnified, but the stars did not.) Yet Galileo had to try to persuade the sceptics that the evidence provided by the telescope was relevant to the problem in hand, and supported the Copernican view.

So, for the Aristotelian it was *irrational* to accept the evidence vouchsafed by Galileo's telescope, and hence we can understand why traditionalists should have refused to look through the instrument at the moons of Jupiter.[77] Why should they waste *their* time looking at evidence that could not be relevant to the point at issue? In fact, according to Feyerabend, it was Galileo's clerical critics who were rational (according to their system of rationality[78]), not Galileo. On the basis of this kind of argument, then, Feyerabend wishes to claim that 'progressive' shifts in theoretical science, involving paradigm changes, entail what are in effect acts of irrationality. (Here, of course, I am thinking of paradigms in the sense of Kuhn$_1$, rather than Kuhn$_2$.) For a person is required to reach out from one knowledge system, so to speak, and grasp (or somehow create) another one that is quite at odds with the first in its most fundamental principles.

A number of important consequences flow from Feyerabend's argument, if it be accepted. If his thesis is correct — that progress in science is linked with acts of irrationality — then it may be said that no particular science, no particular form of knowledge, no particular methodology, no particular way of thinking, can claim any kind of privileged status. And this has important social consequences. Feyerabend presents his case as follows, considering the interactions between modern, scientifically based society and other, older and less sophisticated cultural systems:

The rise of modern science coincides with the suppression of non-Western tribes by Western invaders. The tribes are not only physically suppressed, they also lose their intellectual independence and are forced to adopt the bloodthirsty religion of brotherly love — Christianity. The most intelligent members get an extra bonus: they are introduced into the mysteries of Western Rationalism and its peak — Western science: Occasionally this leads to an almost unbearable tension with tradition ([as in] Haiti). In most cases the tradition disappears without the trace of an argument, one simply becomes a slave both in body and in mind. Today this development is gradually reversed — with great reluctance, to be sure, but it is reversed. Freedom is regained, old traditions are rediscovered, both among the minorities in Western countries and among large populations in non-Western continents. *But science still reigns supreme.* It reigns supreme because its practiioners are *unable to understand*, and *unwilling to condone*, different ideologies, because they use this power just as their ancestors used their power to force Christianity on the peoples they encountered during their conquests. Thus, while an American can now choose the religion he likes, he is still not permitted to demand that his children learn magic rather than science at school. There is a separation between state and church, there is no separation between state and science.

And yet science has no greater authority than any other form of life. Its aims are certainly not more important than are the aims that guide the lives in a religious community or in a tribe that is united by a myth. At any rate, they have no business restricting the lives, the thoughts, the education of the members of a free society where everyone should have a chance to make up his own mind and to live in accordance with the social beliefs he finds most acceptable. The separation between state and church must therefore be complemented by the separation between state and science.[79]

So, in accordance with such arguments, Feyerabend believes that the privileged position of science within the educational curriculum should be abandoned. 'Special creationism' should be taught along with evolutionary biology. Voodoo, witchcraft, astrology, acupuncture and moxibustion should all find a place in the curriculum; or, as the jargon of the curriculum planners has it, they should be offered as 'electives'. There should be complete freedom of choice as to the 'knowledge system' that one espouses, for there is no particular one that can rightfully claim a special place in the educational sun. All this seems to mesh rather well with the libertarianism and pluralism of contemporary Californian culture, where Feyerabend's ideas have come to their full flower.

It might be thought that the whole of society would collapse if Feyerabend's ideas were taken seriously and actually implemented. Needless to say, he thinks otherwise:

There will always be people who prefer being scientists to being masters of their fate and who gladly submit to the meanest kind of (intellectual and institutional) slavery provided they are paid well and provided also there are some people around who examine their work and sing their praises.[80]

In other words (I suppose), science and technology will carry on, some people being silly enough to keep them going, and there will be no great loss if others indulge themselves in non-scientific cultural forms; for all are equally 'valid' (or 'invalid').

How seriously are Feyerabend's arguments to be taken? Presumably not very, since he represents himself as a Dadaist (who is not to be taken seriously), and an exponent of irrationalism. If we accept that Feyerabend favours irrationalism (by the usual twentieth-century Western norms) then there isn't much point in drawing attention to what may seem (from our commonplace perspective) to be errors in his argument. He is, so to speak, in a 'heads he wins/tails you lose' position.

Personally, however, I think that Feyerabend does have a serious argument, and he is right in suggesting that there are many forms of rationality (or ways of reasoning) other than our own, some of which may be highly efficaceous within certain contexts or limits.[81] I also agree that when a new conceptual system has been constructed this system may appear 'irrational' from the perspective of a person still situated within the old frame of reference. (Thus Copernicanism may well appear irrational to one brought up under the aegis of the Aristotelian world view.) In addition, I would agree that 'progress' in science may well be enhanced by a freedom of thought and expression, and a willingness to entertain ideas that are not in accord with the received view, or the *status quo* of the prevailing paradigm. (This, in fact, accords with Popper's recommendation that one should always try to propose and test as many hypotheses as possible.)

However, I cannot accept that the argument in relation to Galileo's *Dialogue Concerning the Two Chief World Systems* used by Feyerabend is adequate to show that science is an altogether irrational enterprise. Galileo's *Dialogue* is essentially a work of scientific 'apologetics', rather than an account of scientific research as actually conducted. It falls within the scope of Reichenbach's 'context of justification' rather than the 'context of discovery'. For in the *Dialogue*, Galileo describes ideas that he had reached long before the book was published. And he was seeking to *persuade* people of the views that he had himself attained by his own complex pattern of reasoning and experimentation, rather than writing a direct account of his *own* original, intellectual thought proceesses. So while it may be true that for an Aristotelian to leap directly to Copernicanism requires an intellectual jump that one might, with some justification, refer to as 'irrational', this does not mean that Galileo's own intellectual pathway was itself irrational, complex though it undoubtedly was. For Galileo's intellectual journey did not involve a leap directly from Aristotelianism to Copernicanism and to the new kinematics. As can be seen if one examines his intellectual biography in detail,[82] Galileo appears anything but irrational in his mental activity. So by taking Galileo's 'apologetic' *Dialogue* as an indicator of his actual intellectual progress Feyerabend is indulging in something of an historiographical sleight of hand.

In my view, therefore, the social theories that Feyerabend educes from his historiographical investigations do not rest upon an adequate historical base. However, I do not suppose that he would find this argument specially persuasive. He rejects out of hand the possibility of making a distinction between the context of discovery and the context of justification. And this, he might want to say, is really the main point. His argument is destructive of positivism in that he suggests that there are no 'basic statements' to which independent observers may have recourse and on the strength of which they can unambiguously determine whether a theory is true or false. There is, it seems, more to theory acceptance or rejection than straightforward rational decision. A whole host of social and psychosocial factors may play a part. In that sense, Feyerabend's argument seems to have strength, using the Galileo texts as he does. Also, we should recognise that his advocacy of methodological pluralism and relativism in science has, in its own terms, some sense. Agreeing with the later Wittgenstein that no correspondence theory of truth is satisfactory, Feyerabend maintains that theories cannot be compared from some privileged vantage point where 'truth' is known (there being no 'basic statements'). However, he holds, two theories can be compared from the perspective of a third, thus avoiding commitment to either of the two being compared. Hence, a proliferation of theories may assist the 'progress' of science, even if all is relative and we have no recognisably progressive direction for science. Thus the argument would seem to run.

Actually, discussion of Feyerabend's work has usually followed a line of argument somewhat different from that of the previous paragraphs, and attention has been focussed chiefly on his so-called thesis of incommensurability. We have seen that this problem emerged in the writings of Kuhn and that in the first edition of *The Structure of Scientific Revolutions*, at least, the view was held that the thinking of a person working under the aegis of one paradigm was 'incommensurable' with that of a person working under another. The two scientists would, so to speak, 'talk past' one another, because the very terms used would have different meanings according to which paradigm was adopted. And totally different interpretations would be placed upon the same pieces of empirical evidence. As we have seen, in his later work Kuhn has drawn back from this thesis of radical incommensurability. But Feyerabend, it appears, would endorse it with enthusiasm. Indeed, he believes that 'progress' occurs in science when theoretical shifts take place from one frame of reference (paradigm) to another. And he seems to think well enough of science to regard this favourably, his relativism and irrationalism notwithstanding.

We find, therefore, that Feyerabend gives a good deal of attention to the incommensurability thesis in his *Science in a Free Society*, where he writes as follows:

> If ... theories are commensurable, ... we simply have an addition to knowledge [when scientific 'advance' occurs]. It is different with incommensurable theories. For we certainly cannot assume that two incommensurable theories deal with one and the same objective state of affairs (to make the assumption we would have to assume that both at least *refer* to the same objective situation. But how can we assert that 'they both' refer to the same situation when 'they both' never make sense together? Besides, statements about what does and what does not refer [*sic*] can be checked only if the things referred to are described properly, but then our problem arises again with renewed force.) Hence, unless we want to assume that they deal with nothing at all we must admit that they deal with different [conceptual] worlds and that the change (from one world to another) has been brought about by a switch from one theory to another. Of course, we cannot say that the switch was *caused* by the change... [But] we no longer assume an objective world that remains unaffected by our epistemic[83] activities, except when moving within the confines of a particular point of view. We concede that our epistemic activities may have a decisive influence even upon the most solid piece of cosmological furniture — they may make gods disappear and replace them by heaps of atoms in empty space.[84]

But is it the case that the positions within two competing paradigms *are* radically incommensurable? Is the situation *really* analogous to that which exists, for example, betwixt the theist and the atheist? *Is* a Kuhnian paradigm the kind of 'entity' that can be the subject of a quasi-religious adherence? *Can* allegiance to one paradigm be shifted to another only by some kind of religious conversion? To take an example from the recent history of science, is part of the reason that some geologists refuse to adopt the theory of continental drift because they cannot understand the *meaning* of the new theory from the perspective of the old? And is it an act of *irrationality* to propose a continental drift model when one has been brought up according to the old point-of-view? Or is it that the old-time geologists (actually mostly Russians at the time of writing) simply fail to accept the new theory because the evidence in its favour does not seem sufficient, or because they have been socialised (educated) in particular ways of thinking about and seeing the world? In other words, are we talking about a psychological/sociological question, which may have bearing on epistemological issues, or are we talking about the radically incommensurable meanings of terms when they are employed by scientists who hold to different theories or paradigms?

Such questions are not easy to answer, yet they are basic to Feyerabend's (and Kuhn$_1$'s) incommensurability thesis. They may be approached from the point of view of either the historian or the epistemologist. In *Against Method* Feyerabend's argument is chiefly historical, and he brings forward some attractive arguments in favour of the incommensurability thesis (including some most interesting discussions from Greek mythology which we do not have space to discuss here[85]). Yet, as I have suggested above, counter-arguments from the history of science may be found. It was, I believe, not an act of *irrationality* for Lavoisier to relocate phlogiston in the caloric of the new oxygen theory of combustion.[86] On the contrary, it was a very sensible thing to do, for it eliminated the need for the *ad hoc* hypothesis of phlogiston having a negative weight. The oxygen theory had (in Popperian language) a much better empirical content than the phlogiston theory. Lavoisier knew the meanings of terms in both the old theory and the new. So, too, did his opponents, though some of them objected to the new nomenclature that Lavoisier and his supporters suggested, on the grounds that if they adopted it they were, in that act, effectively committing themselves to the new chemistry.

But we may be dissatisfied with arguments from history, for who can really know what went on in a man's mind during the past? So can we fall back on an epistemological analysis of the problems involved? This is, naturally, the philosopher's approach. And in fact a good deal of attention has been directed this way, and the problem is still under active discussion. For example, in a paper published in 1973[87] Hartry Field has considered the problem of the denotation of theoretical terms and (denying the incommensurability thesis) has suggested that as a result of a scientific revolution a term may alter its 'partial denotation': [T]he set of things that is partially denoted after ... [the scientific revolution] is a proper subset of the set of things it partially denoted before [the revolution].'[88] Thus, any given term has a certain imprecision in its meaning, so that it may *partially* denote more than one kind of thing or more than one concept. Allowing this 'semantic haze', then, one can use the same term in two different theories, and with enough meaning in common to evade the 'paradox' of incommensurability. So in principle a person should be able to familiarise himself or herself with both kinds of meanings, though in practice this might be difficult on occasions.

In another paper published in 1975, Arthur Fine approaches the problem by settling for a rather restricted account of truth. Truth, he thinks, in matters empirical, is always approximate, not exact. Thus he is prepared to accept that the Kuhnian and Feyerabendian cases of incommensurability are *bona fide* cases of indeterminacy with respect to the sameness or difference of reference (that is, as to what the terms actually refer to). Nevertheless, Fine suggests, theories can hopefully 'pick out a portion of the world that satisfies the core theoretical principles at least approximately'.[90] Consequently, the 'worlds' of different theories may overlap at least to the degree that 'there is a correlation between the terms of the theories that makes correlated terms co- referential in each world of the region of overlap'.[91] Fine points out that because later theories develop from earlier ones there is more than sufficient reason to suppose that there is an overlap of meaning. (The discourse of a phlogistonist, we may suggest, would be intelligible to Lavoisier, for he had formerly held the phlogistic theory of combustion. Also, it would be incumbent on Lavoisier to make himself intelligible to his opponents, otherwise he would never achieve any support for his new theories.)

However, Fine's position may be thought to beg a number of questions. It would seem to suppose that there are, in fact, satisfactory methods for making 'logical and evidential comparisons' — which involves having some kind of neutral reference point, or area of agreement, for comparing variant systems of logic, or for appraising the empirical success or failure theories. I suggest, then, that there are problems still to be resolved satisfactorily in the matter of incommensurability; and this area is one of active inquiry in contemporary philosophy of science — for which we must acknowledge the important contributions of Kuhn and Feyerabend. But in view of the shifting ground in this corner of philosophical territory, we shall not seek to pursue these matters further at this juncture.

Some sociology of knowledge theorists

In several places in this book I have indicated that the philosopher of science would do well to examine science in its social aspects before embarking on any epistemological analysis. And indeed writers such as Kuhn and Feyerabend have been doing just this. But their work is not altogether typical of what has been done in this direction, and we can, with advantage, look at some of the other contributors to the field of sociology of knowledge, in so far as it has had reference to attempted descriptions of science and the characterisation of scientific knowledge.

The area of inquiry known as sociology of knowledge has chiefly German origins, growing from the work of writers such as Marx,

Nietzsche and Max Scheler, and in France from the work of Emile Durkheim and Marcel Mauss. There have also been inputs from areas such as Freudian psychology. But without tracing matters so far back, one can recognise the writings of Karl Mannheim (1893–1947) as of special significance in achieving recognition for the principle that knowledge is necessarily formed within the contexts of particular historical and social situations, and is thereby shaped by such historical and social contexts.

Mannheim did not mean simply that the social context influences how ideas come about, contingently and historically. His thesis was more radical, namely that social relationships influence the very *form* of thought. So epistemology itself is the product of social formations and varies accordingly from one epoch to the next.[92] Such views set Mannheim apart from his positivist contemporaries, who were seeking to establish a secure, lasting and solid basis for knowledge in science; and ultimately, as we have seen, they sought to ground it in logic — which was considered to be invariant — and in some incontestable bedrock of empirical experience, either of sensation or of things. Even so, Mannheim himself was not entirely liberated from the positivist attitude. For while he maintained that, in general terms, knowledge was socially and historically determined, an exception could be made for mathematics and the natural sciences[93] which were supposedly able to be free from what he termed 'existential determination'. It is the gradual erosion of this position that has characterised the development of sociology of knowledge in the period after the Second World War.

If the positivists could have had their way, the methods of the sciences would generate certain and secure knowledge. This had been the hope that had been held out by positivism since its earliest days. For example, when Comte's science had reached its third 'positive' stage, with the proper codification of scientific laws, the uncertainties of earlier 'theological' and 'metaphysical' eras would be gone and a secure basis for a 'scientific' form of society would be established. By concentrating attention on the empirically-determinable laws of nature, Comte hoped to achieve the elimination of the subjective elements of knowledge, producing thereby a condition of pure objectivity. This hope sustained later positivists, even though some of them, as in the case of Mach, sought to establish certainty in science through phenomenalism — which is not *prima facie* compatible with the supposition that man can acquire 'objective' (or inter-subjectively compatible) knowledge. Then in the twentieth century the logical empiricists, with the help of notions such as 'protocol sentences' and 'thing languages', aimed for scientific objectivity based on the applications of the results of logical analysis to the products of the empirical sciences. And even Popper, whom one might represent as a crypto-positivist (although he would, I think, reject the positivist label), has chosen the words 'objective knowledge' for the title of one of his most important books, and in his 'World 3' doctrine he seems to contemplate with enthusiasm the 'existence' of a 'world' of 'objective knowledge' which mankind may create, and also gain access to, through the methods of the empirical sciences. It is, for Popper, the existence of a privileged method for science that supposedly guarantees the objectivity of the product.

But suppose one were to take Mannheim really seriously, making no exemptions for mathematics and the natural sciences (perhaps not even for logic either). Suppose one were to agree that all knowledge is socially mediated, and thereby 'infected' with its historical, cultural and linguistic determinants. Obviously, the inclusion of mathematics and science within the ambit of the sociology of knowledge would sound the death-knell of positivism; all knowledge would be seen to be relativistic, transient, subjective, ... To be sure, this might seem to entail such uninviting paradoxes that one would be extremely reluctant to proceed in this direction. For it might be suggested that the thesis of a thoroughgoing sociology of knowledge (namely, that all knowledge is necessarily socially determined — that what *counts* as knowledge is mediated in or laid down by the society within which that knowledge is generated[94]) could be applied to the thesis itself — which need not therefore be taken too seriously. Is it not the case, then, that an unswerving adherence to the sociology-of-knowledge thesis involves one in a situation of anarchic relativism? This is certainly arguable, but it does not follow thereby that we can simply ignore the social and cultural factors in metascientific analysis. Indeed, any analysis of the dynamics of science is more or less forced to consider them.

There are, I suggest, several reasons why the sociology of knowledge has now come to occupy such a prominent position in metascientific dis-cussions. One is the growing 'power' of sociology itself as an academic discipline in the period since the Second World War. That social phenomenon in itself partly accounts for the matter.

A second relevant factor is, I suggest, to be found in the internal history of philosophy, and particularly in the trends of thought that are to be found in the 'later' Wittgenstein. To revise some points we have mentioned previously,[95] there were very significant differences between the Wittgenstein of the *Tractatus* and that of the posthumous *Philo-sophical Investigations*[96] (though there were perhaps more similarities than was at one time supposed). Wittgenstein came to doubt his account of language, as given in the *Tractatus*, according to the 'picture theory'. Rejecting his earlier view of language as having a kind of logical essence, which was structurally similar to a 'logical' world, he came to think that the important thing to consider is the way language is actually *used* in human life if we are to gauge its meanings.[97] For example, if I say 'Business is business', the meaning of this sentence (which on the face of it looks like a perfect tautology) cannot possibly be apprehended by examining its grammatical structure. Clearly it is not sensible to say that the essential logic of this sentence somehow parallels the 'logical' structure of the world. But see how the sentence is used in *practice*; then you will find out its meaning. Of course, the meaning of a word varies greatly according to its context. Again one finds out just how it does so by practice; there is no preferred or 'essential' meaning.

Now if we accept this argument, then we see that we do not find out anything about the 'logical' structure of the world by examining the structure of language (as the *Tractatus* might have led one to suppose)

So if the examination of language does not reveal reality as it is 'in itself', we can say that in a sense it *creates* reality for us — or what we regard as real.[98] And language, of course, is a social product. So the theses developed by Wittgenstein in the *Philosophical Investigations* had considerable affinities with those developed at about the same time by sociologists of knowledge.

Another important factor that has influenced the development of the sociology-of-knowledge approach to metascience has been the very considerable amount of empirical work that has, since the 1950s, been put into the study of science (or the scientific community) as a social system. Before the Second World War, there were certain historical studies that sought to show the way in which social factors shaped the course of the historical development of science.[99] And in the work of J.D.Bernal (1901–1971) there was a pioneering investigation of the social relations of science that is still highly regarded.[100] Since the war, there have been innumerable studies of the social aspects of science, and some journals such as *Minerva*, *Impact of Science on Society*, and *Social Studies of Science* have been founded with the specific intention of catering for interests in this area. I do not propose to give a general survey of this large field of literature[101] but will concentrate only on the findings that have emerged concerning the way in which scientific knowledge is 'processed' by the scientific community.

One thing can be said straight away that is almost entirely uncontroversial today, namely that science does not operate in some kind of cultural vacuum. In his autobiographical *Prelude*, William Wordsworth remembered his student days at Trinity, Cambridge, where he

> ...could behold
> The antechapel where the statue stood
> Of Newton with his prism and silent face,
> The marble index of a mind for ever
> Voyaging through strange seas of Thought, alone.[102]

But this was the romantic poet's vision of a scientist, not the reality. Even if it may have had some validity for Newton[103] it certainly has minimal applicability to today's conditions of 'big science'[104] where the boundary lines between science, applied science, and technology are so hazy that many commentators maintain that it is quite impossible to make any satisfactory distinctions between the three. And since technology forms the very basis of modern society, whether it be capitalist or socialist, it is obvious that science cannot be detached from its social context. The question, however, is whether the social dimension of science influences the very *form* of scientific knowledge — the very form of the equations of physical theory for example.

That this may be so seems more plausible when we consider that there is also the structure of the social community of science itself to be taken into account; and it is this *community* of science that authors such as those named in note 101 have specifically examined. One must first consider the processes whereby scientists are educated — which is in itself a very selective procedure such that certain ideas are granted

special status and approbation whereas others are ignored or treated
with disdain. Speaking in Kuhnian terms for a moment, the whole process
of scientific education for the would-be research scientist is one of accul-
turation towards the acceptance of a particular paradigm. Or if we like to
use the language of Michael Polanyi (1891–1976),[105] we can speak about
the 'tacit knowledge' of the scientist. By a process akin to that which
occurred in old-time apprenticeships, scientists gradually learn how to
conduct themselves within the scientific community. They learn what
kind of practices are acceptable (and what are not); they learn how to
carry out experimental or theoretical investigations successfully; they
come to know what kinds of problems are potentially solvable and
worthy of examination; and (Polanyi would have us believe) they learn
to have a kind of sixth sense as to what work is sound and reliable and
what is shoddy. Polanyi's arguments were presented from a somewhat
élitist standpoint (only scientists know what is good or bad science), and
were used in favour of arguments for the social autonomy of science, and
hence for favours to be conferred upon the scientific community by the
wider social system.[106] I would not wish to follow Polanyi in these latter
arguments, but his claims that scientists have commitments to the social
system within which they work, and that they have self-regulating
systems whereby they can 'maintain standards', seems to me to be uncontroversial. And the 'tacit knowledge' thesis (though somewhat obscure)
also has plausibility.

The notions of paradigms and tacit knowledge may seem more
plausible when one considers the system of 'social control' that operates
within the scientific community, chiefly through the mechanism of the
'peer-group review system'. All scientific work is 'vetted' by referees
before publication in scientific journals, [107] and even after publication the
process of assessment and appraisal continues, through reviews, annual
reports, etc., and the attempted replication of experiments where this
seems to be necessary. By this lengthy and complex process of 'sieving',
'certified knowledge' is gradually established, on which the scientific
community comes to feel it can place reliance.[108] To be sure, this social
process of sieving material to produce 'certified knowledge' is known to
be imperfect.[109] But that it occurs is certainly not in question. This point
being granted, its consequences for epistemology need to be considered.

Sociologists of knowledge such as Berger and Luckmann, who are not
specifically concerned with the applications of their work to science,
have emphasised that the world, *as it is known to us*, is a socially constructed reality.[110] Such authors are chiefly concerned with the social
reality of everyday life — the world as it is known and appears to us
through our day-to-day 'negotiations' with other people. They argue that
we gradually formulate a whole cosmology against which our social
interactions take place, and according to which they are 'legitimated'.
The social system also 'legitimates' certain ideas and seeks to exclude
others. Questions of power play an important role in all this.

Such arguments would seem to suggest, in a general way, that *all* we
know is necessarily socially mediated: there is no way in which we can

transcend our social circumstances and know the world as it is in itself.[111] The point made is somewhat similar to that made by Kant, long before. Kant, we have seen, maintained that we cannot transcend the categories according to which we think, and somehow apprehend the noumenal realm. The argument of the sociologist of knowledge is analogous in that the possibility of noumenal knowledge is denied. However, there are, on this view, no fixed categories of the understanding. There are 'frames' by which, or according to which, we see the world and seek to make our way in it; but the 'frames' are socially determined and they can vary from one social system or sub-system to the next, or from one epoch to the next.

The position of the sociologist of knowledge is obviously highly relativist from an epistemological point of view. There is no privileged or preferred frame which can claim superiority over all others; so all knowledge is relative. We can have no access to 'truth' in any kind of objective or absolute sense. The important point to consider, then, is whether this account of knowledge can find application to the form of knowledge that is generated by science. Or is it somehow possible that science — by its special experimental methods, by its peer-group review procedures, by its utilisation of mathematics and logic, and so on — can transcend the epistemological relativism that the sociology-of-knowledge thesis would seem to require, achieving thereby a condition of 'objective knowledge', such as that to which Popper would aspire?

One thing is certain. No matter what one's response to the epistemological question just posed may be, there is no reason why the system of science — the scientific community — should be exempt from sociological analysis. It may have appeared to Mannheim, writing as he did in the positivist era, that such an exemption was permissible. But the many able sociological investigations that have been made of science and the scientific community in the past twenty or thirty years amply demonstrate that there is no need for a 'hands-off' attitude towards science. Nevertheless, while it may be a fairly straightforward matter to examine the way the social system of science functions, such inquiries do not in themselves solve our epistemological quandaries.

Well, what *counts* as scientific knowledge is, it would seem, what the scientific community sanctions, through its various journals, reviews, text-books, and so on. So there is at least a *prima facie* argument that there is a social component to scientific knowledge, for it emerges from within a social context, as do other forms of knowledge. But is scientific knowledge 'nothing but' a social construct? Most of us might think otherwise; but before closing our minds on the question let us consider two recent texts that bring us rather close to a 'nothing but' thesis: the work of David Bloor in his *Knowledge and Social Imagery*[112] and of Bruno Latour and Steve Woolgar in their *Laboratory Life*.[113]

Bloor calls his thesis the 'strong programme' in the sociology of knowledge, according to which knowledge is not 'true belief', but 'whatever men take to be knowledge'.[114] He gives four characteristics of his 'strong programme':

1. It ... [is] causal, that is, concerned with the conditions which bring about beliefs or states of knowledge ...

2. It ... [is] impartial with respect to truth and falsity, rationality or irrationality, success or failure. Both sides of these dichotomies ... require explanation.

3. It ... [is] symmetrical in its style of explanation. The same types of cause ... explain ... [both] true and false beliefs.

4. It ... [is] reflexive ... [I]ts patterns of explanation have to apply to sociology itself.[115]

Bloor is not a full-blown 'nothing but' theorist, for he adds to the first criterion of his program the caution that 'there will be other types of causes apart from social ones which will cooperate in bringing about belief.' This allows for an empirical input to knowledge, rather than it being merely and solely a product of social negotiations and interests, the outcome of power-play, the 'Matthew effect' (such that well-published authors find it easy to get their work published), or whatever.

Nevertheless, when we read further into Bloor's book, we find that his program is indeed a 'strong' one, for he includes *mathematics* within his compass as well as science. This may surely startle the unsuspecting reader, for, one may think, is not Pythagoras's theorem (for example) objectively true, quite independent of any social negotiations that may or may not have gone into or accompanied its discovery? No doubt, some social toing-and-froing went into the discovery of Pythagoras's theorem way back in antiquity; but once it was discovered, that was it — the square of the hypotenuse of a right-angled triangle equals the sum of the squares of the other two sides. This always has been so, always will be so, no matter whether man has or has not known about this fact. It is an objective truth, not a social construct. Pythagoras's theorem is firmly situated in Popper's 'World 3' and is not going to be displaced therefrom by trendy discussions in the sociology of knowledge!

But Bloor's thesis cannot be dismissed with a wave of the hand, and it is to his credit that he takes the bull by the horns, so to speak, and considers the *prima facie* very unpromising case of mathematics, rather than (say) intelligence testing, evolutionary biology, or quantum theory. To do this, however, we notice immediately that he turns his back on the labours of philosophers such as Frege and Russell, who worked so hard to display the logical basis of mathematics. Bloor returns to the middle years of the nineteenth century and the 'empiricist' mathematics of John Stuart Mill,[116] in the process mounting an effective critique of the rhetoric used by Frege in his campaign for logicism and the 'objective' character of mathematical knowledge.

To make matters clearer, let us consider one of the particular cases that Bloor adduces in favour of the view that mathematics is an empirical enterprise, susceptible to the analyses offered by the sociologist of knowledge.[117] The Greeks saw an intimate connection between arithmetic and geometry. Thus they thought in terms of 'square numbers', 'triangular numbers', 'oblong numbers' etc:

Figure 47

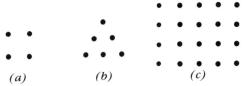

(a)   (b)   (c)

(As Bloor rightly points out, for Frege, in a very different mathematical tradition, an oblong number would be as ridiculous as an oblong concept.)

Consider, now, the problem of expressing the square root of two while thinking of numbers in this Pythagorean way. Such a problem immediately arises, upon the discovery of Pythagoras's theorem, when one considers the triangle of Figure 48.

Figure 48

On the Greek 'geometrical' view of arithmetic, any number ought to be expressible as the ratio of two other whole numbers. So let us see what happens if we try to do this for the square root of two.

Let $p/q = (2)^{1/2}$;
then $p^2 = 2q^2$;
so p^2 is even;
so p is even;
and q is odd (assuming that p/q had first been simplified so that any common factors such as 2 had been removed).

Now, if p is even, it can be written as $2n$.
So $p^2 = 4n^2$
$\quad = 2q^2$.
Therefore $q^2 = 2n^2$.
So q^2 is even;
so q is even;
and p is odd.

This *reductio ad absurdum* implied that the square root of two was not a number in the sense understood by the Pythagoreans for it could not be expressed as the ratio of two integers, p/q. Yet, as we have said, it appeared *geometrically*, when Pythagoras's theorem was applied to the right-angled triangle of Figure 48.

So the link between geometry and arithmetic seemed to be fractured by the discovery of Pythagoras's theorem.[118] For a number, on the Pythagorean view was either even or odd — but not both, as the foregoing analysis might seem to require. It is no surprise, therefore, that there is a legend that the Pythagoreans regarded their discovery of numbers such as the square root of two as something of a scandal and tried to keep it a secret. And thereafter the Greeks sought to establish geometry independently of arithmetic and gave geometry primacy.

The relevance of all this to the point under discussion may not be immediately obvious. But Bloor points out that the Pythagoreans' mathematics was intimately related to their total world view, which was in its turn related to the social formations of the society within which they lived. For example, the following table of opposites held special significance for them:

Male/Female
Light/Darkness
Good/Bad
Odd/Even
Square/Oblong, etc.

Given such a preference for seeing the world in terms of polar opposites, it is not surprising that the Pythagoreans wished to keep numbers like the square root of two out of their mathematics, But, suggests Bloor, in a different culture, not so concerned with cosmic opposites, it might not have been a problem. If one laid emphasis on the way night shades into day, good into bad; if one belonged to a culture of compromisers, blenders, mixers and mediators, and always emphasised the inter-mingling of things, then perhaps 'irrational' numbers might have been accepted much more readily and there would have been no attempt to 'suppress' them. So, Bloor argues, the mathematics that emerges at any given epoch may be highly beholden to particular cultural cir-cumstances. In fact, the general thesis of Bloor's book is that what passes for knowledge within any given society is a kind of transfigured conception of that society.[119] Social images provide the epistemological frames. That is why his book is called *Knowledge and Social Imagery*. Society, being too complex to be grasped as a whole, is comprehended with the help of some kind of ideology (an idea that would have appealed to Marx and to Mannheim). And then the ideology becomes transferred to the knowledge system that is generated by that society. One can easily think of other examples. The social struggle of Victorian England found its way into Darwin's theory of evolution by natural selection. The hie-rarchical structure of eighteenth-century Swedish society provided a model for Linnaeus's hierarchical taxonomy of the animal, vegetable and mineral *king*doms. Also, it has been suggested that the cultural freedom and social relativism towards moral and political questions in Switzerland when Einstein was a young man played their part when Einstein formulated his theory of relativity.[120]

As to the mathematical examples on which Bloor particularly rests his case, it is worth mentioning that Wittgenstein had been setting the scene for such arguments some years before in his *Remarks on the Foundations of Mathematics*,[121] where he argued forcefully that our knowledge of mathematics arises in use and is shaped accordingly.[122] This view accorded with the views of the later Wittgenstein of the *Philo-sophical Investigations*.[123]

But are Bloor (or Wittgenstein) saying that mathematics (and *a fortiori* science) are 'nothing but' social constructs? Assuredly not. Bloor, in fact, seems to adopt a view towards knowledge that one might line up with that of the pragmatists; and the coherence theory of truth also seems to be important to him:

> The indicator of truth that we actually use is that the theory works. We are satisfied if we achieve a smoothly operating theoretical view of the world. The indicator of error is the failure to establish and maintain this working relationship of successful prediction. One way of putting this point would be to say that there is one sort of correspondence that we do indeed use. This is not the correspondence of the theory with reality but the correspondence of the theory with itself. Experience as interpreted by the theory is monitored for *such internal consistency as is felt important*. The process of judging a theory is an internal one. *It is not internal in the sense of being detached from reality*, for obviously the theory is connected to it by the way we designate objects, and label and identify substances and events. But once the connections have been established the whole system has to maintain a degree of coherence: *one part must conform to another*.[124]

As we have seen,[125] neither pragmatism nor coherence theory can provide an adequate criterion of truth (where truth is construed as correspondence between the world and our ideas concerning the world). However, as a sociologist of knowledge, this would be a perfectly acceptable conlusion to Bloor. Knowledge, for the advocate of the 'strong programme', is a social construct; it need have no pretensions to 'Truth', or correspondence with 'The World'.

In a later publication,[126] Bloor has extended his critique to logic, as well as mathematics, building further on ideas to be found in the work of the later Wittgenstein. Bloor raises the question of whether the usual axioms and modes of reasoning of deductive logic are self-evident to all humans at all times and places — as if logical principles are somehow laid up in some Platonic heaven, so to speak. Or should they be regarded as the product of particular social formations, or contingent circumstances, and therefore subject to change and lacking logical necessity?

For example, it is widely held that the 'rules of logic' show that from a logical contradiction as a premiss one can deduce any proposition whatsoever.

Thus, if:
  (i) (p and q) implies p ..... an assumed logical principle
  (ii) (p and q) implies q ..... an assumed logical principle
  (iii) p implies (p or q) ..... an assumed logical principle
  (iv) (p or q), and (not-p), implies q ..... the so-called disjunctive
                                                               syllogism
  (v) p and not-p ..... an assumed logical contradiction
Then:
By (i) and (v) we can conclude p ..... (vi)
By (ii) and (v) we can conclude (not-p) ..... (vii)
By (iii) and (vi) we can conclude (p or q) ..... (viii)
By (iv), (vii) and (viii) we can conclude q ..... (ix)

But one does *not*, Bloor insists, have to take this kind of thing as if the principles of logical reasoning were inscribed on golden tablets in some logical heaven. It will be noted that the assumed logical contradiction refers only to ps, not qs. Yet the conclusion is q. Whence did this q arise? It is evident that it slipped in when we stated (iii) that: p implies 'p or q'. But has this q really any proper place in our thinking at this point? Given

p, are we really justified in saying more than that 'p implies p'? Or again, if we assert 'p or q' and 'not-p', can we validly infer q? Do we not have to know that, for a proper disjunction, the truth of one term does exclude the truth of the other? Actually, in practice whether or not this 'holds' will depend on our interests and circumstances, and the 'relevance' of p to q. For example, suggests Bloor, if p and q are two roads home — and the only two — then if we don't choose one we must choose the other. But if, say, there are three roads, though p and q are still alternatives it does not follow that the rejection of p implies the utilization of q. For this, p and q must exhaust the possiblities. (Of course, if we have no interest in going home at all at that point then our logical apparatus will not be brought into play at all at that juncture; it will be *use*less.)

This little example suggests that there needs to be flexibility of interpretation when *using* logic. Thus, considering the argument outlined above, some people will insist that q can be deduced from the conjunction of p and not-p; others will reject such a bizarre conclusion. So what counts as 'logical' or 'deductively valid' seems, on this view, to be influenced by some kinds of social factors such as needs, interests, or socially-determined and contingent criteria of relevance. This must surely be unwelcome to those accustomed to a 'Platonic' view of logic.

Also, one might add that with respect to a great many matters of fundamental human concern, one does have to entertain 'logical contradictions': that is, a combination of polar opposites, such as 'growth' and 'decay', 'stability' and 'change', 'freedom' and 'discipline'. In each case, *both* poles need to be pursued simultaneously, with the perception that the well-being of an individual or a society will not be attained by emphasis on one to the exclusion of the other. The one does not exclude the other logically and must not be allowed to do so practically. One should seek to attain freedom and non-freedom at one and the same time.

To all this, logicians would counter that one is conflating two different things. A logical system (axiom system), they would say, has to be internally self-consistent though one can choose different axioms, provided they do not involve contradictions. To be sure, different people may indeed have different intuitions about what is or is not logical. But those who, for some reason, choose to intuit contradictions, and define those into an axiom system, will not be able to construct an axiom system that functions in a coherent and consistent manner. Bloor,[127] following Wittgenstein,[128] sees the matter in a different light. If we try to think 'p and not-p', it is not the inherent meaning of 'not' that makes the thought unthinkable. It is the way we commonly *use* the word 'not' that determines the meaning; the logical properties of the expressions used in logic arise from the meanings we give to language in use. That is to say, in order to organise our thoughts about the world and establish coherent patterns of behaviour, we choose to exclude expressions such as 'p and not-p'. But that tells us nothing about the way the world is 'in itself'.

Indeed, it may be that the world is *not* 'logical' at the level of microphysics. We find it difficult to understand how an electron, say, can have two incompatible features simultaneously (i.e., particle and wave), thereby apparently contravening our usual exclusion of 'contradictions' from our thinking. But if this exclusion is, at bedrock, merely a social practice, then it should cause no surprise if it sometimes is inappropriate, particularly in areas that are remote from our usual day-to-day concerns, in relation to which our logical practices have been historically formulated. Thus the sociologist of knowledge faces the conundrums of microphysics with equanimity!

Even so, one may suspect that the reader will find the relativism of one such as Bloor somewhat unpalatable. Yet Bloor's position as a sociologist of knowledge is by no means the most extreme that is to be found in the recent metascientific literature. He certainly does not imagine that there is no empirical input into scientific knowledge. Yet this could well be the conclusion that the reader might draw from the recent book of Latour and Woolgar: *Laboratory Life*.[129] Here we find a 'strong programme' of quite unsurpassed strength and power — moreover, one that is of special interest in that it has been applied in the context of a detailed 'field-study' of a particular scientific institution, rather than through a stiff bout of armchair philosophising. The researchers worked among the 'natives' of the Salk Institute for Biological Studies, in California, observing minutely what went on around them in the laboratories during the prosecution of an important period of research that ultimately led to the award of a Nobel prize. The approach used was like that of the anthropologist visiting some primitive tribe and observing its rituals. To blend with the surroundings as much as possible, one of the investigators (Latour) found himself employment as a laboratory assistant while he was making his sociological investigations.

The tangible product of a research laboratory is, of course, a series of scientific papers, which are highly esteemed as such by the research scientists.[130] But how do the 'facts' described in such publications come into being? To study this matter, Latour and Woolgar constructed a five-point scale to classify statements made verbally or in writing:[131]

Type 5. 'Taken-for-granted' facts — which everyone accepts and which are therefore rarely mentioned in laboratory talk.

Type 4. Uncontroversial facts, but which are, nevertheless usually made explicit — the accepted knowledge disseminated in text-books.

Type 3. Statements containing statements about statements — expressed with qualifications or 'modalities'. (For example, 'Oxytocin is *generally assumed* to be produced by neurosecretory cells of the paraventricular nuclei'.[132])

Type 2. Statements presented as knowledge claims, rather than alleged facts. ('There is a large body of evidence to support the concept of a control of the pituitary by the brain.') Here we have tentative suggestions and ideas for further research.

Type 1. Conjectures or speculations forming part of private discussions, but sometimes included at the end of a paper.

Now, from their observations in the Salk Laboratory, Latour and Woolgar concluded that the many activities involved in the research that they observed made sense if they were construed as being directed towards the gradual elevations of statements in such a 'knowledge hierarchy'. A fact, then, would be a 'Type 5' statement, with no attached qualifications or modalities and no attributed authorship. Or a 'Type 2' statement could gradually become elevated to 'Type 3' over a period of time, through appearance in successive publications. Sometimes, of course, statements could descend the hierarchy and drop away altogether, or never get beyond the first stage.

To see all this at work, Latour and Woolgar noted how the numerous processes of 'social negotiation' occurred within the laboratory, and on a wider stage through the processes of refereeing and publication. Questions of social status and standing were of obvious importance. A person's reputation had a good deal to do with the status accorded to his or her utterances; and the prestige of the journal in which a paper was published had a significant bearing on the reliance that was given to its findings. The 'anthropologists'' investigations made it clear that what was or was not taken to be a 'fact' was very much the outcome of intense processes of social negotiation, as much as the concrete results of experimental investigations. Power and authority were factors of considerable significance.

Also, most interestingly, what started out as tentative statements about hypothetical entities gradually gave way to statements about concrete objects. Indeed, statements gradually came to be superseded by objects. Whereas in the first instance the statements were real and the objects conjectural, gradually the position became reversed: the object becomes the reality accounting for why the statement was made about it in the first place. In this way, Latour and Woolgar have sought to give a sociological account of how we come to believe that there is a one-to-one correlation between language and the world. Thus we have a sociological explanation of our simple-minded(?) belief in the correspondence theory of truth. So, Latour and Woolgar would have it, what we regard as the 'real world' is a consequence of scientific work, rather than the cause of what we think: 'Scientific activity is not "about nature," it is a fierce fight to *construct reality*.'[133]

This is an extraordinary conclusion; yet it is by no means without an *element* of truth. It is certainly consistent with the trend that much metascientific literature has taken in recent years, and can be seen to mesh with the kinds of arguments of writers such as Kuhn and Feyerabend that we discussed earlier in this chapter. But there are one or two fairly obvious points that one might wish to make. One is that although Latour and Woolgar amply demonstrate that there is a very strong social component to what is commonly regarded as scientific knowledge, they do so by downgrading the role of observation in a way that philosophers reared within the empiricist tradition find very difficult to accept. If we recall once again the aphorism of Kant, 'Thoughts without content are empty, intuitions without concepts are blind',[134] it would seem that *Laboratory Life* is concerned with 'thoughts without content'; and as such it might appear unacceptable as a picture of science.

In fact, however, the authors of *Laboratory Life* do concede, or even emphasise, the role of experimentation. Yet they do so in a way that is quite unorthodox so far as the tradition of Anglo-Saxon philosophy of science is concerned. Following ideas initially formulated by the French philosopher Gaston Bachelard (1884–1962),[135] Latour and Woolgar emphasise the view that scientific phenomena are *'constituted'*[136] by the pieces of apparatus used in the laboratory, which are themselves the products of *human* construction — and hence influenced by all the social processes that go on within the scientific community. For it is a matter of decision as to whether a piece of apparatus is to be regarded as reliable or unreliable. Such *decision* may, for example, rest upon the particular reputation of an instrument maker or an instrument manufacturing company. Here, then, emphasis is placed upon the 'social' attributes of scientific instrumentation, rather than (for example) the 'coherence' of observations and theories made with, or based upon, the use of such instruments. Thus, although it is certainly true that Latour and Woolgar do not disregard the experimental activity of scientists — far from it — they construe it within the frame of reference furnished by the exponents of sociology of knowledge, so that any claimed 'objectivity' for science is firmly denied.

A second point I'd like to make is that *Laboratory Life* isn't strictly a sample of metascience in the usual sense — or there are considerable problems involved if it is so regarded. For it is represented by its authors as a specimen of an anthropological investigation of a rather unusual kind. But if this is so, it is a scientific investigation, not one of metascience. So, if the conclusion reached is that scientific knowledge is 'nothing but' a social construct, we may hoist Latour and Woolgar with their own collective petard and say that their findings tell us nothing about the real world of science, but only about the results of *their* social negotiations with other sociologists/anthropologists interested in science; and so on. We should thus be in such a vicious circle (much as seems to be implied in the fourth characteristic of Bloor's 'strong program') that the whole exercise would seem to be self-defeating.

We should perhaps conclude with a cautionary note about the sociology-of-knowledge program. How far may the social origin of knowledge succeed in leaving its mark? I raised the question earlier (p.345) as to whether the very form of the equations of physical theory or those describing the laws of nature, may be socially determined. I failed to answer it at that juncture, but we cannot leave it like that. It is, perhaps, all very well for Latour and Woolgar to regard some of the complex chemicals that may (or may not) be secreted in minute amounts from the brain as — for all that we can really tell — no more than 'social constructs'. But are they choosing a special case to suit their epistemological preferences?

Suppose we consider well-known laws of classical physics such as the ideal gas law, $PV = nRT$; or Snell's law of refraction, $\sin i/\sin r$ = constant. Is the form of these laws a social product? Well, yes and no! I believe that such equations do describe, more or less accurately, the way processes occur in nature. But consider the gas law. It is not exact. There are other equations such as that of van der Waals, $(P + a/V^2)(V - b) = nRT$, which fit the observational data more exactly. None of the available equations is perfect, however, though others can be devised which fit the data more precisely The more 'accurate' equations, however, lack any theoretical base in terms of a model for the structure of the gas. These so-called 'Virial equations' are purely empirical. It becomes, then, a matter of choice as to which equation one chooses according to the practical needs of the particular technical problem in hand at any given time. This allows a kind of social determination of the equation, in *use*.

But, the reader may think, that I am still running away from the issue. Can the sociologist of knowledge point to any kind of formal relationship between some physical equations and the social formations that have generated them? For example, what, if anything, has the *form* of Snell's law, as successfully used today, got to do with the seventeenth-century Dutch society in which Snell worked? Candidly, I think the answer has to be 'nothing'. On the other hand, I do think that the form of Darwin's theory had something to do with the Victorian social milieu within which it was generated.[137] Or the researches of Donald MacKenzie, for example, have shown that the form of the equations used in statistics were shaped by the social purposes of the men that devised them.[138] But statistics is a branch of mathematics, and as such it is a human construct, and thus naturally revealing of its social origins. I am not aware of any historical investigations that have shown that the particular *form* of certain *laws* of physics can be causally connected to the social milieux within which they were generated, though there are published studies that reveal fascinating relationships between physicists' work and their social interests, or between their cognitive styles and their social formations.[139] It is imprudent, however, to extrapolate the results of such investigations beyond what the historical evidence will allow.

I would say, nevertheless, that the work of sociologists of knowledge is of the greatest interest and importance for our understanding of science, and more broadly for a satisfactory epistemology. Latour and Woolgar have, I fear, somewhat overreached themselves, though they would probably argue that they move in a progressive spiral rather than a vicious circle. But writers such as Bloor, somewhat more cautious, can certainly get a lot more philosophical mileage out of the 'strong programme'. At the time of writing, the issues concerned are still under very active discussion, and in an historical text such as that which is presented here, it would be ill-advised to attempt to predict what any final consensus will be. I hope, however, that I have been able to carry readers to the point where some of the issues currently under discussion may be better understood. That has been my limited objective, but more than enough to tax my readers' patience, and not to be prolonged here.

NOTES

1   See I Lakatos ed *Problems in the Philosophy of Mathematics* North-Holland Amsterdam 1967; I
    Lakatos ed *The Problem of Inductive Logic* North-Holland Amsterdam 1968; I Lakatos & A E
    Musgrave eds *Problems in the Philosophy of Science* North-Holland Amsterdam 1968; and I Lakatos
    & A E Musgrave *Criticism and the Growth of Knowledge* Cambridge University Press Cambridge
    1970

2   For a bibliography of Kuhn's writings, see G Gutting ed *Paradigms and Revolutions: Appraisals and
    Applications of Thomas Kuhn's Philosophy of Science* University of Notre Dame Press Notre Dame
    1980 pp 321–4

3   T S Kuhn *The Copernican Revolution: Planetary Astronomy in the Development of Western
    Thought* Harvard University Press Cambridge Mass 1957

4   T S Kuhn *et al* eds *Sources for History of Quantum Physics: An Inventory and Report* American Phi-
    losophical Society Philadelphia 1966; and T S Kuhn *Black-Body Theory and the Quantum Dis-
    continuity, 1894–1912* Oxford University Press London 1978

5   T S Kuhn *The Structure of Scientific Revolutions* University of Chicago Press Chicago 1962; 2nd ed
    enlarged 1970 (It is interesting to note, by the way, that the series in which Kuhn's book first
    appeared — the *International Encyclopedia of Unified Science* – was that in which the logical
    positivists of the Vienna Circle [or those that went to America] had presented their views. The
    project of a 'unified science' was one to which Comte, the founder of positivism, would undoubtedly
    have given his blessing. So Kuhn was somewhat like a Trojan Horse within the positivist citadel. For
    The Structure of Scientific Revolutions was, I believe, one of the works that marked the end of the
    positivist supremacy in metascience.)

6   T S Kuhn 'Second Thoughts on Paradigms' in F Suppe ed *The Structure of Scientific Theories*
    University of Illinois Press Urbana Chicago & London 1974 pp 459–82. A valuable exposition of
    Kuhn's system is: G Doppelt 'Kuhn's Epistemological Relativism: An Interpretation and Defence'
    Inquiry 1978 Vol 21 33–86. This author maintains that it is scientific problems, data and standards
    that are 'incommensurable', not scientific 'meanings'. But see pp 340–42.

9   Gaius Plinius Secundus (*c* 23–79 AD) was author of the celebrated *Natural History* in thirty-seven
    books, which provided the great compendium of ancient knowledge and lore that was widely used
    through the Middle Ages.

10 T S Kuhn *op cit* (note 5 1962) p 17

11 *Ibid* p 10

12 *Ibid* p x

13 The reader will note that we have had occasion to refer, albeit in passing, to all of these in the course
    of this book.

14 A L Lavoisier (1743–1794) was the chief architect of the so-called eighteenth-century chemical
    revolution and founder of the oxygen theory of combustion.

15 It is notorious that Kuhn's own usage of the word paradigm is rather flexible. In a well-known paper,
    Margaret Masterman identified no less than twenty-one usages in *The Structure of Scientific
    Revolutions*, namely: 1. as a universally recognised scientific achievement; 2. as a myth; 3. as a
    'philosophy' or constellation of questions; 4. as a text-book, or classic work; 5. as a whole tradition,
    and in some sense a model; 6. as a scientific achievement; 7. as an analogy; 8. as a successful meta-
    physical speculation; 9. as an accepted device in common law; 10. as a source of conceptual and
    instrumental tools; 11. as a standard illustration; 12. as a device, or type of instrumentation; 13. as an
    anomalous pack of cards; 14. as a machine-tool factory; 15. as a *Gestalt* figure which can be seen two
    ways; 16. as a set of political institutions; 17. as a 'standard' applied to quasi-metaphysics; 18. as an
    organising principle which can govern perception itself; 19. as a general epistemological viewpoint;
    20. as a new way of seeing; 21. as something which defines a broad sweep of reality. (M Masterman
    'The Nature of a Paradigm' in Lakatos & Musgrave eds *op cit* [note 1] pp 59–89) If then, dear reader,
    you form an identikit picture of all this, you have Kuhn's notion of a paradigm, at least as it was in
    1962!

16 Kuhn *op cit* (note 5 1962) p 10, Chs 2–4 & *passim*

17 *Ibid* p 24

18 *Ibid* pp 36–9 & *passim*

19 *Ibid* Ch 7

20 *Ibid* pp 121–2. The *Gestalt* theory supposes that things are perceived as wholes, rather than as sums
    of parts. In the classic illustration of this, a figure of a cube may be viewed as projecting either
    forward, or backwards, but not as something in between. Thus:

Figure 49

The mind seems to switch from one interpretation
to the other in a fraction of a second.

21 *Ibid* p 102

22 T S Kuhn 'Reflections on my Critics' in Lakatos & Musgrave eds *op cit* (note 1) pp 231–78 (at p 264)

23 See, for example: A A Hallam *A Revolution in the Earth Sciences: From Continental Drift to Plate Tectonics* Clarendon Press Oxford 1973; H G McCann *Chemistry Transformed: The Paradigmatic Shift from Phlogiston to Oxygen* Ablex Norward New Jersey 1978

24 K R Popper 'Normal Science and its Dangers' in Lakatos & Musgrave eds *op cit* (note 1) pp 51–8 (at p 53)

25 See above, p 320

26 Kuhn *op cit* (note 5 1970) p 202

27 *Ibid* p 175

28 *Ibid*

29 It is my suggestion, not Kuhn's.

30 Kuhn *op cit* (note 6) p 465

31 *Ibid* p 470

32 *Ibid* p 471

33 See above, p 197

34 Kuhn *op cit* (note 5 1970) p 199

35 *Ibid* p 185

36 I Lakatos 'Proofs and Refutations' *British Journal for the Philosophy of Science* 1963–64 Vol 14 pp 1–25 120–39 221–45 269–342

37 I Lakatos 'Falsification and the Methodology of Scientific Research Programmes' in Lakatos & Musgrave eds *op cit* (note 1) pp 91–195. The ideas in this paper were adumbrated by Lakatos in his paper 'Criticism and the Methodology of Scientific Research Programmes' *Proceedings of the Aristotelian Society* 1968 Vol 69 pp 149–86. It is perhaps important to note here that Lakatos did not use the word 'methodology' in the time-honoured sense. Far from being a set of procedures carried out in scientific investigations, or suggestions as to *how* one might best proceed for successful scientific research, Lakatos regards a methodology of science as a 'set of...rules for the *appraisal* of ready, articulated theories'. See his paper 'History of Science and its Rational Reconstructions' in C Howson ed *Method and Appraisal in the Physical Sciences: The Critical Background to Modern Science, 1800–1905* Cambridge University Press Cambridge 1976 pp 1–39 (at p 2)

38 For further details of Lakatos's life and work, see: P K Feyerabend 'Imre Lakatos' *British Journal for the Philosophy of Science* 1975 Vol 26 pp 1–18; and R S Cohen P K Feyerabend & M W Wartowsky eds *Essays in Memory of Imre Lakatos* Reidel Dordrecht 1976

39 Lakatos *op cit* (note 37) p 132

40 *Ibid* p 133

41 *Ibid*

42 *Ibid* p 135

43 *Figure 50*

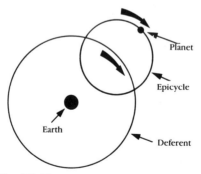

Planet

Epicycle

Earth

Deferent

44 Lakatos *op cit* (note 37) pp 116–20

45 *Ibid* pp 118 & 134 (Lakatos, it may be noted, makes a distinction between 'progressive theoretical problemshifts' and 'progressive empirical problemshifts'.)

46 *Ibid* p 118

47 *Ibid* pp 138–40. Prout's hypothesis suggested that atoms of the different elements were built up of hydrogen atoms as the fundamental building blocks.

48 *Ibid* pp 135–6

49 A E Musgrave 'Method or Madness?' in Cohen Feyerabend & Wartowsky eds *op cit* (note 38) pp 457–91

50 This would, I think, have been agreeable to Lakatos himself, given that he used the Newtonian example in his 'Falsification and the Methodology of Scientific Research Programmes'.

51 S Newcombe 'Gravitation' in *Encyclopaedia Britannica* 11th ed Cambridge University Press London 1910 Vol 12 pp 384–5

52 W Berkson 'Lakatos One and Lakatos Two: An Appreciation' in Cohen Feyerabend & Wartowsky eds *op cit* (note 38) pp 39–54 (at p 52)

53 Lakatos *op cit* (note 37) p 135

54 Musgrave *op cit* (note 49) p 469

55 *Ibid* p 470

56 I Lakatos 'Replies to Critics' in R C Buck & R S Cohen eds *PSA 1970: In Memory of Rudolf Carnap* Reidel Dordrecht 1971 pp 174–82 (at p 174)

57 Musgrave *op cit* (note 49) p 481

58 P K Feyerabend *Against Method: Outlines of an Anarchistic Theory of Knowledge* New Left Books London and Humanities Press Atlantic Highlands 1975

59 P K Feyerabend *Science in A Free Society* New Left Books London 1978

60 P K Feyerabend 'Explanation, Reduction and Empiricism' in H Feigl & G Maxwell eds *Minnesota Studies in the Philosophy of Science* 1962 Vol 3 pp 28–97

61 P K Feyerabend 'How to be a Good Empiricist: A Plea for Tolerance in Matters Epistemological' in B Baumrin ed *Philosophy of Science: The Delaware Seminar* 1963 Vol 2 pp 3–39; reprinted in P H Nidditch ed *The Philosophy of Science* Oxford University Press Oxford 1968 pp 12–39

62 For further discussion of this issue, see pp 340–42

63 P K Feyerabend 'Origin of the Ideas in this Essay' in his *op cit* (note 59) pp 107–22

64 *Ibid* p 109

65 Feyerabend *op cit* (note 59) p 109. For a discussion of Ehrenhaft's views and their relation to physical orthodoxy, see G Holton *The Scientific Imagination: Case Studies* Cambridge University Press Cambridge 1978 pp 25–83

66 Feyerabend *op cit* (note 59) p 115

67 *Ibid* p 116

68 *Ibid* p 118

69 *Ibid* p 119

70 Feyerabend *op cit* (note 58) p 17

71 Closely aligned with Surrealism, the Dadaist movement in art of the 1920s and 1930s advocated a cheeky, irreverent attitude towards artistic activity, taking nothing seriously, and regarding nothing as sacred. A characteristic Dadaist touch, for example, would be to portray the Mona Lisa with a moustache and beard. The following picture is also representative of work of this genre:

Figure 51

From W S Rubin *Dada and Surrealist Art* Thames & Hudson London 1969 p 231 (reprinted 1978)
It is *this* kind of approach that Feyerabend would like to see prevail in philosophy of science! (We

may note, from his brief intellectual autobiography published in his *Science in A Free Society* [*op cit* (note 59) p 120] that Feyerabend studied Dadaism extensively during the Second World War. Now he has come back to it.)

72 Feyerabend *op cit* (note 58) p 21

73 *Ibid* p 81

74 *Ibid* p 83. The passage is from Galileo Galilei, *Dialogue Concerning the Two Chief World Systems — Ptolemaic & Copernican* trans Stillman Drake foreword Albert Einstein University of California Press Berkeley Los Angeles & London 2nd ed 1967 p 249. It will be recalled that this was one of Galileo's 'apologetic' works, in which he sought to persuade his readers of the ideas that he had himself arrived at previously. The text was written in the form of a discussion between a natural philosopher (Salviati — or Galileo himself), Sagredo (an intelligent layman) and Simplicio (an Aristotelian).

75 At least, it seems to us today that his arguments were cogent and persuasive. We know, however, that many scholars of Galileo's day found difficulty in following and accepting his reasoning. Some, of course, did not want to listen.

76 For example, the natural motion of bodies in the superlunary realm was thought to be circular, unlike what one finds in the daily experiences of the sublunary realm.

77 In his well-known play, *The Life of Galilei* (1938–39), the Marxist playwrite Berthold Brecht represents the traditionalist Aristotelians as benighted obscurantists. But according to the argument that Feyerabend seeks to present, they were entirely rational, by their own lights.

78 'Rational', we may recall, means 'endowed with reason'. And 'to reason' means 'to think in a connected, sensible, or logical manner'. The problem, however, is what *counts* as connected, sensible or logical. Here, one man's meat may be another man's poison; and it is because of this that Feyerabend thinks it possible to drive a wedge into the seemingly solid structure of rational science.

79 Feyerabend *op cit* (note 58) p 299

80 *Ibid* pp 299–300

81 A classic illustration of alternative systems of thinking to those of Western science is provided by R Horton 'African Traditional Thought and Western Science' *Africa* 1967 Vol 37 pp 87–155

82 See S Drake *Galileo at Work: His Scientific Biography* Chicago University Press Chicago 1978

83 Epistemic = 'relating to knowledge'. I take it that Feyerabend means here the activities we undertake in order to try to gain knowledge of the world.

84 Feyerabend *op cit* (note 59) p 70. It should be noted that the term 'refer' is one with technical significance in philosophy of science. It is a fundamental question in philosophy of science whether theoretical entities such as 'electron', 'gene', etc *really* exist. The question is whether such terms *refer* to anything, or whether they are but convenient fictions as an instumentalist philosophy of science would claim.

85 See Feyerabend *op cit* (note 58) Ch 17

86 There are many accounts of the history of Lavoisier's chemical revolution. See, for example, H Guerlac *Antoine-Laurent Lavoisier: Chemist and Revolutionary* Scribner New York 1975

87 H Field 'Theory Change and the Indeterminacy of Reference' *The Journal of Philosophy* 1978 Vol 70 pp462–81

88 *Ibid* p 479

89 A Fine 'How to Compare Theories: Reference and Change' *Nous* 1975 Vol 9 pp 17–32

90 *Ibid* p 28

91 *Ibid*

92 K Mannheim *Ideology and Utopia: An Introduction to the Sociology of Knowledge* Routledge & Kegan Paul London 1936 p 261 & *passim*

93 *Ibid* pp 147–8; K Mannheim *Essays on the Sociology of Knowledge* ed Paul Kesckemetic Routledge & Kegan Paul London 1952 p 35

94 For example, educational syllabuses are framed within particular social systems. What is accounted important goes into the syllabus; what is thought irrelevant or inconsequential goes out. And what goes in is what children learn, and are led to believe is important. Of course, the system is very imperfect, and generally there is an 'unofficial' curriculum within a school, running parallel with or (sometimes) counter to the 'official' one. Nevertheless, knowledge is gained within an education system, whether it be 'official' or 'unofficial'.

95 See p 229 above.

96 L Wittgenstein *Philosophische Untersuchungen/Philosophical Investigations* trans G E M Anscombe Blackwell Oxford 1953 (2nd ed 1958)

97 *Ibid* (1958) pp 6 10 14 & *passim*

98 We may refer the reader here back to the example given above on p 227, concerning a simple language and its determination of the way in which one necessarily thinks about, or 'sees', the world. This position, only present in embryo in the *Tractatus*, is developed very much more fully in the *Philosophical Investigations*.

99 The classic example is that in which the Russian Marxist writer, Boris Hessen, sought to show that the work of Newton's *Principia* was fundamentally shaped by questions of socio-economic significance: 'The Social and Economic Roots of Newton's *'Principia'* in N Bukharin *et al* eds *Science at the Cross Roads: Papers presented to the International Congress of the History of Science and*

Technology held in London from June 29th to July 3rd, 1931 by the Delegates of the USSR Kniga London nd pp 1–62. Another very well known study is R K Merton *Science, Technology and Society in Seventeenth-Century England* Fertig New York 1970 (first published in *Osiris* Vol 4 1938)

100   J D Bernal *The Social Function of Science* Routledge London 1939

101   See, for example: B Barber *Science and the Social Order* Macmillan New York 1952; M Polanyi *Personal Knowledge: Towards a Post-Critical Philosophy* Routledge & Kegan Paul London 1958; W O Hagstrom *The Scientific Community* Basic Books New York 1965; J M Ziman *Public Knowledge: An Essay Concerning the Social Dimension of Science* Cambridge University Press Cambridge 1968; J R Ravetz *Scientific Knowledge and its Social Problems* Clarendon Press Oxford 1971; R K Merton *The Sociology of Science: Theoretical and Empirical Investigations* University of Chicago Press Chicago 1973; I I Mitroff *The Subjective Side of Science* Elsevier Amsterdam 1974; B Dixon *What is Science For?* Penguin Harmondsworth 1976; I Spiegel-Rösing & D J de Solla Price eds *Science, Technology and Society: A Cross-Disciplinary Perspective* Sage London 1977; J M Ziman *Reliable Knowledge* Cambridge University Press Cambridge 1978; M Mulkay *Science and the Sociology of Knowledge* Allen & Unwin London 1979; A Brannigan *The Social Basis of Scientific Discoveries* Cambridge University Press Cambridge 1981; B Barnes & D Edge eds *Science in Context: Readings in the Sociology of Science* Open University Press Milton Keynes 1982; K D Knorr-Cetina & M Mulkay eds *Science Observed: Perspectives on the Social Studies of Science* Sage London 1983

102   W Wordsworth ed J C Maxwell *The Prelude: A Parallel Text* Penguin Harmondsworth 1971 p 103

103   In fact, it was not by any means wholly apt. Newton was *educated* in science to some degree by Isaac Barrow, read the work of his contemporaries, and was inordinately concerned with his standing and reputation.

104   Cf D J de Solla Price *Little Science, Big Science* Columbia University Press New York 1963

105   Polanyi *op cit* (note 101). See also: M Grene ed *Knowing and Being: Essays by Michael Polanyi* Chicago University Press, Chicago 1969 Part 3.

106   M Polanyi 'The Republic of Science' *Minerva* 1962 Vol 1 pp 54–73

107   These procedures have been well described by J M Ziman *op cit* (note 101) Ch 6.

108   To a degree, this tends to undermine the problems that we discussed above (p 201) in connection with the 'Duhem/Quine thesis'. However, the fact that large bodies of scientific knowledge are, at any given time, regarded as solid and secure does not remove the *logical* difficulty identified by Duhem and Quine. (But some sociologists of knowledge appear not to set much store by the traditional canons of logic.)

109   One of the most obvious deficiencies is the so-called 'Matthew effect', such that those scientists who have already published a number of papers are likely to find it easier to get their work published than do those with no established reputation, even though the work of the beginners may be superior. See R K Merton 'The Matthew Effect in Science' *Science* 1968 Vol 159 pp 56–63. (Cf The Biblical adage: 'For he who has, to him shall more be given and richly given, but whoever has not, from him shall be taken even what he has.') See also B Broad & N Wade *Betrayers of the Truth: Fraud and Deceit in the Halls of Science* Simon & Schuster New York 1982, which reveals that the amount of deceit in modern science is considerably higher than that which is customarily acknowledged.

110   P L Berger & T Luckman *The Social Construction of Reality: A Treatise in the Sociology of Knowledge* Penguin Harmondsworth 1967. One can, of course, change one's social attachments and thereby gain a different view of the world, though the process can be painful — perhaps analogous to what occurs in a religious conversion. But one cannot see the world in a wholly neutral way, free from any cultural bias.

112   D Bloor *Knowledge and Social Imagery* Routledge & Kegan Paul London 1976

113   B Latour & S Woolgar *Laboratory Life: The Social Construction of Scientific Facts* Sage Beverly Hills 1979

114   Bloor *op cit* (note 112) p 2. (A 'weaker' view of the influence of social factors in science would simply be that ideas are generated in society and their initial form is influenced accordingly; but they are then judged according to objective criteria, thereby shaking off their earlier 'social' features. A somewhat 'stronger' approach would recognise the social character of the process of appraisal of ideas and hence would emphasise that all knowledge has an essential social *component* at least.)

115   *Ibid* pp 4–5

116   *Ibid* p 77 & *passim*. Cf also, p 154 above.

117   *Ibid* pp 105–11

118   For a suggestion as to how the theorem was actually discovered, see J Bronowski *The Ascent of Man* British Broadcasting Corporation London 1967 pp 158–60.

119   This hypothesis, which has been subsequently utilised very considerably in the work of social anthropologists, was first presented in fully-formulated fashion in: E Durkheim & M Mauss 'De Quelques Formes Primitives de Classification: Contribution à l'étude des Répresentations Collectives' *Année Sociologique* 1901–2 Vol 6 pp 1–72. See also: E Durkheim & M Mauss *Primitive Classification* translated...and edited...by R Needham Cohen & West London 1963

120   L S Feuer *Einstein and the Generations of Science* Basic Books New York 1974 p 58 & *passim* (The first section of Feuer's book is entitled 'The social roots of Einstein's theory of relativity'.)

121   L Wittgenstein *Remarks on the Foundations of Mathematics* MIT Press Cambridge Mass 1983. 'The mathematician', Wittgenstein remarked (p 99) 'is an inventor, not a discoverer.'

122 *Ibid* p 37

123 For further discussion of Wittgenstein and the sociology of mathematics, see D Bloor 'Wittgenstein and Mannheim on the Sociology of Mathematics' *Studies in History and Philosophy of Science* 1973 Vol 2 pp 173–91; and D L Phillips *Wittgenstein and Scientific Knowledge: A Sociological Perspective* Macmillan London 1977 Ch 6

124 Bloor *op cit* (note 112) p 33 (emphasis added)

125 See above, p 187 & p 247

126 D Bloor *Wittgenstein: A Social Theory of Knowledge* Macmillan London 1983

127 *Ibid* p 122

128 C Diamond (ed) *Wittgenstein's Lectures on the Foundations of Mathematics Cambridge 1939 from the Notes of R G Bosanquet, Norman Malcolm, Rush Rhees and Yorick Smythies* Cornell University Press Ithaca 1976 177–81; L Wittgenstein *Remarks on the Foundations of Mathematics* 3rd edn Blackwell Oxford 1978 pp 394–401 and *passim*.

129 See note 113

130 The 'cost' of a paper in the years 1975–76 was apparently about $45,000! (*Op cit* [note 113] p 73)

131 *Ibid* p 79

132 If the words 'generally assumed' are deleted one would have a Type 4 statement. There are, of course, very many different types of 'modalities', and it is part of the would-be scientist's task to learn to recognise their significance. This skill is an important part of the scientist's 'tacit knowledge' requirement (in Polanyi's sense).

133 Latour & Woolgar *op cit* (note 113) p 243

134 See above p 120

135 See G Bachelard *Le Matérialisme Rationnel* Presses Universitaires de France Paris 1953

136 Latour & Woolgar *op cit* (note 113) p 64

137 See, for example, L Laudan 'The Pseudo-science of Science?' *Philosophy of the Social Sciences* 1981 Vol 11 pp 173–98; and the reply by Bloor 'The Strengths of the Strong Programme' *ibid* pp 199–213; M Hollis & S Lukes eds *Rationality and Relativism* Blackwell Oxford 1982; K D Knorr-Cetina & M Mulkay eds *Science Observed: Perspectives on the Social Study of Science* Sage London 1983

138 This turn of phrase suggests that metascientific knowledge is itself the outcome of social negotiation. But this point should, I think, already have become sufficiently clear in the course of our exposition.

139 See, for example: B Wynne 'Physics and Psychics: Science, Symbolic Action, and Social Control in Late Victorian England' in B Barnes & S Shapin eds *Natural Order: Historical Studies of Scientific Culture* Sage Publications Beverly Hills & London 1979 pp 167–86; A Pickering 'Interests and Analogies' in B Barnes & D Edge eds *Science in Context: Readings in the Sociology of Science* MIT Press Cambridge (Mass) 1982 125–46; K Caneva 'What Shall We Do with the Monster? Electromagnetism and the Psychosociology of Knowledge' in E Mendelsohn & Y Elkana eds *Sciences and Cultures* Reidel Dordrecht 1981 101–31

CHAPTER 10

CONCLUDING
REMARKS

Our remaining task is to attempt to draw together some of the diverse and tangled threads that have made up the preceding narrative, to draw attention to some of the omissions in what has been said, to attempt to give some brief appraisal of a few of the most fundamental areas of metascientific controversy, and to offer some hints as to possible future directions of inquiry.

The reader will be well aware by now that the principal peg on which I have chosen to hang my account has been a model of the process of scientific inquiry which I have called the 'arch of knowledge'. According to this model, by induction from the world of observed 'facts' (phenomena, or data) one rises to scientific 'principles'; and from these principles deductions are made to other 'facts', which can be tested experimentally, so that the whole 'structure' achieves a certain strength and security. This model, which might also be referred to as a hypothetico-deductive description of science, has been remarkably resilient in the history of 'Western' science, and in tracing its history and its numerous historical variants, we have been able to handle a good deal of the historical metascientific literature with considerable economy of thought. And even in the modern period the model is of value as a general picture of the way some people see the 'structure' of science, even though, as we have seen in the recent publications of writers such as Paul Feyerabend, the arch now seems to be collapsing in a heap. Or perhaps there is a 'structure' to science, but it is so inchoate that no simple-minded picture such as that of an arch may be deemed satisfactory.

Yet even though the 'arch' model seems to carry us through our historical account satisfactorily for many centuries — certainly well into the nineteenth — it must be recognised that it has many inadequacies. To be sure, the deductive approach to science, as it occurs in the 'descending' leg of our arch, is of the greatest importance and continues to find a place in metascientific discussions until the present day. Yet, as we have seen, in the work of Wittgenstein the very notion of universal deductive certainty has been brought into question. As for the 'ascending' side of the structure, it is less clearly defined than its descending counterpart. To ascend may involve the careful collection of data, the classifying of this information, the formulation of hypotheses, the use of so-called canons of induction, thinking with the help of models and analogies, a kind of 'logic' of induction (abduction), sampling procedures and inductive generalisations therefrom, and much else besides. So although one can learn a great deal about the history of metascience by considering the history of the 'arch of knowledge', to proceed in this way, and this way alone, does involve a considerable degree of over-simplification. For we are trying to use a single model to depict a number of distinct activities. On the other hand, the use of a simple guiding thread to help find one's way through the maze of history certainly has advantages, and is therefore not to be decried.

In considering the history of metascience with the help of the model of the 'arch of knowledge', emphasis has been directed towards the procedures of induction and deduction, and hence towards the inductive methods of the natural historian such as Bacon and the deductive methods of the geometer or the mathematically-inclined natural philosopher such as Descartes. Unfortunately, this nice tidy picture can lead to some caricature of figures such as Bacon and Descartes; also it may tend to turn attention away from other modes of scientific inquiry that are of no small importance within the total structure. So it is worth emphasising, for example, that much of science — such as geology — is particularly concerned with *historical* thinking, which is substantially different from the mathematical/experimental investigation of nature that one begins to see being developed in the work of Galileo, and which has subsequently come to form such a large portion of science as a whole.

We have also said very little about probability theory, and probabilistic modes of explanation — which are so important in such branches of science such as statistical thermodynamics, or (less esoterically) in the Darwinian theory. Even so, historical and probabilistic thinking can, to some degree, be accommodated within the 'arch' model. For the geologist certainly formulates hypotheses, and seeks to test their con-sequences by observations in the field or in the laboratory. So the 'arch-of-knowledge' model is by no means entirely inadequate for a simple description of such areas of inquiry. On the other hand, the description of science given by Reichenbach, the metascientist we have considered who has given the most attention to probabilistic thinking, does not fit the 'arch-of-knowledge' model all that well. What he has called the 'context of justification' in science contains elements of both deductive and inductive reasoning. As I have suggested, a rock-climber's route con-struction might offer a better model of Reichenbachian science than does an arch!

A more serious criticism might be that our account has simply left out of consideration large areas of metascience that are of such importance that their omission involves grievous distortion of the 'true' picture. That any historian can ever present a complete and 'true' picture I would, of course, dispute. Nevertheless it must be acknowledged that there are some notable omissions from our account. For example, rather little has been said about theories of explanation, about theories of measurement, about Marxist accounts of the scientific movement, and much else besides. Also, we have, on occasions, taken up a topic such as the problem of induction or the theory of abduction and, while looking at it in a little detail when it first appeared, have tended to neglect it thereafter, so that the subsequent histories of such topics have sometimes tended to be short-changed. All I can say on this is that if we were to pursue each and every topic that we have encountered to its conclusion there would be no possibility at all of bringing this work to a close; so by sheer necessity I have often merely been able to introduce a topic or a problem and invite the reader to investigate its further ramifications alone. Nevertheless, I hope it has been possible to introduce some of the major themes successfully and indicate the chief areas of interest as they have developed over the years.

With these caveats, I should now like to recapitulate some of the major arguments that we have encountered and express some views of my own in relation to what we have been describing and discussing. First, as to methodology, Feyerabend is probably right. There is no certain and secure method which, if carefully followed, will enable one to acquire certain and secure scientific knowledge. Ideas, hunches, hypotheses, can be drawn from any manner of sources, in no rigorously characterisable way, and yet science can progress all the better because of this 'anarchistic' component within its structure. But it does not follow, thereby, that certain carefully controlled procedures (such as one finds codified in Mill's Canons of induction) are irrelevant to science. There is, I suggest, a constant fruitful union in science between 'disciplined' and 'undisciplined' elements. Positivists such as Comte tended to emphasise the ordered, disciplined aspect. Methodological anarchists such as Feyerabend would have it all the other way. But neither of them is exclusively right, I believe, either descriptively or prescriptively. On the other hand, neither is wholly wrong either. There has, through the history of metascience, been a kind of oscillation or dialectical interplay between the opposed elements of 'order' and 'anarchy' — one, perhaps, gaining the upper hand at one period, the other at another. We are, at the moment, I believe, passing through a period where the anarchistic element is in the ascendency. And if Bloor is right, this metascientific view of the world may indeed be a reflection of the social circumstances of our time: knowledge, Bloor tells us, is shaped by social imagery. But while freely granting this point, I do not think that a description of science that presents it as a wholly irrational enterprise can be accounted satisfactory, and in discussing Feyerabend's work I suggested that his historically based arguments did not stand up to close scrutiny. They involve, I believe, an element of historiographical sleight of hand. On the other hand, I must agree with Feyerabend that there can be different forms of rationality within different societies, and that what is regarded as rational in one time and place is not necessarily seen in the same light in another.

One of the chief points to which we have attended has been the question of knowledge — what its constituents are, and how they are acquired. I think it can fairly be said that an epistemological program that rests on an empirical basis alone has to be deemed inadequate. Kant's arguments, back in the eighteenth century, establish this well enough. What we regard as knowledge is always a product, or combination, of both 'objective' and 'subjective' elements. The mind certainly brings something to bear on the world in the processes of cognition. And Kant was surely right in his arguments to the conclusion that we can never know the noumenal realm — the domain of 'things in themselves'. On the other hand, the notion of there being fixed and definite categories of the understanding, with which everyone is endowed, and which do not alter from one historical epoch to the next, cannot be sustained. Even neo-Kantians such as Whewell rightly perceived that the 'spectacles' through which we see the world, and formulate our ideas about it, alter from one generation to the next. The sociologists of knowledge can, therefore, with a little verbal elasticity or semantic licence, be referred to as neo-Kantians. For they claim that our cognitive apparatus is a product of the social formations within which it is nurtured. And our 'knowing' of the world is necessarily shaped by the social system within which we function: we cannot have access to noumenal knowledge. For everything that effectively *counts* as knowledge has to be allowed through the lock-gates of the relevant social formation. Private, incommunicable 'knowledge', even if there were any such thing, is irrelevant to the human knowledge system, which is in its essence social. This position seems to be to be essentially correct, so much so that I suggest that we may feel it appropriate to coin a new 'ism' to give emphasis to the point. Thus let us use the work 'sociologism' to represent the position held by those who wish to emphasise the social component of knowledge and develop a metascience accordingly.[1]

However, even granting the quasi-Kantian arguments of exponents of sociologism, we need not be led thereby towards a state of confused scepticism or assume that knowledge is *nothing but* that which emerges from social negotiation, power struggles, Matthew effects or whatever.[2] Yet this is the danger. As I hinted in the previous chapter, writers such as Latour and Woolgar have come perilously close to this. While acknowledging the importance of instrumentation, they have subsumed even this under the 'social umbrella'. So while we may freely acknowledge the constant influence of social negotiation in scientific research, this does not mean (in my view) that there is no objective component to knowledge whatsoever. The scientists at the Salk Laboratory were doing experiments that yielded information with an empirical aspect. Certainly, this information was talked about and was the subject of intense 'social negotiation'. The theories (papers) that emerged from the Laboratory (at $45,000 a time!) were, to be sure, moulded and shaped by social forces, and no doubt the observations were theory-laden. But the research papers contained *more* than the products of social negotiation *alone*. They did not, we can assume, produce knowledge that was absolute, and true for all time; but the 'knowledge' contained in the papers was certainly related, albeit indirectly, to the physical world as well as the social. I would suppose that Latour and Woolgar would acknowledge this to be so; but they seem (to me) to err by giving attention only to the subjective (social) component of knowledge, and to the virtual exclusion of its objective (empirical) aspects.[3] Even so, one must grant the main thrust of the exponents of sociology of knowledge: it is impossible to transcend one's social condition to arrive at some kind of wholly objective cognitive state. And so it is perfectly proper, if one so wishes, to underscore the social features of human knowledge, including scientific knowledge. And maybe it is appropriate for sociologism to have its say now, after years of empiricism and logicism being in the ascendant.

Incidentally, it is worth of remark that William Whewell, in saying long ago that science is a process in which 'fundamental ideas' are gradually made clear, can perhaps be represented as an early theorist of the sociology-of-knowledge variety. For according to his understanding it was through dialectical discussion that scientists were supposed to effect a clarification of the 'fundamental ideas'. Indeed, if we wanted to stretch further an already somewhat overextended analogy, we might propose Plato as a sociologist of knowledge! For it was, after all, through the processes of dialectical discussion that the philosopher tyros were to come to an understanding of the world — grasping the Ideas in the transcendent world of forms. However, for Plato there was an 'absolute' truth residing in the world of forms, waiting to be apprehended by the skilled dialectician, who was not thought to 'construct' knowledge (Ideas). And for Whewell, once the fundamental ideas had been satis-factorily established, there was no more uncertainty in the matter: truth had been discovered. In that sense, there was a positivist component to Whewell's position that was characteristic of his age, though he is not usually regarded as a representative of the positivist school.

What, then, should be our attitude towards the enormous positivist component of metascience, in all its many manifestations? It has, to the present, been by the most important overall contributor to metascientific doctrine. In positivism — except in its instrumentalist version — there was always the presumption that empirical science would provide the basis for secure knowledge, a claim with a political message attached. For ultimately one might have a basis for a stable and contented social system founded upon a scientific sociology, or science of society.

The positivist nirvana was to be reached with the help of the rigorous application of empirical methods (perhaps underwritten by a phenomenalist epistemology, as in the case of Mach). Or it might be attained with the help of the results of modern logic. For example, as we have seen, Carnap sought to tie (or 'hook') the theories of science to their empirical bases with unbreakable logical connections. But just as the political dimension of the positivist program was over-ambitious, so too was its plan for the construction of a unified science based on the union of logic(ism) and empiricism. For even if it were possible to give a thoroughgoing logical reconstruction of a scientific theory or group of theories, this would be quite misleading as a description of science as actually practised. The 'static' rational reconstructions of science produced by the logical positivists were artificial structures, unre-cognisable to those who actually practised science, and consequently of little use for the unification of science in any practical sense. Even the 'arch of knowledge' was more like science as practised in a dynamic sense than were the one-legged static deductive formalisms of the positivists' rational reconstructions of scientific theories.

So it might be argued that the union of logicism and empiricism had a somewhat baneful and distorting influence on metascience, particularly if the role of that enterprise was to present a picture of science that was something like science itself, rather than a logicised caricature or a normative/prescriptive exercise. It is interesting, therefore, that since Kuhn's day practising scientists have treated metascience with considerably more respect than it was accorded in the heyday of logical positivism. For by recognising the importance of the dynamic aspects of science, with attempts made to give some account of theory *change*, and the processes whereby scientific knowledge is mediated in the community, metascientists have begun to make their work more generally intelligible (and therefore acceptable) to scientists. For example, Kuhn's work was widely acclaimed almost immediately after it was published. Scientists could recognise themselves when described by Kuhn. When clothed in Carnap's logical apparatus, metascience seemed to be something arcane and irrelevant.

Nevertheless, if scientific theories claim to be logical in their structure, and scientists claim to be logical in their thinking, it seems not unreasonable to try to present the logical structures of theories, and to display the precise relationships between theory and observation and between theories of different degrees or levels of generality. The logical positivists encountered all sorts of problems in their efforts to do this, but the effort was by no means entirely wasted. The 'received-view' structure of a 'completed' scientific theory was at least plausible and attempts to give rational reconstructions of theories could sometimes reveal logical loopholes in those theories.[4] However, as has been said, the logical positivists could not do justice to the dynamics of the scientific enterprise, and the social processes whereby science achieves its status and standing within the community.

One of the most interesting debates in the history of the philosophy of science has been that concerned with the issue of realism and instrumentalism. The instrumentalist view of scientific theories seems to have an austere air of self-denial about itself. The thoroughgoing instrumentalist makes no claims that scientific theories find out anything about the 'real world'. The realist, on the other hand, claims first that there is a real world, and second that we can gain knowledge of it through our scientific investigations. However, it is not altogether easy to find examples of hard-line instrumentalists, and, as Giedymin has shown fairly recently,[5] some persons commonly represented as instrumentalists do not really deserve that label. In any case, I think it will be granted that the instrumentalist view, or a thoroughgoing version thereof, is not very plausible. The successful coordination of observations by means of a theory does seem *prima facie* evidence for the truth of that theory. We know, well enough, that evidence in favour of a theory is not proof of the truth of that theory. But that does not mean that we are obliged to be wholly agnostic on the question of the truth of successful scientific theories. What is needed, of course, is an open-mindedness on the question of truth. Successful coordination of observations, successful coherence, and successful prediction are *indications* of the truth of theoretical claims, even though they fail the acid test of strict deducibility. The truth of scientific theories cannot be *deduced* from successful inductions. But logical certainty in science has always been somewhat too strict a requirement. It was the deductive prong of Hume's fork that led to all the pother about induction.

But if a thoroughgoing instrumentalism appears unreasonable, given the known successes of science, what of the pragmatist doctrine? Unfortunately, a scientific theory's practical successes do not *prove* its truth, as the history of science bears ample witness. But they certainly lend support, and provide no warrant for wholesale scepticism at the metascientific level. Yet it must be remembered that what *counts* as practical success is (if the sociologists of knowledge are right) a question that is socially mediated and determined.

We have therefore the real world, our observations of this world, our experiments carried out on the real world, which experiments and observations are conducted through the medium of our cognitive apparatuses. And our cognitive apparatuses are themselves socially moulded, language playing a particularly important role in all this. Possibly we have some innate knowledge, or at least certain innate propensities — for example, to learn and to use language. Given this complex of factors, it is clearly imprudent to suppose that science can yield certain knowledge of the truth about anything. But this should be regarded as a counsel of prudence and a guard against intellectual arrogance — the arrogance of scientism. It should not be construed as a recipe for intellectual scepticism, anarchy or solipsism, or the proposition that science offers a fabric of ignorance rather than knowledge. It certainly has had remarkable theoretical and practical achievements. To take just a trite example, theory is used in moon shots. This theory is vindicated with the successful arrival of the astronauts at their destinations. This, in itself, seems to be an adequate rebuttal of wholesale scepticism; and most people regard it as such. There seems no reason to doubt that science does gradually find out about the world — not as à Kantian 'thing-in-itself', but a real world nonetheless; and *some* reliance can be placed on scientific knowledge! We know more today than we did five centuries ago. To suppose otherwise seems to me to be quite perverse.

As for the role of logic, it is clearly used in scientific reasoning, but to suppose that logic and the world are 'mirrors' of one another, as Wittgenstein formerly supposed, is a dubious proposition; although what we regard as logically acceptable is, I suggest, a joint product of the way things behave in the world and the way we see the world, as mediated by the language we use and the society in which we live. Yet logic, though important in science, has perhaps wielded an influence that has been too strong in metascience. The philosophical snares that have emerged from time to time in metascience have often, I believe, been a result of paying more attention to logic than common sense. And the incredible philosophical cathedrals that were constructed during the heyday of logical positivism were the product of a logicism run riot. The point I'd make here is that undue attention to logical factors can cause the metascientist to form a distorted picture of the object of his or her vision: science. And often the allure of logic may lead the metascientist into seeking to lay down rules for the scientist to follow that seem quite unnatural and unnecessary.

One further point is worth making. Our account of the history of metascience has shown that the divergence between would-be descriptions of science and science itself has often been quite extraordinary. For example, Descartes' theory of science and his actual practice were worlds apart. Or Popper's claim that the essence of good science lies in scientists trying to *falsify* their own ideas has been regarded as absurd by many scientists (though admittedly some have claimed to find Popper's methodological suggestions beneficial for their work). And Feyerabend's suggestion that the very notion of there being a procedure that can be characterised by the term 'scientific method' is likewise rejected by most scientists. We have, then, this extraordinary (but not unusual) rift between science and metascience.

Why should this be so? It certainly seems rather strange. One suggestion is that metascience often serves as part of the rhetorical armoury of science, whether wittingly or unwittingly. Descartes' *Discourse* was intended to display the supposed solid metaphysical grounding of science. His methodological pronouncements were intended to legitimate the new Cartesian philosophy. Or Popper's demarcation criterion was intended to exclude enterprises such as Freudian psychoanalysis, which might be seeking to claim scientific status, and which Popper wished to see kept out of the domain of science. His demarcation principle, therefore, would serve to sustain the purity of science. The verification principle of the Vienna Circle theorists may readily be seen in the same light.

Of course, this must seem paradoxical. It is hardly to be expected that writings of metascientists, intended to legitimate the work of scientists, should be so mistaken in their descriptions of science. Yet this may well be the case. When scientists take a day off from their science to do metascience, their reasons for doing so need to be scrutinised. If they are in the business of attempting to legitimate their accustomed scientific work, then their descriptions of it may become more distorted. And when professional metascientists are at work, since they are often remote from day-to-day scientific activities, and may have been professionally trained as historians or philosophers rather than scientists, or are many years away from their original scientific training, we can sometimes find some rather curious products. Some, like Popper, may be interested in the legitimation of science. Others, like Feyerabend, may have diametrically opposed intentions. All in all, the relationship between science, metascience, and 'the truth' about science is highly confused and confusing. And it would be presumptious to suppose that works such as the present one, occupying a niche at the level of meta-metascience, will really sort everything out satis-factorily.

But all this does not mean that we should simply disregard what metas-cientists may have to tell us. Certainly, metascientists have their own various axes to grind. Even so, there is no doubt that the history of metascience is a subject of the highest interest. It reveals some of mankind's most distinguished intellects wrestling with problems of very considerable difficulty and complexity; and in tracing this history we find out a great deal about the Western intellectual tradition. Philosophy of science is, moreover, a subject that has by no means reached its terminus. As science continues to develop, so also may we expect to see the continuing development of ideas about the nature of science (even if these ideas may often best be seen as part of the process in which science seeks legitimation). Only fairly recently, for example, have the social aspects of scientific knowledge been sufficiently recognised. It may be that in the future we shall see other equally important dimensions introduced or given greater prominence, though at the moment it is not easy to see exactly what these might be. Possible candidates are the dimensions of morality and value. Maybe soon we shall need to have chapters on the ethics of science and the axiology[6] of science, even in elementary works on metascience such as the present one (which, as I say, is really a sample of meta-metascience). Indeed developments in this direction are already beginning to manifest themselves.

However, all this is looking to the future, whereas the attention of this book has been directed to the past. It is more than sufficient for our present purposes that we seek to understand developments in metascience to our own age. The hope is that some general understanding of its intricate history may assist in its future development. But for those who wish to pursue the subject at a more elevated level, it is worth emphasising the role that metascience often plays — the legitimation of science, while ostensibly engaged in clarifications or critiques of what scientists do and say. So those who may wish to climb a step or two higher, above the first few that I have sought to shape and smooth here, may well be advised to keep this thought in mind as they make their further ascent.

FINIS!

NOTES

1 The word 'sociologism' is not new. But mostly, in previous usage, it has had a derogatory sense. (See, for example, K R Popper *The Open Society and its Enemies Volume II The High Tide of Prophecy: Hegel, Marx and the Aftermath* Routledge & Kegan Paul London 4th ed 1962 p 208 Ch 23 & *passim.*) No derogatory sense is intended here.

2 Kant, of course, would have found any such 'nothing but' view an utter anathema.

3 Sociologists of knowledge themselves tend to regard 'objectivity' as residing in something social, whereas personal experience is held to be subjective.

4 A particularly good instance of this was offered by the Darwinian theory. There is a large literature on its structure and it soon became apparent that the theory, as presented by Darwin, was not strictly logical. Eventually, after a good many attempts, a 'satisfactory' logical structure was developed by Mary Williams. But in order to do this she had to depart from Darwin's original theory to quite a large degree. See M B Williams 'Deducing the Consequences of Evolution: A Mathematical Model' *Journal of Theoretical Biology* 1970 Vol 29 pp 343–85; and for my own discussion of the problem see *Darwinian Impacts: An Introduction to the Darwinian Revolution* New South Wales University Press Kensington, Open University Press Milton Keynes and Humanities Press Atlantic Highlands 1980 pp 117–25.

5 J Giedymin 'Instrumentalism and its Critique: A Reappraisal', in R S Cohen P K Feyerabend & M W Wartowsky eds *Essays in Honour of Imre Lakatos* Reidel Dordrecht 1976 pp 179–207

6 Axiology is the branch of philosophy that attends to theories of value. (Greek *axios* 'worth' or 'worthy'.) A recent book which discusses values as an essential component for understanding philosophy of science is: L Laudan *Science and Values: The Aims of Science and their Role in Scientific Debate* California University Press Berkeley Los Angeles & London 1984. But Laudan is, of course, discussing cognitive values, methodological norms and rules, not problems in ethics.

BIBLIOGRAPHY
SUGGESTIONS FOR
FURTHER READING

The secondary literature on the topics discussed in this book is immense, and the following suggestions must be regarded merely as indicative of the field rather than comprehensive or in any way definitive. Reading of relevant primary sources is, of course, always to be recommended, but for reasons of space such texts are not listed here, since they are generally easy to locate with the help of library catalogues. The lists of holdings published by the British Library, the Bibliothèque Nationale and the Library of Congress are particularly useful for this purpose, and are held by most large libraries. Also, for reasons of space, I have largely omitted the vast number of journal articles in the secondary literature. However, relevant articles may readily be located in publications such as the *Isis Critical Bibliography*, issued annually by the History of Science Society, the *Social Sciences Index*, the *Social Sciences and Humanities Index*, *The Philosopher's Index: An International Index to Philosophical Periodicals*, or from the bibliographies to be found in the various books listed below. A comprehensive bibliography of writings on scientific method, up to Mach, was published by Laurens Laudan in 1968: 'Theories of Scientific Method from Plato to Mach: A Bibliographical Review' *History of Science* Vol 7 1968 pp1–63. This lists virtually all the books, journal articles and doctoral dissertations directly related to the theme of the present book, to 1968 (from Plato to Mach, that is).

I shall list first a few introductory texts on logic and philosophy of science, some general histories of philosophy, and some books of an encyclopaedic character that are relevant to our theme. These last, in particular, are well furnished with references. There follow details of the few works known to me that cover broadly the same ground as does *The Arch of Knowledge*, namely the general history of the philosophy of science. I shall then list some works that discuss more than one of the authors dealt with in the present book. Finally, works deaing with individual authors, periods or schools will be listed, the order of presentation being that which has been used in *The Arch of Knowledge*. For authors who are not dealt with individually in any book with which I am familiar, I have had recourse to doctoral dissertations, bound xerographic reproductions of which may be obtained from University Microfilms, Ann Arbor, Michigan, U.S.A. In the case of Robert Hooke only, I shall refer the reader to journal articles rather than books or theses, there being none of these, to my knowledge, dealing exclusively with Hooke's philosophy of science.

Anthologies of primary sources are not included in the present bibliography, although they frequently contain introductions with useful expositions and appraisals of the writings of specific authors (see, for example R.E.Butts ed *William Whewell's Theory of Scientific Method* University of Pittsburgh Press Pittsburgh 1969). Such books may, however, readily be located by the use of bibliographic tools such as those mentioned above.

INTRODUCTORY TEXTS ON LOGIC AND PHILOSOPHY OF SCIENCE

Ackerman R *The Philosophy of Science: An Introduction* Pegasus New York 1970

Chalmers A F *What is This Thing Called Science? An Account of the Nature and Status of Science and its Methods* 2nd ed University of Queensland Press 1982

Cohen M R & Nagel E *An Introduction to Logic and Scientific Method* Routledge London 1934

Frank P *Philosophy of Science: The Link Between Science and Philosophy* Prentice-Hall Englewood Cliffs 1957

Hempel C G *Philosophy of Natural Science* Prentice-Hall Englewood Cliffs 1966

Hodges W *Logic* Penguin Harmondsworth 1977

Lambert K & Brittan G G *An Introduction to the Philosophy of Science* Prentice-Hall Englewood Cliffs 1970

Medawar P B *Induction and Intuition in Scientific Thought* Methuen London 1969

Richards S *Philosophy and Sociology of Science: An Introduction* Blackwell Oxford 1983

Theobold D W *An Introduction to the Philosophy of Science* Methuen London 1968

Toulmin S E *The Philosophy of Science: An Introduction* Hutchinson London 1953

Trusted J *The Logic of Scientific Inference: An Introduction* Macmillan London 1979

Trusted J *An Introduction to the Philosophy of Knowledge* Macmillan London 1981

HISTORIES OF PHILOSOPHY

Copleston F C A *A History of Philosophy* 10 vols Burns & Oates and Search Press London 1951– (The 10th volume is still in preparation at the time of writing.)

Kneale W & Kneale M *The Development of Logic* Clarendon Press Oxford 1962

Kolakowski L *Positivist Philosophy from Hume to the Vienna Circle* Penguin Harmondsworth 1972

O'Connor D J ed *A Critical History of Western Philosophy* 3rd ed Free Press New York 1979

Stumpf S E *Socrates to Sartre: A History of Philosophy* McGraw-Hill New York 1966

Warnock G J *English Philosophy Since 1900* Oxford University Press London 1969

ENCYCLOPAEDIAS, DICTIONARIES, ETC.

Bynum W F Browne E J & Porter R eds *Dictionary of the History of Science* Macmillan London 1981

Edwards P ed *Encyclopedia of Philosophy* 8 vols Collier-Macmillan New York 1967

Flew A ed *A Dictionary of Philosophy* Pan London and Macmillan London 1979

Gillispie C C ed *Dictionary of Scientific Biography* 13 vols Scribner New York 1970–76

Grooten J & G J Steenbergen *New Encyclopedia of Philosophy* Philosophical Library New York 1972

Runes D D ed *Dictionary of Philosophy: Ancient — Medieval — Modern* Littlefield Adams & Co Paterson 1956

Wiener P P ed *Dictionary of the History of Ideas: Studies in Selected Pivotal Ideas* 5 vols Scribner New York 1973–74

HISTORIES OF PHILOSOPHY OF SCIENCE

Fowler W *The Development of Scientific Method* Pergamon Oxford 1962

Harré R *The Philosophies of Science: An Introductory Survey* Oxford University Press 1972

Losee J *A Historical Introduction to the Philosophy of Science* 2nd ed Oxford University Press 1980

Smith V E *Science and Philosophy* Bruce Milwaukee 1955

Westaway F W *Scientific Method: Its Philosophy and Practice: Its Philosophical Basis and its Modes of Application* 4th ed Blackie London 1931

WORKS DEALING WITH MORE THAN ONE AUTHOR

Bennett J F *Locke, Berkeley, Hume: Central Themes* Clarendon Pres Oxford 1971

Blake R M Ducasse C J & Madden E H *Theories of Scientific Method: The Renaissance Through the Nineteenth Century* University of Washington Press Seattle 1960

Brown H I *Perception, Theory and Commitment: The New Philosophy of Science* Precedent Publishing Ltd Chicago 1977

Buchdahl G *Metaphysics and the Philosophy of Science: The Classical Origins Descartes to Kant* Blackwell Oxford 1969

Burtt E A *The Metaphysical Foundations of Modern Physical Science: A Historical and Critical Essay* Routledge London 1924

Butts R E & Davis J W eds *The Methodological Heritage of Newton* Blackwell Oxford 1970

Crombie A C ed *Scientific Change: Historical Studies in the Intellectual, Social and Technical Conditions for Scientific Discovery and Technical Invention, from Antiquity to the Present* Heinemann London 1963

Frank P *Modern Science and its Philosophy* Harvard University Press Cambridge 1949

Garforth F W *The Scope of Philosophy: An Introductory Study Book* Longman London 1971

Giere R N & Westfall R S eds *Foundations of Scientific Method: The Nineteenth Century* Indiana University Press Bloomington 1973

Hacking I ed *Scientific Revolutions* Oxford University Press Oxford 1981

Koyré A *From the Closed World to the Infinite Universe* John Hopkins University Press Baltimore 1957

Laudan L *Science and Hypothesis: Historical Essays on Scientific Methodology* Reidel Dordrecht 1981

Morris C R *Locke, Berkeley, Hume* Oxford University Press London 1931

Stove D *Popper and After: Four Modern Irrationalists* Pergamon Oxford 1982

J R Weinberg *Ockham, Descartes, and Hume: Self-Knowledge Substance, and Causality* University of Wisconsin Press Madison 1977

WORKS DEALING WITH INDIVIDUAL AUTHORS, PERIODS OR SCHOOLS
PLATO

Crombie I M *An Examination of Plato's Doctrines* 2 vols Routledge London 1962–63

Findlay J N *Plato and Platonism: An Introduction* Times Books New York 1978

Hare R M *Plato* Oxford University Press Oxford 1982
Ross W D *Plato's Theory of Ideas* Clarendon Press Oxford 1951
Taylor A E *Plato: The Man and his Work* Methuen London 1926
Vesey G *Plato's Theory of Forms* Open University Press Milton Keynes 1980

ARISTOTLE
Adler M J *Aristotle for Everybody: A Difficult Thought Made Easy* Macmillan New York 1978
Barnes J *Aristotle* Oxford University Press Oxford 1982
Barnes J Schofield M & Sorabji R eds *Articles on Aristotle: 1. Science* Duckworth London 1975
Farrigton B *Aristotle: Founder of Scientific Philosophy* Weidenfeld & Nicolson London 1965
Ferguson J *Aristotle* Twayne Boston 1972
Lloyd G E R *Aristotle: The Growth and Structure of his Thought* Cambridge University Press Cambridge 1968
Randall J H Jr *Aristotle* Columbia University Press New York 1960
Ross W D *Aristotle* Methuen London 1923
Taylor A E *Aristotle* revised ed Dover New York 1955

MEDIAEVAL AUTHORS
Boehner P *Medieval Logic: An Outline of its Development from 1250 to c.1400* Manchester University Press Manchester 1952
Carré M H *Realists and Nominalists* Oxford University Press London 1946
Crombie A C *Augustine to Galileo* 2 vols Mercury Books London 1961
Crombie A C *Robert Grosseteste and the Origins of Exerimental Science: 1100–1700* Clarendon Press Oxford 1953
McEvoy J *The Philosophy of Robert Grosseteste* Oxford University Press Oxford 1982
Moody E A *The Logic of William of Ockham* Russell & Russell New York 1965
Moody E A *Studies in Medieval Philosophy, Science, and Logic: Selected Papers, 1933–1969* University of California Press Berkeley 1975
Moody E A *Truth and Consequence in Medieval Logic* North-Holland Amsterdam 1953
Weinberg J R *Abstraction, Relation, and Induction* University of Wisconsin Press Madison 1965
Weinberg J R *A Short History of Medieval Philosophy* Princeton University Press Princeton 1946

RENAISSANCE AUTHORS
Edwards W 'The Logic of Zabarella' Ph D dissertation Columbia University 1960
Gilbert N W *Renaissance Concepts of Method* Columbia University Press New York 1960
Howell W *Logic and Rhetoric in England: 1500–1700* Princeton University Press Princeton 1956
Randall J H *The School of Padua and the Emergence of Modern Science* Antenore Padua 1961
Wallace W A *Prelude to Galileo: Essays on Medieval and Sixteenth-Century Sources of Galileo's Thought* Reidel Dordrecht 1981

GALILEO
Butts R E & Pitt J C eds *New Perspective on Galileo* Reidel Dordrecht 1978
Clavelin M *The Natural Philosophy of Galileo: Essay on the Origins and*

Formation of Classical Mechanics M I T Press Cambridge 1974

Drake S *Galileo Studies* University of Michigan Press Ann Arbor 1970

Drake S *Galileo at Work: His Scientific Biography* University of Chicago Press Chicago 1978

Drake S *Galileo* Oxford University Press Oxford 1980

Finocchiaro M *Galileo and the Art of Reasoning: Rhetorical Foundations of Logic and Scientific Method* Reidel Dordrecht 1980

Geymonat L *Galileo Galilei: A Biography and Inquiry into his Philosophy of Science* McGraw-Hill New York 1965 (1st Italian ed 1957)

Koyré A *Galileo Studies* Harvester Hassocks 1978 (1st French ed 1939)

Shapere D *Galileo: A Philosophical Study* Chicago University Press Chicago & London 1974

Shea W *Galileo's Intellectual Revolution* Science History Publications New York and Macmillan London 1972

BACON

Anderson F H *The Philosophy of Francis Bacon* University of Chicago Press Chicago 1948

Farrington B *Francis Bacon: Philosopher of Industrial Science* Schuman New York 1949

Jardine L *Francis Bacon: Discovery and the Art of Discourse* Cambridge University Press Cambridge 1974

Quinton A *Bacon* Oxford University Press Oxford 1980

Stephens J *Francis Bacon and the Style of Science* University of Chicago Press Chicago 1975

Vickers B ed *Essential Articles for the Study of Francis Bacon* Sidgwick & Jackson London 1968

HOOKE

Hesse M B 'Hooke's Development of Bacon's Science' *Actes du Dixième Congrès International d'Histoire des Sciences: Ithaca 1962* Paris 1964 Vol 1 pp 265–8

Hesse M B 'Hooke's Philosophical Algebra' *Isis* Vol 57 1966 pp 67–83

Oldroyd D R 'Robert Hooke's Methodology of Science as Exemplified in his *"Discourse of Earthquakes"'* *The British Journal for the History of Science* Vol 6 1972 pp 109–30

DESCARTES

Beck L J *The Method of Descartes: A Study of the Regulae* Clarendon Press Oxford 1952

Clarke D M *Descartes' Philosophy of Science* Manchester University Press Manchester 1982

Gaukroger S ed *Descartes: Philosophy, Mathematics and Physics* Harvester Sussex and Barnes & Noble New Jersey 1980

Hooker M ed *Descartes: Critical and Interpretive Essays* Johns Hopkins University Press Baltimore 1978

Kassman A *Cartesian Scepticism* Open University Press Milton Keynes 1976

Keeling S V *Descartes* Oxford University Press 2nd ed London 1968

Kemp Smith N *New Studies in the Philosophy of Descartes: Descartes as Pioneer* Macmillan London 1963

Kenny A *Descartes: A Study of his Philosophy* Random House New York 1968

Sorrell T *Descartes: Reason and Experience* Open University Press Milton Keynes 1982

Vesey G *Descartes: Father of Modern Philosophy* Open University Press Milton Keynes 1971

Williams B *Descartes: The Project of Pure Enquiry* Penguin Harmondsworth 1978

Wilson M D *Descartes* Routledge & Kegan Paul London 1978

PORT-ROYALISTS

Rea L *The Enthusiasts of Port Royal* Methuen London 1912

NEWTON

Cohen I B *Franklin and Newton: An Inquiry into the Speculative Newtonian Experimental Science and Franklin's Work in Electricity as an Example Thereof* American Philosophical Society Philadelphia 1956

Cohen I B *The Newtonian Revolution: With Illustrations of the Transformation of Scientific Ideas* Cambridge University Press Cambridge 1980

Koyré A *Newtonian Studies* University of Chicago Press Chicago 1968

Westfall R S *Never at Rest: A Biography of Isaac Newton,* Cambridge University Press Cambridge 1980

LEIBNIZ

Brown S *Leibniz: Reason and Experience* Open University Press Milton Keynes 1983

MacDonald Ross G *Leibniz* Oxford University Press Oxford 1984

Parkinson G H R *Logic and Reality in Leibniz's Metaphysics* Clarendon Press Oxford 1965

Rescher N *Leibniz: An Introduction to his Philosophy* Blackwell Oxford 1979

Rescher N *The Philosophy of Leibniz* Prentice-Hall Englewood Cliffs 1967

Saw R L *Leibniz* Penguin Harmondsworth 1954

LOCKE

Dunn J *Locke* Oxford University Press Oxford 1984

Khin Zaw S *John Locke: The Foundations of Empiricism* Open University Press Milton Keynes 1976

Mabbott J D *John Locke* Macmillan London 1973

Mandelbaum M *Philosophy, Science and Sense Perception* Johns Hopkins University Press Baltimore 1964

O'Connor D J *John Locke* Penguin Harmondsworth 1952

Tipton I *Locke: Reason and Experience* Open University Press Milton Keynes 1983

Woolhouse R S *Locke's Philosophy of Science and Knowledge: A Consideration of Some Aspects of An Essay Concerning Human Understanding* Blackwell Oxford 1971

Yolton J W *John Locke and the Way of Ideas* Clarendon Press Oxford 1956

Yolton J W *John Locke and the Compass of Human Understanding: A Selective Commentary of the 'Essay'* Cambridge University Press Cambrdige 1970

BERKELEY

Bracken H M *Berkeley* Macmillan London 1974

Brook R J *Berkeley's Philosophy of Science* Nijhoff The Hague 1973

Luce A A *Berkeley's Immaterialism: A Commentary on his A Treatise Concerning the Principles of Human Knowledge* Nelson London 1945

Pitcher G *Berkeley* Routledge & Kegan Paul London

Tipton I C *Berkeley: The Philosophy of Immaterialism* Methuen London 1974
Urmson J O *Berkeley* Oxford University Press Oxford 1982
Vesey G *Berkeley: Reason and Experience* Open University Press Milton Keynes 1982
Warnock G J *Berkeley* Penguin Harmondsworth 1953

HUME

Ayer A J *Hume* Oxford University Press Oxford 1980
Basson A H *David Hume* Penguin Harmondsworth 1958
Cockburn D and Bourne G *Hume: Reason and Experience* Open University Press Milton Keynes 1983
Flew A *Hume's Philosophy of Belief* Humanities Press New York 1961
Hanfling O *Cause and Effect* Open University Press Milton Keynes 1973
Passmore J A *Hume's Intentions* Cambridge University Press Cambridge 1952
Penelhum T *Hume* Macmillan London 1975
Stove D C *Probability and Hume's Inductive Scepticism* Clarendon Press Oxford 1973
Stroud B *Hume* Routledge & Kegan Paul London 1977

KANT

Bird G *Kant's Theory of Knowledge: An Outline of One Central Argument in the Critique of Pure Reason* Humanities Press New York 1962
Brittan G G *Kant's Theory of Science* Princeton University Press Princeton 1978
Cassirer E *Kant's Life and Thought* Yale University Press New Haven & London 1981
de Vleeschauwer H J *The Development of Kantian Thought: The History of a Doctrine* Nelson London 1962
Kemp J *The Philosophy of Kant* Oxford University Press London 1968
Körner S *Kant* Penguin Harmondsworth 1955
Strawson P F *The Bounds of Sense: An Essay on Kant's Critique of Pure Reason* Methuen London 1966
Vesey G *Kant's Copernican Revolution: Speculative Philosophy* Open University Press Milton Keynes 1972
Walker R C S *Kant* Routledge & Kegan Paul London 1978
Walsh W H *Kant's Criticism of Metaphysics* Edinburgh University Press Edinburgh 1975
Werkmeister W H *Kant: The Architectonic and Development of his Philosophy* Open Court La Salle 1980
Wilkerson T E *Kant's Critique of Pure Reason: A Commentary for Students* Clarendon Press Oxford 1976
Wolff R P ed *Kant: A Collection of Critical Essays* Macmillan London 1968

HERSCHEL

Jain C L 'Methodology and Epistemology: An Examination of Sir John Frederick William Herschel's Philosophy of Science with Reference to his Theory of Knowledge' Ph D dissertation University of Indiana 1975

WHEWELL

Walsh H T 'The Philosophy of William Whewell' Ph D dissertation University of Michigan 1960
Butts R E ed *William Whewell's Theory of Scientific Method* University of Pittsburgh Press Pittsburgh 1968

MILL

Anschutz R P *The Philosophy of J S Mill* Clarendon Press Oxford 1953
Britton K *John Stuart Mill* Penguin Harmondsworth 1953
Jackson R *An Examination of the Deductive Logic of John Stuart Mill* Oxford University Press London 1953
McCloskey H J *John Stuart Mill: A Critical Study* Macmillan London 1971
Ryan A *The Philosophy of John Stuart Mill* Macmillan London 1970

COMTE

Evans-Pritchard E E *The Sociology of Comte: An Appreciation* Manchester University Press Manchester 1970
Lewes G H *Comte's Philosophy of the Sciences: Being an Exposition of the Principles of the Cours de Philosophie Positive of Auguste Comte* Bell London 1883
Mill J S *Auguste Comte and Positivism* Michigan University Press Ann Arbor 1961
Simon W M *European Positivism in the Nineteenth Century: an Essay in Intellectual History* Cornell University Press Ithaca 1963

MACH

Bradley J *Mach's Philosophy of Science* Athlone Press London 1971
Blackmore J T *Ernst Mach: His Work, Life, and Influence* University of California Press Berkeley 1972
Cohen R S & Seeger R J eds *Ernst Mach: Physicist and Philosopher* Humanities Press New York 1970
Weinberg C B *Mach's Empirio-Pragmatism in Physical Science* Albee Press New York 1937

PRAGMATISTS

Ayer A J *The Origins of Pragmatism: Studies in the Philosophy of Charles Sanders Peirce and William James* Macmillan London 1968
Davis W H *Peirce's Epistemology* Nijhoff the Hague 1972
Fann K T *Peirce's Theory of Abduction* Nijhoff The Hague 1970
Gallie W B *Peirce and Pragmatism* Penguin Harmondsworth 1952
Hookway C *Peirce* Routledge & Kegan Paul London 1985
Kannegiesser H J *Knowledge and Science* Macmillan Melbourne 1977
Moore E C *American Pragmatism: Peirce, James and Dewey* Columbia University Press New York 1961
Moore E C & Robin R S eds *Studies in the Philosophy of Charles Sanders Peirce* University of Massachusetts Press Amherst 1964
Morris C *The Pragmatic Movement in American Philosophy* Brasiller New York 1970
Reilly F E *Charles Peirce's Theory of Scientific Method* Fordham University Press New York 1970
Rescher N *Peirce's Philosophy of Science: Critical Studies in his Theory of Induction and Scientific Method* University of Notre Dame Press Notre Dame and London 1978
Scheffler I *Four Pragmatists: A Critical Introduction to Peirce, James, Mead and Dewey* Routledge & Kegan Paul London 1974
Sebeock J A & Uniker Sebeock J *You Know My Method: A Juxtaposition of Charles S. Peirce & Sherlock Holmes* Gaslight Publications Indianapolis 1979
Skagestat P *The Road of Inquiry: Charles Peirce's Pragmatic Realism* Columbia University Press New York 1981

POINCARE

Dantzig T *Henri Poincaré, Critic of Crisis: Reflections on his Universe of Discourse* Scribner New York 1954

Giedymin J *Science and Convention: Essays on Henri Poincaré's Philosophy of Science and the Conventionalist Tradition* Pergamon Oxford 1982

DUHEM

Harding S G ed *Can Theories be Refuted? Essays on the Duhem-Quine Thesis* Reidel Boston 1976 (This book does not deal with Duhem's work in general, but is classified here for convenience.)

Jaki S *The Life and Work of Pierre Duhem* Nijhoff The Hague 1984

Lowinger A *The Methodology of Pierre Duhem* Columbia University Press New York 1941

FREGE

Currie G *Frege: An Introduction to his Philosophy* Harvester Sussex and Barnes & Noble New Jersey 1982

Dummett M *Frege: The Philosophy of Language* Duckworth London 1973

Sluga H D *Gottlob Frege* Routledge & Kegan Paul London 1980

RUSSELL

Ayer A J *Russell* Fontana/Collins London 1972

Brown S *Realism and Logical Analysis* Open University Press Milton Keynes 1976

Eames E R *Bertrand Russell's Theory of Knowledge* Allen & Unwin London 1969

Sainsbury R M *Russell* Routledge & Kegan Paul London 1979

Schilpp P A ed *The Philosophy of Bertrand Russell* 4th ed Open Court La Salle 1971

Watling J *Bertrand Russell* Oliver & Boyd Edinburgh 1970

WITTGENSTEIN

Anscombe G E M *An Introduction to Wittgenstein's Tractatus* Hutchinson London 1959

Bartley III W W *Wittgenstein* Quartet Books London 1974

Bloor D *Wittgenstein: A Social Theory of Knowledge* Macmillan London 1983

Fogelin R J *Wittgenstein* Routledge & Kegan Paul London 1976

Janik A & Toulmin S E *Wittgenstein's Vienna* Weidenfeld & Nicolson London 1973

Kenny A *Wittgenstein* Penguin Harmondsworth 1975

Mounce H O *Wittgenstein's Tractatus: An Introduction* Blackwell Oxford 1981

Parkinson G H R *Saying and Showing, An Introduction to Wittgenstein's Tractatus Logic-Philosophicus* Open University Press Milton Keynes 1976

Pears D *Wittgenstein* Fontana/Collins Glasgow 1971

Phillips D L *Wittgenstein and Scientific Knowledge: A Sociological Perspective* Macmillan London 1977

LOGICAL POSITIVISM AND THE VIENNA CIRCLE

Bergman G *The Metaphysics of Logical Positivism* Greenwood Press Westport 1978

Brown S *Verification and Meaning* Open University Press Milton Keynes 1976

Hanfling O *Logical Positivism* Blackwell Oxford 1981

Joergensen J *The Development of Logical Empiricism* Universty of Chicago Press Chicago 1951

Kraft V *The Vienna Circle: The Origin of Neo-Positivism — A Chapter in the History of Recent Philosophy* Greenwood New YorK 1969

Suppe F 'The Search for Philosophic Understanding of Scientific Theories' in Suppe F ed *The Structure of Scientific Theories* 2nd ed University of Illinois Press Urbana 1977 pp 1–232

Urmson J O *Philosophical Analysis: Its Development Between the Two Wars* Clarendon Press Oxford 1956

Weinberg J R *An Examination of Logical Positivism* Routledge London 1936

CARNAP

Buck R C & Cohen R S eds *P.S.A. 1970: In Memory of Rudolf Carnap* Reidel Dordrecht 1971

Hintikka J ed *Rudolf Carnap, Logical Empiricist: Materials and Perspectives* Reidel Dordrecht 1975

Runggaldier E *Carnap's Early Conventionalism: An Inquiry into the Historical Background of the Vienna Circle* Rodopi Amsterdam 1984

Schilpp P A ed *The Philosophy of Rudolf Carnap* Open Court La Salle and Cambridge University Press London 1963

REICHENBACH

Salmon W C ed *Hans Reichenbach: Logical Empiricist* Reidel Dordrecht 1979

EINSTEIN

Bernstein J *Einstein* Fontana/Collins Glasgow 1973

Infeld L *Albert Einstein: His Work and its Influence on the World* Scribner New York

Miller A I *Albert Einstein's Special Theory of Relativity: Emergence (1905) and Early Interpretation (1905–1911)* Addison-Wesley Reading Mass 1981

Pais A *Subtle is the Lord...: The Science and Life of Albert Einstein* Clarendon Press Oxford 1982

Peyenson L *The Young Einstein: The Advent of Relativity* Adam Hilger Bristol & Boston 1985

Schilpp P A ed *Albert Einstein: Philosopher-Scientist* 3rd ed Open Court La Salle 1970

Schwartz J & McGuinness M *Einstein for Beginners* Writers & Readers London 1979

Williams L P ed *Relativity Theory: Its Origins and Impact on Modern Thought* Wiley New York 1968

BRIDGMAN

Benjamin A C *Operationism* Thomas Springfield 1955

EDDINGTON

Douglas A V *The Life of Arthur Stanley Eddington* Nelson London 1956

Stebbing L S *Philosophy and the Physicists* Penguin Harmondsworth 1944

Witt-Hansen J *Exposition and Critique of the Conceptions of Eddington Concerning the Philosophy of Physical Science* Gads Copenhagen 1958

Yolton J W *The Philosophy of Science of A.S. Eddington* Nijhoff The Hague 1960

MODELLISTS

Hesse M B *Models and Analogies in Science* Sheed & Ward London 1963
Leatherdale W H *The Role of Analogy, Model and Metaphor in Science* North-Holland Amsterdam and American Elsevier New York 1974

POPPER

Ackerman R N *The Philosophy of Karl Popper* University of Massachusetts Press Amherst 1976
Burke T E *The Philosophy of Popper* Manchester University Press Manchester 1983
Johansson I *A Critique of Karl Popper's Methodology* Scandinavian University Books Stockholm 1975
Magee B *Popper* Fontana/Collins Glasgow 1973
O'Hear A *Karl Popper* Routledge & Kegan Paul London 1980
Schilpp P A ed *The Philosophy of Karl R. Popper* Open Court La Salle 1974

LAKATOS

Cohen R S Feyerabend P K & Wartofsky M W eds *Essays in Memory of Imre Lakatos* Reidel Dordrecht

KUHN

Barnes B *T.S.Kuhn and Social Science* Macmillan London 1982
Gutting G ed *Paradigms and Revolutions: Appraisals and Applications of Thomas Kuhn's Philosophy of Science* University of Notre Dame Press Notre Dame 1980

FEYERABEND

Guneratne R D *Science, Understanding and Truth: A Study of the Meaning Variance Problem in the Philosophy of Science* Ministry of Higher Education Sri Lanka Colombo 1980

SOCIOLOGISTS OF KNOWLEDGE

Brannigan A *The Social Basis of Scientific Discoveries* Cambridge University Press Cambridge 1981
Merton R K & Gaston J eds *The Sociology of Science in Europe* Southern Illinois University Press Carbondale 1977
Mulkay M *Science and the Sociology of Knowledge* Allen & Unwin London 1977
Social Studies of Science 1981 Vol 11 No 1 (for a discussion of the 'strong program' in the sociology of knowledge)

INDEX